Moon Rhythms in Nature

Moon Rhythms in Nature

How Lunar Cycles Affect Living Organisms

Klaus Peter Endres
Wolfgang Schad

Floris Books

Translated by Christian von Arnim

First published in German in 1997 under the title
Biologie des Mondes. Mondperiodik und Lebensrhythmen
by S. Hirzel Verlag, Stuttgart, Leipzig
First published in English in 2002 by Floris Books

© 1997 S. Hirzel Verlag, Stuttgart
Translation © 2002 Floris Books, Edinburgh

All rights reserved. No part of this publication may
be reproduced without the prior permission of
Floris Books, 15 Harrison Gardens, Edinburgh.

British Library CIP Data available

ISBN 0-86315-360-7

Printed in Great Britain
by Bell & Bain, Glasgow

Contents

Preface	7
Introduction	9
How widespread are lunar rhythms in the living world? 14; Specialist terms and issues 19	
1. The Moon	23
How do lunar eclipses occur? 25; The Moon in the sky 28; Other lunar rhythms 30	
2. The Tides	35
High and low tides across the Earth 35; The causes of tides 41	
3. Life in Tidal Coastal Areas	47
The tidal rim of the North Sea 47; Crabs, snails, diatoms and a turbellarian 50; The Atlantic shore 58; The marine midge, Clunio 59	
4. A Worm, Fish, Squid and Corals from the Pacific Ocean	63
The palolo worm 63; The grunion 71; The pearly nautilus 74; The origin of the Moon 79; Corals and the Moon 81	
5. The Causal Question	87
A sea snail, lunar periods and the Earth's magnetic field 89; Endogenous rhythms? 92; Exogenous rhythms? 95; Life indivisible 97; Uniformity or multiplicity in temporal behaviour 100	
6. Moon Rhythms in Non-Marine Organisms	107
Freshwater animals 107; Land animals 109; Land plants 111; The evolutionary aspect of biological rhythms 112	

7. Moon Rhythms in Human Beings 117
*Sensitivity of the eyes to Sun and Moon rhythms 117;
Kidneys, lungs and Moon rhythms 123; Birth and
Death 126*

8. The Rhythmical Spectrum of Human Beings 133
*Biological rhythms 134; Biographical rhythms 137;
Conclusion 141*

9. The Quality of Time 147
The riddle of time 148; Time integration in life 151

Species catalogue 155

References 219

Index of authors 277

Index of species 289

Subject index 299

Preface

Chronobiology established itself as a new branch of science in the second half of the twentieth century. Complementing our knowledge of organisms as spatial phenomena, chronobiology is also used to study their forms over time, mostly in the context of the rhythmical structure of their physiological functions. In doing so, it is concerned with the whole spectrum of rhythmical life processes, ranging from milliseconds and seconds via minutes and hourly rhythms, days and months, to the year and beyond. The results of this research are of relevance for many practice-based branches of science such as pharmacology, medical therapies, occupational medicine, education and ecology. In addition, there is special interest in the contribution which chronobiology can make to the present interdisciplinary debate about the concept of time itself. Organisms live in the *quality* of time. That is why they can tell us such a lot about it.

In this book we concentrate on lunar rhythms, and thus on the field of chronobiological research which has given us access to a new scientific aspect relating to the Moon and our knowledge of it. The steady demythologizing of our nearest celestial neighbour during the last four hundred years reached its climax when human beings landed on the Moon for the first time — a technical triumph of applied sciences. Astronomical and physical investigation of the Moon was followed by petrochemical investigation of Moon rock. This confirmed the previously established picture of the Moon as a giant cinder, covered by craters and dust, subject to bombardment by meteorites. In contrast, we also experience the Moon as an expressly romantic, mysterious body. Both interpretations of the Moon — the physical and emotional claims on it — are one-sided. Both sides are complemented and integrated by chronobiology, which meets the requirements of scientific methodology while at the same time revealing the objective connection of the Moon with the living world.

Moon Rhythms in Nature is an introduction to the chronobiology of lunar rhythms within the biosphere of the Earth for the reader without specialist knowledge. The text is structured so that no previous specialist knowledge is assumed. Any specialist terms that are used are explained so it is advisable to read the book in sequence.

The detailed species catalogue provides an overview, and also contains all bibliographical references which have not been quoted in the text. The index of species at the end should be used to find a specific entry in the species catalogue.

Readers may be surprised by the range of interaction between the Earth and the Moon in the biosphere of the Earth, and yet there is nothing strange about it. It would be even more surprising if life on Earth reacted every day to the Sun but not to our nearest companion in space. This book deals with these questions to the extent that it offers the scientifically interested reader references to the original research.

This research into species, pursued over many years, has consistently been supported by individual sponsors. They are due our thanks, as are the student assistants who were also involved, Eva-Marie Fischer, Georgia Borchert, Kathrin Ahrens and Sören Schmidt. We thank Thomas Schmidt for reading the chapter on astronomy and Jochen Bockemühl for reading the biological section. The Karl Schweisfurth Foundation is due thanks for financial support for key stages of the book.

Klaus Peter Endres
Wolfgang Schad

Introduction

Aristotle, the first naturalist to rely solely on the experience of his senses and thinking, reported as early as the fourth century BC on a Moon rhythm in organisms, using the example of the Mediterranean sea urchin. In his description of the anatomical structure of the animal — its jaws are still called 'Aristotle's lantern' by zoologists today — he mentions that the ovaries swell at the time of the Full Moon [1]. This can also be confirmed for the Diadem sea urchin (*Diadema setosum*) (Figure 1, Plate 13) from the Red Sea near Suez in 1921. (For these and many other species, compare the index of species which refers to the corresponding place in the catalogue of species with the respective bibliographical reference.)

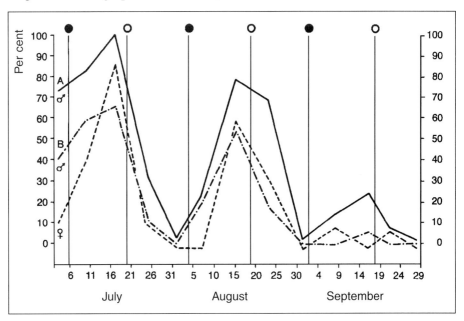

Figure 1. Diadem sea urchin (Diadema setosum). *Percentage of animals with mature gametes (eggs or sperm) in per cent of numbers caught during the summer of 1921 near Suez. Line A ♂: males with only mature sperm (without spermatocytes). Line B ♂: males, which spontaneously spawned during the study. Line ♀: females with exclusively mature eggs. Full Moon: white circle. New Moon: black circle [13].*

In Cicero [2] and Pliny [3] we find references to the quantity of oysters in the sea increasing and decreasing with the phases of the Moon. The reproductive activity of the oyster (*Ostrea edulis*) (Plate 6) largely at Full and New Moon can also be confirmed in the summer months on the Dutch and English coasts.

When Columbus discovered America on October 11, 1492, about one hour before moonrise he and his companions saw many lights flickering like candles, in the water off the coast of the Bahamas. In 1935 the English marine biologist, Crawshay [4], stated that the cause of this light is likely to have been the Atlantic luminescent worm (*Odontosyllis spec.*). At about the time of the last lunar quarter, the females swim to the surface of the sea in the hours after sunset and emit flashes of light. The males, whose attention has been caught in this way, then also swim to the surface glowing with phosphorescence. Once in the vicinity of the females, both eject their reproductive cells and the brief process is quickly at an end.

However, it was the palolo worm (*Eunice viridis*) in the South Pacific which first created a general awareness of lunar rhythms in living beings. The first reports by Stair and Gray [5, 6] tell of a worm living in the coral reefs surrounding the Samoan Islands which always swarmed at night when the Moon was in its third quarter in the early summer months — October/November in the southern hemisphere. The island inhabitants have always known about this lunar rhythm and still wait for it today to catch the mass of worms as they swim to the surface and eat them as a delicacy. Numerous studies at the turn of the last century and in the 1950s confirmed the regularity of the palolo's behaviour. Many hundreds of other species with lunar periodicity living today have been discovered in the meantime in all the oceans. However, dependence on the phases of the Moon is not just characteristic of marine organisms, as some cases closer to home illustrate.

Everyone knows that the growth speed of the first shoots of spring is dependent on the weather and, above all, temperature. In the botanic gardens in Padua, northern Italy, measurements of this temperature-dependence revealed that the growth speed of some blossoming plants was particularly temperature-dependent at New Moon; the same applied at Full Moon, but not to the same extent. At Full Moon, the plants concerned grew more steadily, according to their own laws and without such dependence on the weather. These plants were: the wood anemone (*Anemone nemorosa*); bulbous corydalis (*Corydalis cava*); ground elder (*Aegopodium podagraria*); creeping bellflower (*Campanula rapuncu-*

loides); a type of Jerusalem-oak (*Chenopodium botrys*); and, tuberous comfrey (*Symphytum tuberosum*). For monocotyledons such as snowdrops (*Galanthus nivalis*) and wild garlic (*Allium ursinum*), a dependence on external temperature relating to lunar periods could not be established, and with broad beans (*Vicia faba*) there was only a weak link [7]. However, plants germinating with two cotyledons (dicotyledons) reacted more sensitively to the phases of the Moon.

In the same way the grape vine (*Vitis vinifera*) produces a better grape harvest, taken as a statistical mean, if the New Moon phase occurs in the first half of June during blossoming (based on data covering 425 years).

Plant seeds need humidity to germinate and, when the seeds of the common bean (*Phaseolus vulgaris*) swell during germination, the intake of water is much greater every seven days. The maximum intakes occur during the Full and New Moon and with the waxing and waning Half Moon. Bean seeds separate easily into two halves — the cotyledons and are thus dicotyledonous.

Among European animals, two types of worm also display this kind of dependence on lunar phases. In clean, rapidly flowing forest and meadow streams, we commonly find small, black flatworms, which creep along like snails on the underside of stones lying in the water. The most frequent species, the river planaria (*Euplanaria gonocephala*) loves darkness — like all its relatives. If its sensitivity to light is tested, it becomes evident that planaria prefer darker places during Full Moon than New Moon. These turbellarians are least sensitive to light with a waxing and waning Moon.

The familiar earthworm loosens, aerates and digests the soil and is therefore an irreplaceable aid for all higher plant life. In 1881 Darwin wrote enthusiastically about it in his last work, where he asserted that humankind would never have been able to rise to high levels of agriculture without the earthworm [8]. In 1957 it was discovered that the activity and rest periods of the earthworm are governed by four rhythms: the solar day (24.0 hours), the lunar day (24.8 hours), and the half and whole synodical lunar month (14.75 and 29.5 days respectively). Its movement is greatest in the evening and at night, as well as about three hours before the lowest position of the Moon below the horizon each day, during Full and particularly New Moon. Then there is an annual rhythm. It therefore participates in the whole delicate interaction among Sun, Earth and Moon in the way that it lives.

In 1971 it was discovered that the honey bee *(Apis mellifera)* also

Figure 2. European river eel (Anguilla anguilla) *[1081].*

knows the Moon. The Carnica strain, much kept in Europe, displayed the greatest flight activity on days of the Full Moon when the honeycomb was in a north-south alignment, and the flight hole pointed north. In an east-west alignment, with the flight hole facing east, the flight maximum occurred at New Moon. In all events there was a monthly, or lunar, rhythm of activity (29.5 days) throughout the summer. Before the start of winter rest, it then changed into a half-monthly or semi-lunar rhythm (14.7 days). The Moroccan bee strain, on the other hand, displayed a semi-lunar activity rhythm in summer which reached its maximum partly at Full Moon and partly at New Moon, with interspersed neutral lunar periods. When brought from Morocco to Frankfurt by plane, these bees changed to a monthly rhythm. We can see here very subtle patterns of rhythms which allow us to become more familiar with the highly differentiated behaviours of these special insects.

Among more the complex vertebrates, the European eel (*Anguilla anguilla*) (Figure 2) is a good example of behaviour linked to the periodicity of the Moon. Once it has eaten its fill in freshwater and has grown to full sexual maturity, it travels down the rivers back to the sea where its lays its eggs and dies in mid-ocean — in the Sargasso Sea in the middle of the Atlantic. The still transparent young *glass eels* or *elvers* swim to the coast and back into freshwater. The catch records of

Figure 3. The European nightjar (Caprimulgus europaeus) *about to land [1082].*

eel fishermen on the Upper Rhine from 1938 to 1950 show that eel migrations are dependent on the water level, temperature, time of year and day, and the phases of the Moon. The highest catches were recorded during the autumn when the water was rising or high, in cloudy weather which was not too cold, and at night when the Moon was in its third quarter. The same conditions were shown to apply on the island of Rügen in the Baltic and on the River Bann in County Wexford, Ireland.

The reproductive behaviour of the nocturnal European nightjar (*Caprimulgus europaeus*) (Figure 3) is also connected with the phases of the Moon. Two eggs — never more — are laid directly on to heather at night during the third quarter of the Moon. The young birds hatch after sixteen to eighteen days, when the Moon is waxing, so that they can be fed sufficiently during the time when the Moon approaches its greatest brightness.

Eighteen days after they have hatched they are fully fledged. It is clear that the maturing of the gametes in the gonads at the start of the reproductive cycle and thus mating and ovulation are also co-ordinated with the synodic lunar rhythm. This is also seen in the mating call of the male which only commences late at night when the Moon is full, about 95 minutes after sunset, while at the time of the New Moon his peculiar humming can be heard as early as 15 minutes after sunset — but only if the weather is not too bad.

How widespread are lunar rhythms in the living world?

We have reason to believe that it is not only such unusual animals as the migrating eels and the nightjar, which are in harmony with the rhythm of the Moon. Such rhythms can be found in many organisms if they are measured precisely enough. The American biologist, F.A. Brown Jnr., worked in the laboratory on small individual pieces of carrot root and potato tubers, which were kept under constant illumination, temperature, air pressure, etc. Oxygen consumption as a measure of vital activity regularly reached its maximum in the carrot tissue (*Daucus carota*) in the third quarter, its minimum at Full Moon, while in the potato pieces (*Solanum tuberosum*) half-monthly (14.75 days) and lunar diurnal (24.8 hours) rhythms occurred. The maximum was reached at Full and New Moon, always when the Moon was setting.

In mammals rhythms of activity which ran in parallel with the phases of the Moon were also found, particularly in the laboratory rodents. Golden hamsters (*Mesocricetus auratus*) show a monthly rise and fall of daily running activity. In the white laboratory mouse (*Mus musculus*), the running activity rhythm reaches its maximum shortly before the lowest position of the Moon below the horizon (Moon midnight).

INTRODUCTION   15

   We can also include the well-known Guppy fish (*Lebistes reticulatus*), which is common in home aquariums. This brightly coloured little fish, which originates in the freshwater pools of Venezuela and the Caribbean islands of Barbados and Trinidad, can be kept with such ease even by beginners, and reproduces so easily, that it is an ideal animal for experiments dealing with chromatic sensitivity. When it is illuminated from the side, it places itself at an angle in the water with its back facing the light. The angle of divergence from the vertical is a sure measure of light sensitivity. This fluctuates with the synodical rhythm and is greatest for the yellow light at Full Moon and lowest at New Moon (Figure 4). As early as the 1940s, similar experiments in people produced the same phenomenon for orange light. It may be assumed that this monthly rhythm of light sensitivity is spread throughout vertebrates.
   Next, let us examine how the rhythms of natural light change geographically across the Earth. In the tropics the brief sunsets and sunrises are all the more dramatic for their shortness. What can take half an hour or longer at middle latitudes depending on the weather — the gradual transition from the dark of night to daylight — takes no more than ten minutes at the equator. The blackness of night is replaced with blazing brightness at breathtaking speed. The interplay of colours at dawn and dusk is compressed into a dramatic light display of all colours and shades. Day and night, like their transition periods, are much more sharply accentuated than at middle latitudes. Thus the rhythm of day and night is the dominant rhythm for the whole of the living world there.
   The situation is different with regard to the course of the year. Each day and night are equally long at the equator. The lengthening and shortening of the day throughout the year do not occur. There the yearly rhythm hardly becomes evident; its monotony is only relieved slightly by the alternation of wet and dry seasons. Yet it is quite normal in the tropics to see all the *phases of the year* simultaneously displayed in the trees: while one tree is just greening, another is blossoming, yet another has shed its leaves, and a fourth is just bearing fruit. Indeed, the various states of vegetation can even be seen on the various branches of one and the same tree (Plate 1).
   In place of the seasons, which are so familiar in middle latitudes, it is the Moon which determines the structure of the year in the tropics. In the equatorial regions of the Earth it rises to its highest position in the celestial sphere like the Sun during the day and therefore has a

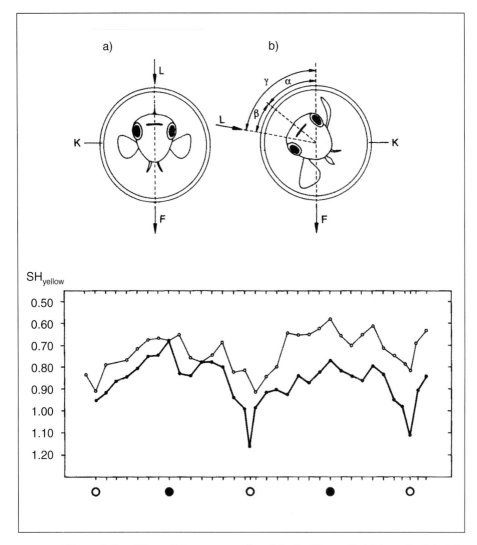

Figure 4. Top. *The well-known aquarium fish, the Guppy* (Lebistes reticulates) *in a glass test tube (K). If light (L) comes from above (a), it turns its back to face upwards. If it comes at angle γ from the side (b), it turns its back by angle α to the side. Its plane of symmetry forms a resultant between light and gravity effect (F), thus precisely indicating its light sensitivity.*

Bottom: The stimulus effect of yellow light can be compared with that of white light of the same intensity by forming the ratio from the reaction to both light sources (spectral brightness stimulation ratio) (SH_{yellow}). Lower curve: male; upper curve: female. The ratio in both animals changes synchronously with the Moon phases, being highest at Full Moon (white circle) and lowest at New Moon (black circle).

Both animals only experienced 12-hour artificial daylight and had never seen natural daylight or moonlight [1999].

much stronger impact at night, particularly at Full Moon, than in middle latitudes. It is not surprising, therefore, that life processes in a number of tropical animals are determined by synodical rhythms. There are, for example, two toads (*Bufo fowleri* and *Bufo melanostictus.* See Plate 16), and a species of frog (*Rana cancrivora*) in Java, which lay their eggs in rhythm with the phases of the Moon. Thus the eggs of the black-spined toad (*Bufo melanostictus*) mature during the waxing Moon, that is during the first and second quarter (increasing Half Moon and Full Moon).

In the animals of the far north, we encounter life cycles which run in parallel with another motion of the Moon. In polar regions the daily rhythm turns into the yearly rhythm — it is day for half the year and night for half the year. (We will explain the astronomical reasons for this in the brief introduction to the Moon in the next chapter.) In northern countries the interrelationship between the synodic Moon and the yearly rhythm plays a significant role in some animals' reproduction. Thus various birds and mammals increase particularly strongly every three to four or ten years respectively. The most well-known animal in this respect is the Norway lemming (*Lemmus lemmus*), a type of large spotted burrowing mouse of the Norwegian fjell landscape. Something similar is found in the snowshoe hare (*Lepus americanus*), the waxwing (*Bombycilla garullus*) and in the capercaillie (*Tetrao urogallus*) and black grouse (*Lyrurus tetrix*) [9, 10, 11, compare also 12].

In the snowshoe hare in Canada, for example, strong population growth occurs at intervals of seven to twelve years with a mean value of 9.6 years. Siivonen and Koskimies [11] link this change in the rate of population growth to the date of the Full Moon, compared with a reference spring date. In Figure 5, the interval of the Full Moon in days is listed over one hundred years, with April 6 as the original reference date. Given that a specific lunar phase must be repeated with the precision of plus/minus one day, a 9.6 year rhythm becomes evident which coincides with the increased reproductive rate of the snowshoe hare, also shown. This is a semi-Metonic cycle; an overall rhythm in which the same lunar phase recurs precisely on the same calendar date. It is assumed that a specific lunar phase at the time of a particular period in the spring, which is determined by sunlight, represents a necessary condition for the full reproductive success of those species with cyclical reproduction. The further the Full Moon is distant from this date, the lower the reproductive rate. On the basis of the fur sales reports of the Hudson Bay Company, conclusions can be drawn about

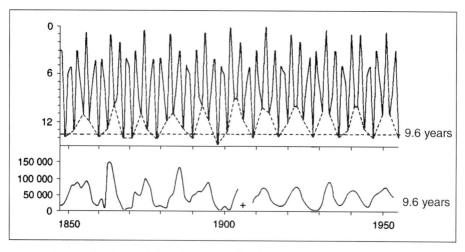

Figure 5. The north American snowshoe hare (Lepus americanus) *displays a 9.6 year cycle of population growth. Upper spiked curve: interval between dates of the Full Moon from April 6 (0 to 12 days) over 100 years. Dotted curve: given that a specific Moon phase must be repeated with the precision of ±1 day, the peak values fluctuate in a 9.6 year rhythm. Bottom curve: the population numbers of this northern species of hare fluctuate in this semi-Metonic cycle [11].*

the population density of other animals as well for a period covering many decades. Thus the statistical analysis also produced 9.6 year rhythms for the following species: coyote, red fox, lynx, pine marten and fishers, mink, muskrat, skunk, wolf, wolverine and Atlantic salmon [13]. However, since the snowshoe hare is an important prey of the lynx or coyote, it is assumed that population fluctuations in these two species and — by secondary effect — also in other animals is due to the differing frequency in population of the snowshoe hare. Nevertheless, other causes such as climatic factors must also be taken into account. Thus a cyclical pattern of about 18 to 19 years can be identified for summer weather (July, August) in the two American states of Illinois and Iowa [14, 15, 1091, 1092, see also 1079]. This corresponds to the orbit of the lunar nodes *(see* Chapter 1).

The population maximum of the waxwing (*Bombycilla garullus*) corresponds to the full Metonic cycle of 18 to 20 years. Given that a specific lunar phase will be repeated with a precision of four days with respect to a spring reference date, the reproductive cycles of species with three to four-year population periodicity can be aligned with the Metonic cycle of the Moon in a similar fashion to the snowshoe hare. (Nevertheless, here too there is a wealth of other explanatory patterns; for the cyclical reproductive rate of Norway lemming compare [1080].)

INTRODUCTION

The evidence relating to the movement of the Moon in organisms with such long-term rhythms is small. Scientific studies focus mostly on daily and monthly lunar influences, above all in the area of coastal tides, as the tides represent the most direct effect on the biosphere of the Moon's position in relation to the Earth and the Sun. Numerous plants and animals on our coasts and in the adjoining shallow seas form part of it. Seaweed, coral, worms, crabs, snails, mussels and fish display their lunar periodicity specifically in their reproductive rhythm.

Specialist terms and issues

The time structure of organisms is studied by CHRONOBIOLOGY. This includes all rhythmical manifestations of the life processes which, apart from the lunar rhythms, also include daily and annual rhythms as well as hourly and minute rhythms — for example, in the digestive system — or even higher frequencies up to the millisecond range which, for example, govern our nervous system. Chronobiology has undergone rapid development in the last five decades and has expanded in several directions. While CHRONOPHYSIOLOGY describes and analyses the structures of rhythmical processes of the organism's various functions and its interrelationship with environmental processes at a physiological level, CHRONOPHARMACOLOGY investigates the chronobiological effect of medicines. Then further differentiation is made in modern medicine between CHRONOPATHOLOGY, CHRONOTOXICOLOGY, CHRONOTHERAPY and CHRONOHYGIENE. Including the human being, we will investigate biological lunar rhythms in the whole of nature. Such rhythms can be divided as follows:

Period (duration of cycle)	Nomenclature
12.4 hours	tidal, semi-diurnal (lunar) (sometimes semi-lundian)
24.0 hours	diurnal or daily rhythm, (solar) diurnal (sometimes dian)
24.8 hours	lundian, (lunar) diurnal
14.75 days	semi-lunar or syzygian lunar
29.5 days	lunar, synodic or monthly rhythm (sometimes trigintan)
1 year	annual, yearly rhythm
6939.9 days = 9 solar years	Metonic cycle

The SYNODIC LUNAR rhythm is the time that the Moon takes to travel around the Earth until it returns to the same place in relation to the Sun; for example from one New Moon to the next New Moon. After 29.5 days the Moon has reached the same phase again. In life processes or biological processes, we can therefore refer to *synodic* or *lunar* rhythms or *monthly* rhythms. In some academic papers the word *trigintan* is also used, based on the Latin word for thirty (*triginta*), with the number thirty referring to the days of one month. On one hand, this frequently describes all rhythms related to the Moon, and on the other describes monthly rhythms in a narrower sense. Here we use the term *lunar* as set out in the table above, that is exclusively for rhythms with a period of about thirty days. The totality of all rhythms (tidal, lundian, semi-lunar, lunar, Metonic) are referred to as RHYTHMS OF THE MOON, LUNAR RHYTHMS or LUNAR PERIODICITY. In general, we will use the terms *rhythm* and *period* (or *periodicity*) for the more or less regular repetition, of living phenomena in organisms and of specific events within so-called *abiotic* systems. In processes with half the cycle period of the lunar rhythm we speak of SEMI-LUNAR rhythms. Semi-lunar and lunar phases of the Moon involve movements of the Moon and Sun (from a geocentric perspective), and should, therefore, strictly be called SOLUNAR or SOL-SEMI-LUNAR or (SEMI) SOLUNAR as the American F.A. Brown has already suggested; however, we will continue to use the commonly used term *lunar*.

We refer to TIDAL rhythms when the period of the rhythms observed in organisms is equal to the period from one high tide to the next. For the North Sea coast and most Atlantic Ocean coasts the time required for this is 12.4 hours, so that there are slightly less than two tides per day for one solar day of 24 hours. Since 24.8 hours pass from one moonrise to the next moonrise, two tides thus occur in the specified maritime regions in one lunar day. With rhythms whose period is equal to that of the lunar day, we speak of LUNDIAN rhythms.

Our summary of the organisms listed in the species catalogue by related groups shows that rhythms related to lunar periods were found in all larger groups of the plant and animal realms. Equally prevalent is geographical spread; in all regions of the world we encounter organisms whose vital processes run in parallel with the rhythms of the Moon.

If we observe organisms in the context of their environment, we see that the organisms' biological rhythms often take place in harmony with rhythmical environmental processes. The temporal order of organisms is embedded in the temporal order of their environment. That, at least, is how it appears in most cases. When seeking causes and reasons, the question then arises; in what way are they reciprocally dependent? Experimental research therefore removes the organisms from their familiar habitat into the laboratory in order to study the independent stability of their biological rhythms under artificially constant environmental conditions. It emerged that the studied organisms frequently maintained their rhythms under these conditions too, but occasionally it was also shown that the biological rhythms ended with the absence of the environmental temporal order. Chapter 5 is devoted to the causes of biological rhythms. If in the descriptive chapters the sample species are generally observed in their native environment, we do on occasion also include the research results relating to plants and animals under constant conditions.

CHAPTER 1

The Moon

The Moon is the Earth's closest neighbour and yet its appearance is clearly subordinate to the Sun as a source of light in its own right. The Moon's light is part of the light of the Sun just as the light of the Earth is, but on Earth sunlight is transformed by the atmospheric mantle of the Earth, thus supporting life. The ozone layer filters out a large part of harmful cosmic radiation. When the Moon is positioned *behind* the Earth and opposite the Sun, the whole of its surface is lit by the Sun, and we see the Full Moon at night. Two weeks earlier or later the Moon is positioned between Sun and Earth, and the sunlit side of the Earth reflects its light on the night side of the Moon lying in shadow. However, we do not see the New Moon because it stands near the Sun in the daytime sky, and the light of the Earth throws too little light on it. The dark disc of the Moon, with its silvery grey shimmer, is only seen when the Moon is still near its New Moon position but is already distant enough from the Sun that it can be seen above the horizon when the Sun is below the horizon: in other words, after sunset in the western night sky after New Moon, or before sunrise in the eastern sky before New Moon. The shadowy disc of the Moon within the brightly illuminated Sickle Moon is always an impressive sight on a clear night. The dark part of the Moon becomes less and less visible the more the Moon moves out of the sunlit half of the Earth; that is, the further it moves away from the New Moon position, waxing through the Half Moon to Full Moon.

In its orbit around the Earth, the Moon shows us a different appearance each day. These changes are called LUNAR PHASES. The sequence of lunar phases from the New Moon via the waxing Half Moon (first quarter), Full Moon (second quarter) and the waning Half Moon (third quarter) to the next New Moon is called one LUNATION. The Moon lent its name to this period: one month. The lunar phases too show us nothing more than the constantly changing positions of Sun, Earth and Moon. Their coming together (Greek *synodos* = session) also gives lunation its other name of SYNODIC MONTH. Its mean value is 29 days,

12 hours, 44 minutes and 2.9 seconds, though the period fluctuates by up to 7 hours. Since the synodic month of about 29.5 days is slightly shorter than the mean calendar month, slightly more than 12 lunations — namely 12.4 — fit into one year. Thus the same date of the following year always falls into a different lunar phase from the previous year.

The Sun visibly changes its daily trajectory over the course of the year. It is higher at midday in the summer than in the winter, and at the equinox it is in the middle position at the celestial equator. The New Moon, being closest to the Sun, follows it in this respect. At Full Moon, in contrast, when it is exactly opposite the Sun, the reverse applies: Full Moon in summer is as low as the Sun in winter; the winter Full Moons are positioned as high as the Sun in summer. This remains the same from year to year.

The daily paths of all the celestial bodies — Sun, Moon, planets, fixed stars, star nebulae and comets — (in the northern hemisphere) from the eastern horizon to their apex in the southern sky *(upper culmination),* down to the western horizon, and on to their lowest position below the northern horizon *(lower culmination),* is the visible result of the Earth's rotation, which we do not perceive because we are engaged in the same motion. If this motion did not exist, the fixed stars in the sky would be at rest, but against their background we would be able to observe the slower counter movement of Sun and Moon in relation to the alternation of day and night. This movement means that they gradually change their position day by day in relation to the fixed stars from east to west. The trajectory of the Sun through the whole celestial orbit takes one year. The time it takes to return to its starting point in relation to the stars is called the SIDEREAL YEAR (Latin *sidus* = star) and comprises 365 days, 6 hours, 9 minutes and 10.7 seconds. Seen from outside, we can also think of this movement as the trajectory of the Earth around the Sun. We see the Moon moving much more rapidly. Its trajectory around the Earth against the background of the stars is 27 days, 7 hours, 43 minutes and 11.5 seconds. Seen from our perspective, this period represents the movement from one star through the whole orbit back to the original star and is thus called a SIDEREAL MONTH. Since the Sun has slowly moved on after one calendar month by about $1/_{12.4}$ of its orbit of the heavens, the Moon reaches its next identical phase (the next New Moon in proximity to the Sun for example) a little later than it reaches the same stellar location, which is why its synodic month is a good 2 days and 5 hours longer than its sidereal one. This means that the (sidereal solar) year only contains 12.4 synodic months but 13.4 sidereal lunar months. The difference between the synodic and sidereal month means that the lunar phases always fall into the subsequent zodiac constellation on their return.

If two celestial bodies reach their closest proximity from our perspective, this 'connection' is called their CONJUNCTION. Their furthest distance apart at opposing positions in the celestial sphere is called OPPOSITION. When there is conjunction between the Sun and Moon we have a New Moon. When both celestial bodies are in opposition (180° apart), we have a Full Moon. When Sun, Moon and Earth are aligned, that is either conjunction and opposition, the configuration is called SYZYGY (Greek *syzygos* = linked, pair). If two celestial bodies are at right angles to one another from our perspective, this is called their QUADRATURE. The Moon in quadrature to the Sun is either a waxing or waning Half Moon (first or third quarter).

How do lunar eclipses occur?

Anyone who has a clear spatial sense will have already wondered why at New Moon each month the Moon does not move in front of the Sun with the precision to produce an eclipse of the Sun. Equally, with an exact monthly opposition of 180° the Earth would have to be positioned on the line between the Full Moon and the Sun causing a lunar eclipse with its shadow. It is self-evident that a solar eclipse can only occur at New Moon and a lunar eclipse at Full Moon. Yet the majority of syzygies occur without eclipses. Clearly the Moon is able to pass above or below the axis, which connects the Earth and the Sun. That would not be possible if the orbital plane of the Moon was the same as the plane on which the Earth moves around the Sun. The orbital planes are inclined to one another at the slight angle of 5° 9'. (The symbol ' represents one minute of arc, a sixtieth of one degree.) The one half of the Moon's orbital plane lies below, the other lies above the orbital plane of the Earth. The orbital plane of the Earth is called the ECLIPTIC and is the same as the Sun's apparent annual path. The line of the ecliptic passes through a band of fixed stars known as the zodiac. The Moon will thus travel through the constellations of the zodiac each month for some time slightly above the ecliptic and for a time slightly below it. At the changeover it is positioned exactly on the ecliptic, in other words, where the orbital planes of the Earth and Moon intersect. These intersections are called LUNAR NODES, and depending on the direction in which the Moon is moving they are referred to as *ascending* (from south to north) and *descending* (from north to south) lunar nodes. Twice a month the Moon is positioned on one of the nodes and thus in the same plane as the Sun. If the Full Moon occurs exactly at one of the nodes, we have a total eclipse of the Moon, if the New Moon occurs at one of the nodes then a total eclipse of the Sun can be seen somewhere on Earth. If the syzygies do not occur directly on, but only close to the node, we only have a partial eclipse.

The position of the two lunar nodes against the background of the stars is not fixed, but changes gradually. A full (sidereal) revolution of a node takes 18.6 years (exactly: 18.5997 years = 18 years, 218 days, 21 hours and 22 minutes, which is 6793.39 days). This slow shift of the Moon's orbit also has a weak effect on the position of the Earth's axis so that the latter produces a slight wobble over the period of 18.6 years. That is why this rhythm is called NUTATION PERIOD (Latin *nutare* = to nod). Since the lunar nodes regress in a westerly direction — counter to the monthly orbit of the Moon — the Moon reaches the same node a little sooner than the same sidereal location. This orbit of the Moon from one ascending node via the descending one to the next ascending one is the DRACONITIC (or NODAL) MONTH which lasts for 27 days, 5 hours, 5 minutes and 35.8 seconds.

When the sphere of the Sun or the Full Moon suddenly turns dark in a cloudless sky, both human beings and animals are overcome with a feeling of something abnormal or sinister. In earlier times people referred to the Fenris wolf which gobbled up the Sun or the Moon, or to the dragon which interposed itself between the Earth and the celestial bodies. An impressive description of such a mood can be found in the German writer Adalbert Stifter's account of the solar eclipse on July 8,1842 [16]. The lunar nodes as locations for possible eclipses were considered to be locations of the dragon. Even today they are still depicted using the snaking symbol ☊ (ascending node = head of dragon) and ☋ (descending node = tail of dragon). Just as the Moon passes through the two lunar nodes once every month, the Sun passes through each lunar node once every year. That is why eclipses only take place twice a year, approximately six months apart. A lunar eclipse can always be seen from all locations on the night side of the Earth (for then the Full Moon is above the horizon), while a solar eclipse can only be seen on the day-side of the Earth in the narrow zone where the shadow of the Moon falls. The zone can be up to 250 km wide, but is usually narrower.

With the western shift of the lunar nodes, the times of the lunar eclipse gradually move forwards through the year. Only after the passage of about 6585.3 days, or 18 years and 10 days, does the Moon return to the same position in relation to the Sun, Earth and node so that solar and lunar eclipses are repeated at this interval. This SAROS CYCLE comprises 233 synodic (6585.32 days) and 242 draconitic (6585.36 days) months.

There is a further peculiarity associated with eclipses which reveals another lunar rhythm. It is a remarkable fact that generally the Full Moon appears to be as large as the Sun's disc. The distances of the Sun and the Moon from the Earth are in such a relationship with regard to their diameters that the smaller celestial body makes up with its proximity what the larger loses through its distance. Thus the disc of the Moon covers the disc of the Sun completely during an exact eclipse of the Sun (total eclipse). However, even when all three celestial bodies are positioned exactly along an axis, the eclipse is not always a total one. As with all celestial events, this too has its rhythmical variations. The Sun can, on occasion, appear larger than the Moon, causing an ANNULAR (Latin *annulus* = ring) rather than a total eclipse to occur. In this event the dark disc of the Moon is surrounded by a shining solar ring. Since the Sun and the Moon have not changed size, it must be their distances from the Earth which fluctuate. This shows us that the Earth does not revolve around the Sun in an exact circle but in an almost circular ellipse. The Sun is positioned in one of its *foci* (first planetary law established by Johannes Kepler in 1605). Thus there is a lesser distance when the Earth is in proximity to the Sun *(perihelion)* and half a year later a greater distance when it is farthest away *(aphelion)*. The apparent diameter of the Sun fluctuates between 31.5' and 32.5'. In the same way, the Moon also orbits the Earth in an almost circular ellipse and is close *(perigee)* or more distant *(apogee)* from the Earth every fourteen days, and its apparent diameter varies between 29.5' and 33.5'. If the New Moon is at a node when the Moon is distant from the Earth, and the Sun is close to the Earth, an annular eclipse of the Sun with the widest ring will occur — a relatively rare event. The monthly rhythm of the Moon from perigee to perigee is called an ANOMALISTIC or APSIDAL MONTH and lasts for 27 days, 13 hours, 18 minutes and 33.1 seconds. This means that it is close in terms of its duration to the sidereal month.

The position of the ellipse of the Earth's trajectory may influence the climate over major ages of the Earth. The greatest solar proximity, the PERIHELION, currently occurs when the Sun is positioned in the constellation of Sagittarius on January 2. The disc of the Sun, which looks 1' larger compared to its diameter on July 4 when it reaches the aphelion of Gemini, also shines more strongly on us and very slightly alleviates the coldness of winter which rules in the northern hemisphere. In addition, in the half year when the Sun is distant from the Earth, the latter moves more slowly through its orbit which means that our summer is slightly longer than the winter. If the perihelion and aphelion were moved by half a year, the temperature decreases would be greater and cause the Earth's surface to cool down more; here the great land masses of the northern hemisphere do not possess the moderating influence of the water masses of the southern hemisphere. The imaginary line between perihelion and aphelion, called LINE OF APSIDES, move over great periods of time. Their full rotation takes 21,000 years. Then there are also further effects from neighbouring planets on the Earth's trajectory. The long-term climate fluctuations of the ice ages and intermediate periods are presumed to be a consequence of this. (Milankovich cycles) [17].

The Moon in the sky

Let us now return to a direct observation of the Moon. If we allow ourselves to experience the direct impression which this satellite makes on us, spontaneously and unencumbered by any knowledge, we experience its peculiar double character. On a night when there is Full Moon, it covers the landscape in a soft, indeterminate twilight with a mild, silky light, which leaves the brighter stars visible. Something without contours, dream-like emanates from it which takes away the harshness of day:

> Filling bush and valley once again
> Silently with shimmering mist
> All my soul is finally released.

Thus the Weimar poet, Goethe, captured this mood in the valley of the River Ilm. The structure of the eye is such that the details of the Moon and the nocturnal landscape become indeterminate in the weak light of the Moon. The elements which allow for sharp and colour vision (the *cones*) are hardly used. We see the surface of the Moon as lighter or darker fields, and with the Half Moon the edge of the shadow as a finely serrated edge, but these irregularities only create some variation in the general indeterminate impression.

However, if we look through a simple pair of binoculars or, indeed, a telescope, a wholly different character appears: covered in numerous ring-like boundaries of hard-edged mountains, the Moon turns into an image of hard, barren desert. In the forsaken sequence of craters we see the mountainous areas of the Moon as light areas (Figure 6). The flat areas, criss-crossed by a few ditches and grooves, appear darker and this gives us the impression of the 'man' or the 'hare' in the Moon. These flat areas were also called *mare* (from Latin 'sea') by seventeenth-century astronomers, but there is no water, no clouds, not even an atmosphere. This means that the Moon does not have the soft, changing profile of a surface created by climatic events on Earth.

The surface of the Moon which is visible to us never changes, for the Moon only ever turns one face towards us. Seen from outside, it does have a rotation of its own but since this own rotation coincides with its orbit around the Earth it always shows us the same face (CAPTURED or SYNCHRONOUS ROTATION). What a contrast to what astronauts see when they look at the Earth: shimmering in blue, full of rotating cloud fields in their captivating atmospheric diversity.

The impression of the Moon as lifeless and rigid can be gained best from the polar regions of the Earth without a telescope. In the Polar night, when the Sun — and also the New Moon — is positioned below the horizon for six months, the Half Moon initially appears above the horizon at the time of the winter solstice. One week later the Full Moon describes a circular path, remaining at the same height in all directions. spirals in directionless uniformity. During this time the Full Moon is visible to the human eye in the sharpest possible clarity and contours. The lowest atmospheric layer, the troposphere, is only eight kilometres high in the Polar regions. The cold and consequent heavy atmospheric masses create an almost permanent high pressure with the clearest weather. The air has extremely low humidity and dust levels. Here we see nothing of an idyllic character; the Moon displays its forsaken character with maximum clarity.

In the tropics, the silvery-mild, velvety impression of the Moon predominates. Here the troposphere is not only heavy with humidity, but is also at sixteen kilometres at its thickest, so that the contours of the Moon become diffuse even when the weather is clear. The Moon rises rapidly and vertically in the east, and at midnight the Full Moon regularly spreads its velvety dome of light from the zenith over the landscape of palms, the nocturnal song of the cicadas and swarming fireflies. However, all the other phases up to, and including, the silver sickle (shortly before or after New Moon), belong to the most impressive images of our closest celestial body carrying the dark disc of the Moon horizontally like a boat and shimmering in the ash-grey light from the Earth.

Other lunar rhythms

The hours of daylight determine the four seasons in the middle latitudes. At the spring and autumn equinoxes the days and nights are equally long. At midsummer the Sun gives the greatest amount of light, at the time of the winter solstice it traverses its lowest orbital plane across the horizon. Sunrise at the spring equinox takes place exactly in the east —that is, at the location in the sky where celestial equator and ecliptic intercept one another. The time which passes until the Sun reaches the same position in the spring of the following year is slightly shorter by about twenty minutes (365 days, 5 hours, 48 minutes and 46 seconds) than the sidereal year. The shorter TROPICAL YEAR in comparison to the sidereal year makes us aware that — just as the orbital plane of the Moon in relation to the ecliptic — the intersection of the ecliptic with the celestial equator is not fixed but travels slowly counter to the annual orbit of the Sun. A full orbit of the vernal equinox through the circle of constellations takes about 26,000 years. This period is called a PLATONIC or COSMIC YEAR. In our age the vernal equinox lies in the constellation of Pisces.

The rise and fall as observed in the Sun during the course of the year takes place with the Moon in about 27 days. The monthly course of the Moon from the vernal equinox to the same point again, or the change from one highest position until it reaches its highest position again, is described as a TROPICAL MONTH. Since the vernal equinox is a little closer to the Moon after one incomplete sidereal orbit, the tropical month at 27 days, 7 hours, 43 minutes and 4.7 seconds is shorter by

Figure 6. The lunar surface through a telescope in the third quarter as can be seen a week after Full Moon in the latter half of the night. North is at the top. Where the Sun's rays strike vertically (as here on the left edge), the surface appears without shadow and flat. Where they strike horizontally as on the border between lunar night and day (at the right), the lunar mountains appear in sculpted clarity. Since the time of Galileo the light parts of the Moon have been called terrae *(= continents), covered in craters, while the* mare *(= seas) are dark areas made up of waterless dust deserts. At the top right of the picture is the large Mare Imbrium, and to the south of it the equally unmistakeable crater Copernicus, which has white rays reaching out from it in all directions for hundreds of kilometres [1083].*

seven seconds than the sidereal lunar month. At midsummer, the New Moon, together with the Sun, has its highest — in northern latitudes, its northernmost — daily path: at the winter solstice it is the Full Moon; in spring it is the waxing Half Moon; and in the autumn the waning Half Moon. During the opposite phase the Moon describes its lowest arc over the horizon in the relevant season. If the Moon is positioned above the celestial equator, its daily trajectory — like that of the Sun in summer, is greater over the horizon than under it. If it is positioned below the celestial equator its daily arc, like that of the winter Sun, is greater under the horizon than over it.

The height of the daily arcs of the lunar arcs thus depends on its position in the zodiac. For about fourteen days it spirals upwards in its daily cycle from the constellation of Sagittarius through Pisces to Gemini where it orbits at the highest level. From the constellation of Gemini via Virgo to Sagittarius it then spirals downwards again over the next two weeks until it reaches its lowest daily arc. The rising and setting points of the Moon are also influenced by its position in the zodiac. Like all celestial bodies, the Moon moves each day from east to west due to the rotation of the Earth, but its counter monthly movement, from west to east, means that it rises later each day on the eastern horizon than on the previous day. It rises on average fifty minutes later each day than on the previous day, and is only delayed by exactly this amount only at the times that it has reached its highest and lowest position in the constellation of Pisces and Sagittarius respectively, when it moves almost in parallel to the celestial equator, and rises and sets fifty minutes later compared to the previous day. However, if the Moon is positioned in the constellation of Pisces (that is at the vernal equinox), its trajectory intersects that of the celestial equator at a vertical angle in middle northern latitudes. It then rises only half an hour later than on the previous day, and we have to wait for it to set almost ninety minutes later than on the day before. Exactly the opposite happens when the Moon has reached the constellation of Virgo. If the monthly orbit of the Moon coincided with the celestial equator itself, the daily delay of fifty minutes would remain the same, but since it moves forward along the line of the zodiac, which is inclined to the celestial equator at an angle of 23.5°, the visible daily arcs grow larger or smaller in the middle latitudes, and the location at which the Moon rises and sets also changes. In the northern hemisphere while the Moon's arcs grow higher, the rising point shifts from south-east via east to north-east, and the setting point moves from south-west via west to north-west. The interval between successive Moonrises decreases, while the interval of Moonsets increases until it reaches the constellation of Pisces. These intervals equalize again when the Moon reaches its highest point in the constellation of Gemini, and then they show the opposite pattern as the Moon moves towards the constellation of Virgo. In the southern hemisphere the whole pattern is reversed.

The highest point in the celestial sphere is called the ZENITH, the lowest point, invisible below the Earth's horizon, the NADIR. Both terms, like many in ancient astronomy, come from Arabic. Between these two *poles* the HORIZON forms a balance, a kind of 'equator.' Each point on Earth has its own zenith, nadir and horizon. At the Earth's poles the line of the horizon is identical with the celestial equator. At the Earth's equator it runs vertically in relation to the latter. The celestial equator runs through zenith and nadir and intersects, as everywhere else on Earth, with the horizon in its eastern and western points. (This does not apply to the poles, for here, as we have already mentioned, the line of the horizon and the celestial equator coincide.) At the equinox, the Sun is positioned on the celestial equator. At the Earth's equator it thus passes exactly through the zenith at midday. Since the trajectory of the Sun, the ecliptic, is inclined against the celestial equator by 23°, the Sun can reach its midday position in the zenith also up to 23° latitude in the north and south. The area between these fields on Earth is the tropics (Greek *tropein* = turning). In Africa, for example, the Sun still reaches the zenith in Aswan in southern Egypt, and in the Kalahari Desert in Botswana.

Since the orbit of the Moon is inclined by 5° against the plane of the ecliptic, the Moon can shift its own *tropics* from 23° – 5° = 18° to 23° + 5° = 28° of geographical latitude, both in the northern and southern hemisphere of the Earth during the regression of the lunar nodes, in a rhythm of 18.6 years. The highest position can be 5° higher than the highest position of the Sun. This means that in extreme cases the Moon can shift its zenith orbital plane beyond the tropics of the Sun as far as Florida, the Canary Isles or Delhi in the north, and correspondingly far in the south to southern Brazil, Natal or Brisbane. There the Moon can be seen directly above the observer every 18.6 years — an impressive spectacle at Full Moon in winter at midnight. In the Arctic the Moon can become circumpolar, that is not setting, from as low a latitude 61° (90° – 28°). As far south as anchorage in Alaska or Tampere in southern Finland, the Sun in the south and Full Moon in the north can both be visible at the same time in the sky at midday in December. While the midnight Sun can be seen in June every year north of the Arctic circle (66° latitude, 90° – 23°) the midday December Moon can only be seen for a few years during the 18.6-year rhythm of the lunar node.

The duration of the revolution of the lunar nodes (nutation period) must not be confused with two other similarly long Moon cycles: the SAROS PERIOD, which lasts a good eighteen years and the METONIC CYCLE which lasts almost exactly nineteen years. Not only do the synodic and draconitic lunar months fit very closely into the period of the Saros cycle, but nineteen nodal solar revolutions of 346.62 days (from lunar node to same lunar node) are also completed in this time. The Metonic cycle lasts almost exactly nineteen years and consists of the whole number coincidence of 235 synodic and 254 sidereal months, which are just 2 hours, 4.5 minutes short of exactly nineteen solar years. After nineteen years the Moon is thus positioned in the same phase in the same stellar location on the same day of the year, but regardless of the position of the lunar nodes.

The exact values for the three long-term lunar rhythms are:

Saros cycle: 6585.32 days = 18 years 10 days 8 hours
Nutation cycle: 6793.39 days = 18 years 218 days 21 hours 22 minutes
Metonic cycle: 6939.78 days = 19 years (less 2 hours 4.5 minutes

In conclusion, we will list here too once more the average periods of the five monthly lunar rhythms we have dealt with:

Synodic month:	$29^d\ 12^h\ 44^m\ 2.9^s$	(29.53059 days)
Anomalistic month:	$27^d\ 13^h\ 18^m\ 33.1^s$	(27.55455 days)
Sidereal month:	$27^d\ 7^h\ 43^m\ 11.5^s$	(27.32166 days)
Tropical month:	$27^d\ 7^h\ 43^m\ 4.7^s$	(27.32158 days)
Draconitic month:	$27^d\ 5^h\ 5^m\ 35.8^s$	(27.21222 days)

CHAPTER 2

The Tides

High and low tides across the Earth

The unique character of the Earth in relation to the Moon and planets is demonstrated particularly by its oceans. Of all the celestial bodies in our solar system it is the only one whose surface is covered with water, which surrounds the continents of the Earth. More than two thirds of our planet are covered by three large oceans and their peripheral seas. Therefore, our water-rich atmosphere makes the Earth shine into the cosmos like a blue star. Life on Earth would not exist without the hydrosphere. After all, it is water in particular which has the capacity to mediate between extremes, a feature of all life. What is firm easily falls into extremes. Continental summers are very hot and continental winters extremely cold. These influences are toned down in those regions to which the influence of the sea extends. The air currents coming off the sea warm up the land in winter and provide cooling in summer. Yet an Earth covered only by water would represent just as much of an extreme. We experience the interaction of the elements more intensely where they touch and interpenetrate directly: on the coast, that active fringe 'where the land ends and the sea begins' [18]. The frontiers of freshwater and salt water, coastal land and substantial air currents, periodic flooding of dry land by the sea *(transgression)* and drought *(regression),* are in constant transformation. In this transitional landscape in particular, decisive stages of evolutionary development of organisms have taken place. Jablonski and Bottjer [19] were able to show that most of the main groups of marine invertebrates have their origins in the shallow zones of coastal waters. It was from here that the high and deep seas were colonized. Schad [20] noted that the coastal areas of brackish water were also the predominant biotope for the macro-evolution of vertebrates. This is where the development of deep sea fish and higher land animals had its origin. *Archaeopteryx* for example, the famous 'archetypal bird' from the Jurassic deposits of the Frankish Alb was the inhabitant of such a transition zone. The evolutionary evidence of feather impressions contained in the fossil record show it to be a bird, as much as other characteristics show it to be a reptile.

The tidal movements of the sea are a noticeable element in the character of coastal zones. The constant change in the rise and fall of the tides, the eternal to and fro of the water on the beach, the constantly regenerating power of the breaking waves, their unpredictable absence and renewed appearance are evidence of an inexhaustible dynamic which appears to have its origins in the oceans.

These movements are not, of course, equally pronounced in all oceans. Caesar was surprised by the great rise and fall of sea level in the English Channel. Such regular differences in water level were unknown to him from the Mediterranean. For us, too, it is always a surprise when we stand on the quay in a port on the Channel Coast and see the water deep below us. It is as if the water had been pumped out of the harbour basin. If we had enough time, we could observe how it gradually rises. It takes more than six hours for the water to reach its highest level and the harbour basin to fill up. Then the water begins to fall again it reaches its lowest point after approximately 6.5 hours, and has thus run through a full cycle of rising and falling.

If we spend several days on the Channel coast or by the North Sea we soon notice that the water does not reach the same place at the same time as before but a little later. Longer-term observation of the water level shows that the period from one rising tide to the next takes on average about 12 hours, 25 minutes. From one day to the next, then, the water is delayed after two tides by an average of fifty minutes. This is the same period by which the meridian passage of the Moon is delayed on average. The meridian is the great circle passing through north and south celestial poles and the observer's zenith. The Sun, for example, crosses it at midday (upper culmination), and at midnight invisibly in the north at the exact opposite point (lower culmination).

If the Moon is positioned in the southwest today at high tide for example, high tide recurs on the next day when it is again in the southwest. There is thus a relationship to the extent that high tide at a particular location always occurs at a specific position of the Moon. This observation has led to so-called *port time,* specified for different ports (sometimes also called *lagging of tides),* and is the time by which the tide follows the passage of the Moon through the meridian at Full or New Moon. With a few small exceptions, high tide in Heligoland on the North Sea always occurs roughly eleven hours and twenty minutes

2. THE TIDES

after the passage of the Moon through the meridian. The tides thus copy the movement of our closest celestial body in their periods.

If we observe continuously the fluctuating water level over a longer period we see that the twice-daily change of high and low tide is subject to many different variations. For example, see Figure 7: the tidal curve at Immingham (east coast of England) for March 1936. The top scale — from right to left — shows the days for a specified period. Directly below the phases and positions of the Moon are shown which are explained underneath the diagram. Q is the moment at which the Moon crosses the celestial equator. In equatorial regions of the Earth, the Moon culminates in the zenith at this time. At its greatest northern or southern declination (divergence from the celestial equator), it reaches its tropic. The scale on the left side of the graph shows water level in metres, (the method used is such that negative numbers are avoided).

We see not only that the water falls and rises approximately twice a day, but also that this motion has a longer-term pattern superimposed on it. The difference between high tide and low tide is particularly great for a time, then becomes less, reaches a minimum, and then begins to grow again. The difference of the water level between high tide and the previous low tide is described as RISE OF TIDE and, conversely, between low tide and previous high tide, as FALL OF TIDE. The mean value of rise of tide and fall of tide within a tide is called TIDAL AMPLITUDE or TIDAL RANGE.

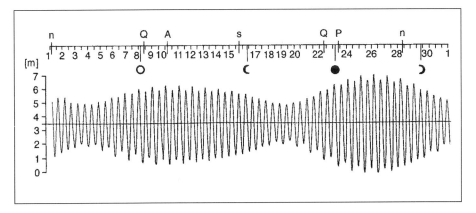

Figure 7. Semi-diurnal tidal forms at Immingham/Grimsby, east coast of England, for March 1936, in metres (m). Full Moon: white circle. New Moon: black circle.
n: greatest northern, s: greatest southern declination of the Moon.
Q Moon crossing equator. A (= apogee): furthest distance from the Earth,
P (= perigee): closest distance to the Earth of the Moon [22].

We can now see that the tidal amplitude varies with a period of fourteen days. Its highest value is called SPRING TIDE, its lowest value NEAP TIDE. Spring tides and neap tides follow one another at approximately weekly intervals. About half a month passes from one spring or neap tide to the next. This is called the half-monthly inequality. In most locations the spring tides take place a number of days after Full and New Moon, and the neap tides about the same time after the quadratures. Long-term evaluation of the tidal curve shows that the period from one spring tide to the next equals half a synodic month. The time-lag in relation to Full and New Moon is described as age of tide. In the pattern of spring and neap tides the tidal rhythms also reproduce the periods of the Moon's movement in relation to the Sun.

In the course of the second spring tide around March 25, even greater values are found for the tidal range than for the earlier spring tide around March 10. An exact analysis reveals a further pattern here, the period of which is the same as that of the anomalistic month. When the Moon moves further away from the Earth (apogee) the tidal range is less (again a few days later). The tidal range is greater (again with the same delay) after perigee. If, as Figure 7 makes clear, the Moon's greatest proximity to the Earth falls at New Moon (or Full Moon), the spring tide is particularly high. This is referred to as MONTHLY INEQUALITY.

Careful reading of the graphic in Figure 7 reveals a further inequality. In a comparison of both the high and low water level we can see alternating rises and falls. The one high water during the day rises higher than the other, and the one after that rises higher again than the previous one. The same applies to the low water level. This alternation appears clearly in phases, becomes less clear, and then becomes apparent again after a certain period. If we investigate this phenomenon more closely it becomes evident that the high water, which follows the upper culmination of the Moon, is the greater one for almost two weeks, and then for the same period the one following the lower culmination becomes the greater (DAILY INEQUALITY).

In order to extend the horizon of our observations, we will use a further three graphs *(see* Figure 8). Let us look, first of all, at the tidal curve of San Francisco, at tides as they occur on the Pacific coast of North America. Here the ocean comes and goes in a semi-diurnal cycle, but the daily inequality which we have already discovered for Immingham is more clearly marked. In other words, the level of the two high tides and low tides respectively falling in a day has risen to such an extent, that at certain times one of the two semi-diurnal tides almost disappears. What is apparent here in its initial stages does in fact happen in the sea off the Philippines — a tidal form with a diurnal period.

The tidal curve for Manila shows phases in which only one high tide and low tide in the day remains. Here the water rises slightly longer than twelve hours and then falls for approximately the same period. However, as we have already seen for the tidal curve of San Francisco, there are also times in which there are two tides in a day, as is common on most European coasts. During these times the tidal range is at a minimum.

A further characteristic of Manila is that the half-monthly inequality, which is so pronounced in Immingham, does not occur here. Here too there is a longer-term superimposition of the semi-diurnal and diurnal water level movements but their period is clearly shorter. We can read them if we look at the tidal curve for Do-Son, in Vietnam. On the coasts of this region the semi-diurnal tidal form appears to have been almost completely suppressed in favour of the diurnal one. As we can see, the daily inequality pulsates not in the rhythm of the synodic month but follows the periodicity of the half tropical month. For the tidal range is at its maximum some time after the greatest northern and southern declination of the Moon, (and here *maximum* means that high tide occurs only once a day). The tidal range is at its minimum some time after the Moon crosses the celestial equator. This can be seen particularly clearly in the tidal curve for San Francisco. Since — as we have already seen above — the higher high water level follows approximately two weeks after the upper, and then approximately two weeks after the lower, culmination of the Moon respectively, the daily inequality turns out to have the period of a full tropical month.

There are other inequalities in addition to the ones mentioned so far, but these ones cover the main tidal features. The tidal curves discussed here appear as a sequence of examples in which there is the transition from one to the next, starting with the conditions in Immingham and ending in Do-Son. In Immingham we have two daily high and low tides, in which the high tides follow the meridian transit of the Moon at an almost constant interval, and the spring tides follow the Full and New Moon. This semi-diurnal tidal form occurs throughout the whole of the North Sea, the Atlantic Ocean excluding the Gulf of Mexico, and in parts of the southern Pacific Ocean. On the Californian coast there are also two daily tides, although with great inequalities. The

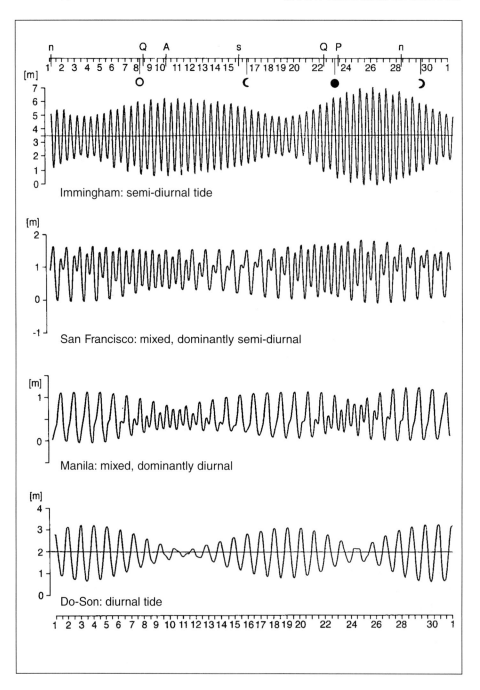

Figure 8. semi-diurnal and diurnal tidal curves also for March 1936 [22]. Key as in Figure 7, p. 37.

maximum inequality occurs at the time of the greatest Moon declination and escalates when the Sun has also reached its greatest declination (= distance from the celestial equator), at the time of the summer and winter solstice. Then one of the two high or low tides may be suppressed completely. This mixed, but dominantly semi-diurnal tidal form predominates in the Pacific Ocean, as well as in Japan and the Arabian Sea. In comparison to the Atlantic and the Pacific, the tides on the coasts of the Indian Ocean take an intermediate position. Semi-durnal and mixed, dominantly semi-diurnal tidal forms are kept in balance here. At maximum declination of the Moon the tidal curve for Manila displays only one high and one low tide per day. Two high tides occur only in the time of the equatorial transit of the Moon. This, too, is a mixed form but a dominantly diurnal form. Other examples of this are the Gulf of Mexico, the Persian Gulf, the Gulf of Thailand and the sea around the Philippines. There is only one high and low tide a day at Do-Son in the Gulf of Tonking. The spring and neap tides here follow the maximum and minimum declination of the Moon respectively. This diurnal tidal form also occurs in the Java Sea and in parts of the Sea of Okhotsk (eastern Siberia). We thus particularly encounter diurnal and mixed, dominantly diurnal tidal forms in the peripheral seas of the Atlantic, Indian and Pacific Oceans.

The causes of tides

What we have so far seen as the rhythm of the tides, repeats in its periods the motions of the Moon in relationship to the Earth and the Sun. This, of course, begs the question of the causal connection between the phenomena of high and low tide and the Moon. At each high tide the water rises and at each low tide it falls again. What forces cause the tidal movements of the oceans? Information about the astronomical, physical, mathematical treatment of this question can be found in the listed literature [21, 22, 23, 24]. We only wish to provide a brief first look at the information provided there. The phenomena of the tides is understood as a consequence of tide-producing forces which can be applied to Earth-Moon and Earth-Sun systems in application of Newton's Law of Gravity. We will look at this relationship only in respect of the Moon, but the same applies to the Sun.

The system of Earth and Moon revolves around the common centre of gravity in such a way that the totality of gravitational forces is balanced by the totality of centrifugal forces. This balance applies to the

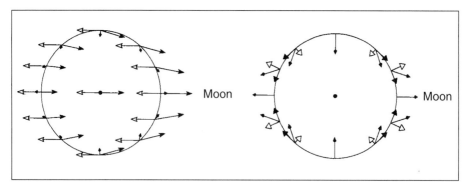

Figure 9. Left: the interaction of the Earth's centrifugal and the Moon's gravitational forces at various locations on Earth. They are balanced at the centre of the Earth. Observe the small resultants as they change across the Earth's surface. Right: the horizontal and vertical components of the tide-producing force resultants across the Earth [23].

totality of the combined masses of Earth and Moon respectively and not to individual points on the surface of the Earth. The centrifugal force is the same for all points on Earth and is applied in the same direction. The gravitational force, on the other hand, depends on the distance of the observation point from the Earth and is directed in each case at the centre of the Moon. The gravitational force is stronger than

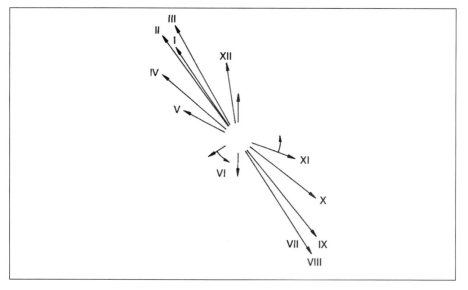

Figure 10. Change of direction and strength of the tidal stream at the Weser light ship at the mouth of the Weser in the course of one tidal cycle measured hourly [1084].

2. *THE TIDES* 43

the centrifugal force on the side of the Earth facing the Moon, while the situation is reversed on the side of the Earth facing away from the Moon. There is thus a residue of forces. These are the tide-producing forces. Figure 9, on the left, shows the resulting forces for some points of the Earth. The tide-producing forces can further the dissected into a horizontal and a vertical component (Figure 9 on right).

The vertical component bears little relevance as a tide-producing force since it only makes up one nine millionth part of gravity, also set at vertical. The horizontal component, however, whose effect runs parallel to the surface of the Earth, is of the same order as other horizontal forces occurring on Earth. Thus the actual cause of the tides is not the vertical *lift* forces but the horizontal *thrust* forces. The primary event of the tides is thus the tidal stream and not the rise of tide.

It is true, of course, that the rise of tide is much more accessible than the tidal stream. For example, the tidal range at spring tide in thee Bay of Fundy, Nova Scotia can be twenty metres or more. In contrast, measurement of the tidal stream requires special measuring instruments, and is made more difficult by other currents which may be caused by such factors as the wind. If these investigations are carried out in mid-ocean, a complicated picture emerges with regard to tidal streams. Owing also to the influence of the Earth's rotation, we find a constant change in the tidal stream depending on amount and direction. Figure 10 shows how the direction of flow and strength of the tidal stream changes in the course of one tide at the mouth of the River Weser in Germany. The tidal stream of one tide changes its direction there in an anti-clockwise direction, rotating in a complete circle, whereby the stream of the incoming and outgoing tide points in opposite directions.

The rise and fall of the ocean water level can be easily observed on the coast, but depth sounding also allows us to measure the rising and falling tide in mid-ocean. The next question is thus about the spatial connection between specific phases of the occurrence of the tides. If a spring tide occurs in a location at a specific time, where else do spring tides and the other phases of the tides occur? Figure 11 shows the results of numerous observations of this kind for the North Sea. The continuous curves are lines which join places where high tide occurs at the same time. These lines are referred to as *cotidal lines*. The figures refer to the number of hours after the Moon's meridian passage at Greenwich. High tide occurs nine hours after the transit of the meridian at all points which are linked by the number 9, for example. In addition, the broken lines show the mean range of spring tide in metres. These lines run roughly at right angles to the cotidal lines.

It is noticeable that the cotidal lines in the North Sea frequently converge on the same point. For example, the cotidal line starting at Edinburgh (two hours) moves towards the same centre as the cotidal lines starting at more southerly points of three hours and four hours, etc. The cotidal lines of two hours, three hours, four hours and so on through zero hours to one hour, rotate around their common centre in an anti-clockwise direction. High tide rotates around the centre in approximately half a day, then a new cycle starts. The centre is called an *amphidromic point* (Greek *amphi* = around; *dromos* = course). All the different phases of a tide are present simultaneously at the various locations around the periphery. Such a system of radiating cotidal lines are called an *amphidromic system.* As the tidal range lines show, the range decreases towards the centre. There are no changes in the centre itself, at the intersection of all lines. If we include the easterly section of the map, we can see how the cotidal line coming from the Atlantic (four hours) divides and enters the North Sea on the one hand via Scotland and, on the other hand, via the English Channel. There is very little to slow down the northern Scottish tidal wave and it enters the North Sea at 80 to 100 km/h making a significant contribution to the tidal events there. However, the Channel wave is significantly restricted by the mainland, producing a considerable rise of tide in some places, but after it enters the much wider North Sea it no longer has any significant effect. Thus the North Sea receives its tidal influences from the North Atlantic.

While the tide-producing forces as a cause of the tides can be calculated exactly with regard to their size, distribution around the Earth and their fluctuation over time, the way that these forces act on the water masses of the oceans has not been completely answered. Several theories have been developed in this respect, but to date no one has succeeded in making a complete prediction of tidal occurrence by purely deductive methods. 'In practical terms this means that the prediction of the occurrence of the tides in the oceans is not fully satisfactory because transforming influences as a result of complicated ground topography cannot be fully included' [22 p.404]. The topography of the ocean floor like that of the shelf regions and thus the mainland, most certainly exerts an influence on the occurrence of the tides. In the totality of tide-producing forces, the tides are able to adapt

2. THE TIDES

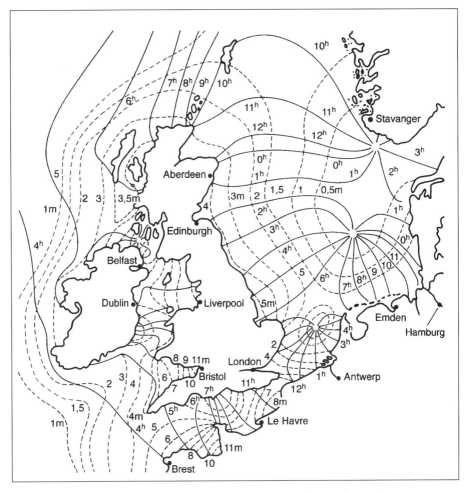

Figure 11. Map showing cotidal lines (lines which join places where high tide occurs at the same time) at intervals of 1 hour (figures are hours after Moon's meridian passage at Greenwich). The dotted lines show places with the same range of spring tide (in metres). Note the circular movement of high tide around the North Sea. Also note the increased range in the Severn Estuary and in the Gulf of St Malo, Brittany.

within the oceanic water masses to the given circumstances provided by the topography of the fixed continents.

Earlier we saw how the world's oceans with their tidal occurrence are differentiated from one another. This also throws light on the development of the Earth's continents. At the end of the Palaeozoic age during the Permian period, all the continents were situated together on one side of the globe. Since then they have been distributed across the

Earth more evenly without having completely lost their original asymmetry. One hemisphere still has more land than the other. Alongside the internal forces of the Earth which essentially opened up the young oceans (Atlantic and Indian), we may also assume that the extent of continental drift was influenced by the modulating influence of the tidal roll in the world's oceans. Thus it is feasible that continental drift was, and still is, influenced by the motion of Earth, Sun and Moon. Thus tidal events have also contributed to the way that the Earth's surface looks today. The interrelationship between both actions is visibly expressed today in the existing correspondence of the tides with the development of sea basins, and thus with the position of the continents across the Earth.

CHAPTER 3

Life in Tidal Coastal Areas

The tidal rim of the North Sea

The landscape at the edge of the sea lives in constant change. The clear path of the wind which can easily escalate to storm strength means that the weather changes constantly, and the rollers crashing in polish the rocky coast and through their backwards and forwards motion, shape the tidal flats. The tides dominate all coastal life. When the tidal flats are uncovered seagulls, oystercatchers, mergansers, ducks and geese search for food. At high tide they rest on the water or on land, or fly away.

Harbour seals (*Phoca vitulina*) on the German and Dutch North Sea coasts behave in exactly the opposite way. At low tide they rest on the sandbanks in the tidal flats and during high tide they embark on their search for food (Plate 3). Young animals no older than thirteen months have horizontal rings at the roots of their incisors; the number of rings approximately reflects their age in months. A four-week-old animal shows one ring, a four-month one four rings, and a thirteen-month animal thirteen rings. Since seals feed on coastal fish the question arises as to what extent periodic fluctuations in the supply of food are involved here. For Atlantic herring (*Clupea harengus*) for example, a lunar rhythm can be shown in relation to North Sea catches.

In what follows, we are, however, particularly interested in the world of the small organisms for which we see the evidence in the ground-down shells of coastal surf.

There are three types of bivalves whose shells are most frequently washed up on the northern coasts of Europe. We find the fine and regularly ribbed cockles (*Cardium edule*) everywhere, then less frequently the bluish-black blue mussel (*Mytilus edulis*) with its shinning mother of pearl interior, and more occasionally still the large soft-shelled clam (*Mya arenaria*) with its plaster white shell (Plate 5). Living blue mussels cannot be missed at low tide. Masses of them are attached with their tough threads, the *byssus*, to the wood and stones of harbours and sea walls, or on coastal rocks in the highest areas to which the water extends (Figure 12). At low tide they are exposed to the air for hours, but the drying wind, hot Sun or fresh rainwater do not damage them in any way, since the two bluish-black, slim, pointed shells fit together so tightly that a tiny little 'sea' is maintained inside them. Once high tide reaches the mussels again, they open up and small incurrent and outcurrent pipes are extended with an intensification of respiration, feeding and metabolic activity. Thus their periods of rest and activity follow the rhythm of the tides, but their reproduction is also determined by the Moon. The microscopically small larvae are ejected into the water, in their natural environment, most frequently at Full Moon, and most rarely at New Moon.

A similar situation exists with the soft-shelled clam. Its larvae, too, are ejected mainly in the warm months from June to September, preferably in the days around Full Moon, but in other respects this bivalve is very different from the blue mussel. Its shell cannot close properly because the two halves do not meet properly at front and back where its foot and the joined incurrent and outcurrent pipes, the *siphons*, stretch far out of the shell. It does not, however, face any great danger of drying out, for the soft-shelled clam lives up to thirty centimetres in the sand of the tidal flats, pushing its equally long siphon, which it uses for feeding itself, up to the water level (Figure 12). While the blue mussel has hardly any interlocking teeth, and thus both halves of the shell are symmetric, the interlocking teeth on the soft-shelled clam — particularly on the left shell — are very asymmetric. As it is very wide, the soft-shelled clam's size gives it a massive and plump effect with rough strips across its surface. What a difference to the pointedly formed, mirror-smooth blue mussel, exposed to the air and gleaming black!

The ribbed cockle lives in the surface silt, that is in the middle between the other two types described above as far as its living environment is concerned (Figure 12). In terms of its whole organization, too, it takes an intermediate position. Foot and siphon are of medium length. It only digs down a little way and as a young animal lives wholly on the surface of the tidal flats, still covered at low tide in the water pools left by the retreating tide (Plate 4). Its shell is covered by a series of regular ribs which start close together at the projecting *umbo* and increasingly spread out. These ribs on the shell are growth spirals in the pattern of natural logarithms. The alternating brown and white transverse stripes display regularly decreasing intervals like the number scale on the logarithmic slide rule. Black and white colouring

3. LIFE IN TIDAL COASTAL AREAS

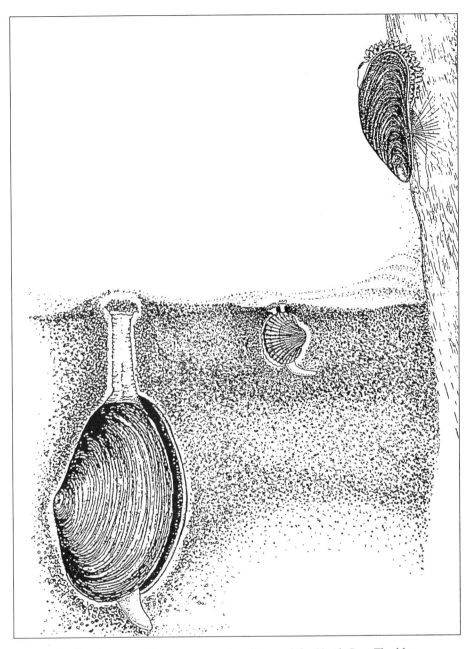

Figure 12. The three most frequent coastal molluscs of the North Sea. The blue mussel (Mytilus edulis) *is firmly anchored in the spray zone with tough byssus threads to resist the waves. In contrast, the soft-shelled clam* (Mya arenaria) *lives up to thirty centimetres deep in the sand of the tidal flats. Between both the ribbed cockle* (Cardium edule) *lives in the surface silt of the rippled tidal flats. Their shells form the main part of the ground shells in the coastal surf.*

interpenetrate equally. The shell is neither too long nor too wide. The interlocking teeth are slightly asymmetric without overly projecting.

Even the small surface patterns of the shell, when looked at with a magnifying glass, display the fine harmonic rhythms which are characteristic of this common mollusc. If the shell is examined under the microscope, the individual growth strips can even be analysed. It appears to be true for this bivalve, at least in certain phases of its life, that a new growth strip is added in the course of one tide and that alongside the tidal rhythms the lunar rhythm of spring and neap tides are reflected in the regular alternation of thicker and thinner growth strips. The fact that even animals, which are kept artificially submerged, show traces of periodic shell formation is an indication that the rhythm governing the ribbed cockle might have a component which is independent of its environment.

The ribbed cockle is positioned in the middle between the high-zoned, tightly closed blue mussel and the soft-shelled clam which feeds deep in the silt. These three most common types of bivalves, each as a part of the whole, provide an image of the way in which the coastal zone represents an independent and balanced interrelated ecological unit through their way of life and organization.

Many species of water fauna in the Baltic have been lost due to the increasing freshwater content in its eastern and northern parts in a sea, which in any case is low-saline. Nevertheless, these three species of bivalve are still represented in coastal zones — for example in the German Baltic coast — albeit in stunted form. Also represented is the Baltic macoma (*Macoma baltica*), a smaller relative of the ribbed cockle (Plate 5).

Let us now keep our eyes open for other living creatures which are influenced by the lunar periods, and investigate the water level at low tide.

Crabs, snails, diatoms and a turbellarian

We often find masses of small, segmented animals, which may be up to 1.5 centimetres in length, in the highest coastal zone, the region of the blue mussels, and usually in damp surf rather than on steep rocks. They jump about like tiny grasshoppers feeding on the organic material which is washed up by the water.

They are called sand hoppers but belong to the crustaceans (amphipoda). The sand hopper (*Talitrus saltator*) (Figure 13) is able to

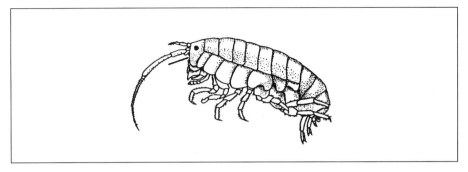

Figure 13. The sand hopper (Talitrus saltator), *here a male, is only 16 millimetres long [1085].*

orient itself excellently by the Sun. If taken to the dry zones of the sand beach it will return by the shortest route possible to the surf zone, no matter whether it does so early in the morning or late in the afternoon. The direction is determined by the sand hopper as a specific angle to the Sun even as this changes throughout the course of the day. At night, the sand hopper orientates itself using the Moon. The surprising thing is that these animals are still able to orientate themselves properly if they have been kept in the dark for ten hours and are then exposed to the natural position of the Moon. Even then they escape in the direction of the wash using the shortest route.

The genus *Orchestia* belongs to the same family as the sand hopper. These amphipoda live hidden under stones in the wash zone and wait for the next high tide. Once this has passed its highest point, they emerge from the dark and search for food in the material which has just been washed up. The rhythm of their activities is determined both by the day-night rhythm (Sun) and the tidal rhythm (Moon), and is also maintained under artificial conditions in the laboratory. The animal itself thus keeps the rhythm of both astronomical cycles.

The periwinkles of the genus *Littorina* (Plate 5) are conspicuous snails of the tidal zone. Four quite similar species share this area in European latitudes. Here the relatively slim common periwinkle (*L. neritoides*) searches and grazes the spray zone for microscopically fine algae, while the wider flat periwinkle (*L. obtusata*) likes to live off the seaweed plants at the low water line. The rock periwinkle (*L. saxatilis*) and the great periwinkle (*L. littorea*) fit in between these in terms of their size and habitat, with the former reaching a little higher in the coastal zone and the latter — living slightly lower — being the largest of the four indigenous species at 2.5 centimetres.

The most pronounced lunar periodicity is shown by the common periwinkle, which lives in the spray zone together with the blue mussel. Its reproductive period lies in the winter half-year between September and April; it lays its eggs every fourteen days at the time of the syzygies, that is at Full and New Moon. The rock periwinkle is also particularly active at this time. This semi-lunar rhythm of activity is maintained for many months even under artificially constant laboratory conditions. In contrast, the great and the flat periwinkle display tidal patterns of activity.

In Britain and France, the strange limpet, *Patella vulgata,* is found on the steep rocky coastline in the highest part of the spray zone. The conical non-spiral shell of this snail is pressed so tightly against the rock at low tide that the edge of the shell forms a tight seal with the rock and fits into all irregularities. At high tide it crawls around at such a distance and for such a time that it can reach its previous spot again at the next low tide. Its vital activity, measured by its oxygen consumption, is highest at high tide and lowest at low tide, thus fluctuating with the tidal rhythm. The small, colourful bivalve, Baltic macoma (*Macoma balthica*), (Plate 5) is found in the silt areas of the tidal zone, which appear at low tide, in other words, in the zone of the ribbed cockle. Mostly a delicate raspberry red, sometimes also coloured with a strong orange or yellow, the associated shells, which have been washed up often cling together for a long time. Their reproduction is closely linked with lunar phases. The microscopically small larvae swarm in later summer, particularly from July to September, at the time of the Full Moon, and do so least at the New Moon. We thus have a lunar rhythm here.

The open tidal flats are a gigantic flower bed irrigated daily by the sea. Myriads of tiny living creatures swarm about on its surface. At the start of the food chain is the green, brown or faint yellow coating of the silt, which can solidify into damp glistening patches. Under the microscope this is revealed to be lower algae. Here we have the green-tinged *Euglena limosa*, the brown luminous algae (*Chromulina psammobia*), a green diatom (*Hantzschia virgata*) and a brown one (*Pleurosigma aestuari*) (see Figure 14). At low tide during the day they come out from under the top grains of sand and silt and feed on the Sunlight by synthesizing starch or similar organic substances from light, carbon dioxide and water (photosynthesis). The solar day (24 hours) and the tidal rhythm or lunar day (12.4 and 24.8 hours respectively) overlap in its rhythm of activity. Since the daily low tide lags behind the one of

Figure 14. Myriads of the microscopically small shuttle-like diatoms (Pleurosigma spec.) *live in the tidal flats [1086].*

the day before by almost an hour and eventually progresses into the dusk period, a jump is recorded every fourteen days when suddenly, after 12.4 hours, on the next morning, the migration into the light starts again and the evening migration no longer takes place. All the photosynthesizing microbes in the tidal flats behave in this way. These micro-organisms have integrated this solar-lunar rhythm to such an extent that they maintain their rhythm independently even under constant laboratory conditions [25].

The first beneficiaries of the food chain referred to above are all those microscopic creatures which cannot photosynthesize but which live on the tiny algae, and have thus linked their own life's rhythm to their behaviour over time. Such activity co-ordinated with the tides is demonstrated, for example, by the turbellarians (*Convoluta roscoffensis*) (Figure 15), which are from the group of flatworms and about five millimetres long, and which have been studied particularly on the coast of Brittany, near Roscoff. If we take a walk at receding tide along the many-branched troughs (channels), we can see strong green colouring on the otherwise light-coloured base. If we stamp our foot on the ground the green colouring of the edge of the channel gradually turns to a light sand colour. On closer investigation, we find a massive occurrence of turbellarians which host numerous single-celled algae (*Platymonas convolutae*) in their tissue. The turbellarians are directly dependent on the matter provided by these algae as the result of photosynthesis because their mouth opening grows shut at an earlier developmental stage, and they can, therefore, no longer ingest any particles of food. The algae, in turn, profit from the products of the animal's metabolism: a perfectly symbiotic relationship.

Now *Convoluta* not only has its own daily tidal rhythm in order to take what we might call its own colony of algae — up to 30,000 green cells can be in each animal — for a walk at low tide in the light, but it also orientates itself by the Moon as far as reproduction is concerned. The eggs are shed at two-weekly intervals in relation to the neap tides, that is around the quadratures of the Moon (waxing and waning Half Moon).

The swimming rhythm of the small isopod, *Eurydice pulchra,* (Figure 15) which darts about in the pools of the open tidal flats searching for food (Figure 16) follows a tidal pattern. Swimming activity escalates for another three or four days after the syzygies (Full and New Moon) and is thus also semi-lunar.

The obvious phenomenon of the open tidal flats includes numerous little casts of sediment which are pushed up from below. If we dig into

3. LIFE IN TIDAL COASTAL AREAS

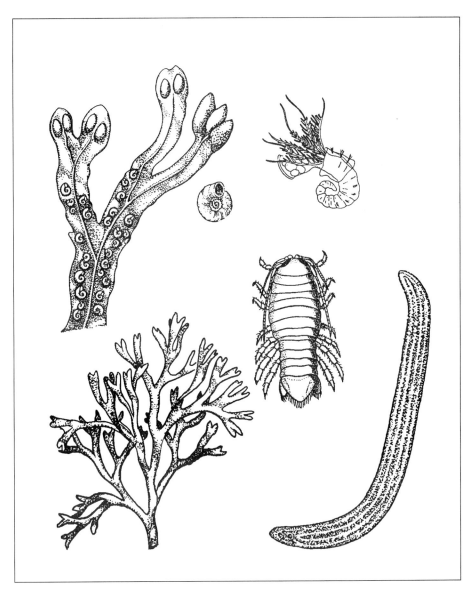

Figure 15. Further classic lunar-periodic inhabitants of western European tidal flats are: top, the post horn worm (Spirobis borealis) *with extended mouth tentacles. The calcified, spiral-shaped tube it inhabits, only a few millimetres long, is often found in great quantities on seaweed (left) [294]. Middle: the isopod* Eurydice pulchra *busily swimming about in tidal flat pools only grows to 4 mm (♂) and 7 millimetres (♀) respectively [1087]. Bottom left: the brown algae,* Dictyota dichotoma, *lives below the surface also at low tide, attached to the rocky sea floor, and grows up to 20 cm in height. Bottom right: the spinach-coloured turbellarian* (Convoluta roscoffensis) *is only 3–5 mm long, but its great numbers can colour the uncovered tidal flats green [410].*

the sand we find a fat annelid which looks similar to the earthworm and is distantly related to it: the lug worm (*Arenicola marina*) (Plate 8). Its metabolic rhythm, which can be determined from its respiratory oxygen consumption, shows a stable tidal rhythm with maximum values at high tide and minimum values at low tide. Its Japanese relative, *Arenicola cristata,* is guided in its reproduction by the lunar phases. This species swarms particularly during the quadratures (waxing and waning Half Moon) from mid-July to mid-September.

A tiny relative of this worm which lives on European coasts behaves in exactly the same way. It does not look like a worm, but like a tiny snail. Its calcareous spiral shell, with a diameter of a few millimetres, is often found attached to the surface of bladderwrack and similarly large seaweed. Inside this shell lives the tiny, hermaphroditic post horn worm (*Spirobis borealis*) (Figure 15). From late summer into the autumn it only releases its larvae in the days around the quadratures of the Moon (neap tides) and always at low tide.

If our route takes us to that part of the tidal flats where there is still water cover in still tidal pools, channels or the surf, we will encounter rich growth of seaweed with even richer animal life which once again correspond to lunar rhythms. There are, first of all, primitive underwater plants of which two species, also occuring in the North Sea, reproduce in line with the lunar periods. Reproduction occurs by directly shedding mobile reproductive cells into the water which either reproduce sexually through copulation (gametes) or which produce the next generation asexually through cell division (zoospores). The swarming of these flagellated cell types occurs with gut weed (*Enteromorpha intestinalis*), a tubular green algae, in the summer months three to five days before Full and New Moon respectively. The brown algae, *Dictyota dichotoma* (Figure 15), has its gametes swarm in the same semi-lunar rhythm but in this case in the period shortly before the syzygies. A closer study of samples on Heligoland found that the brightness of the nocturnal Full Moon triggers the maturing of the sex cells in such a way that they are released on the tenth and twenty-sixth day after natural, or (in the laboratory) artificial, illumination. With the green algae *Halicystis (= Derbesia) ovalis*, which also occurs in Heligoland, the gametes swarm exactly at Full and New Moon.

The tiny plant-like hydroid polyp *Laomedea geniculata* (medusa form = *Obelia*) (Figure 17), only a few centimetres high, can be found underwater on seaweeds and stones like a delicate bearded growth. It reproduces in the hot summer months from July to September by

Figure 16. Water draining away in the channels of the tidal flats, the habitat of Eurydice pulchra.

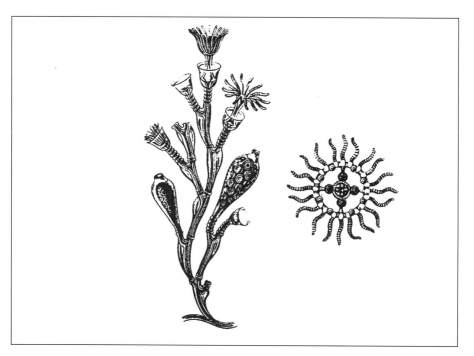

Figure 17. The polyp Laomedea (= Obelia) geniculata *with the free-swimming medusa,* Obelia lucifera *[1088].*

ejecting small, vegetatively produced *embryos* over ten days following the last quarter of the Moon, which grow into free-swimming medusa. These subsequently form sex cells which fertilize one another, and in turn produce *benthic polyps.* They often attach themselves not only to stones, but also to the hard carapace of living crabs, which also display some behaviour related to lunar periodicity.

Of the crabs living in the shallow water, it is the green crab (*Carcinus maenas*) (Plate 11) and two types of shrimp (*Palaemon elegans,* Plate 7, and *Palaemon serratus*), which develop their most intense movement activity in close co-ordination with the tides.

The Atlantic shore

The oyster (*Ostrea edulis*) and the large and small scallop (*Pecten maximus, Pecten opercularis*) (Plate 6) always live below water. Both are common to the warmer French and English coasts. The oyster stands out because of its thick shells, which develop asymmetrically as the lower one attaches the oyster to its base. The scallop shells, with their regular raying out ribs, are also asymmetrically formed with the upper one much flatter than the lower one. By rapidly snapping its shell shut, the scallop creates enough thrust to shoot along underwater. There is a long series of small eyes along the edge of its mantle, which it uses to orient itself in the light. Thus moonlight might be a contributing factor in the reproductive cycle. From early spring to start of summer, these animals lay eggs in a monthly rhythm above all when the Moon fully reflects the light of the Sun. The oyster also prefers to spawn both at Full and New Moon, thus displaying a semi-lunar rhythm for spawning, but only in the period from June to August.

It is also worth mentioning two other coastal inhabitants which orientate themselves by the diurnal rhythm of the Moon, as well as the long-term rhythms mentioned above. We include this lunar day-oriented behaviour even though it was noted in species on the North American east coast. One is a bladderwrack (*Fucus spec.*) and the other the lovely mud dog whelk (*Nassarius obsoletus*) (Plate 5) of the tidal flats zone. Day to day both of these display their greatest respiratory activity when the Moon dips below the horizon in the west, irrespective of phase, while their respiration grows weakest when the Moon rises. This lunar day-oriented rhythm also clearly links some marine animals with the lunar orbit.

The marine midge, Clunio

We will now look in greater detail at the lunar dependence of an insect whose larvae live underwater in a few places on European coasts. This is the marine midge, *Clunio marinus* (Figure 18). It has been studied in detail in Heligoland and on the Baltic, as well as in other coastal regions. Interesting divergences in its development became apparent which were always in complete harmony with its geographical environment, both in respect of the geographical and cosmic circumstances. The marine midge was discovered in 1855 near Dublin on the Irish coast, but it was not until the 1930s that it was found for the first time in Heligoland, Germany (Thienemann, Caspers). They are not found in the mud or sand of the North Sea coast, as this tiny midge is tied almost exclusively to rocky coastlines and lives in European habitats, ranging from the Mediterranean area to the far north of Norway.

First, let us describe the life cycle of the midge occurring in Heligoland. The larvae live for at least two months underwater in algae (particularly the red algae *Rhodochorton floridulum*) on rock banks which at the time of the spring tides (every fifteen days or so) become

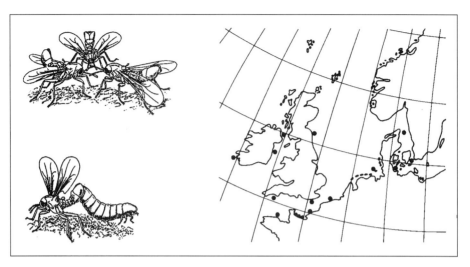

Figure 18. The Heligoland marine midge (Clunio marinus), *only 3 millimetres in size. Top left: Males looking for each other. Bottom left: mating with the wingless female. Right: some north-west European occurrences of various populations of this species.* [33].

dry for a few hours at low tide. The pupa stage lasts for three to five days, then the fully formed midges emerge to live for just a few hours. Without consuming any further nourishment, they mate and lay their eggs. Only the males have wings and by far outnumber the females — there are about ten to twenty males for each of the flightless females. The males actively seek out the females. Immediately after copulation, the female sheds a clump of forty to eighty eggs and instantly dies. Generally, the males also die shortly after but in some cases can continue to live for six to twelve hours.

The mature insects only appear in Heligoland in the summer months (end of April to start of October). The midges hatch and swarm at this time of the year but only at the time of the lowest afternoon ebb, during the spring tides shortly after Full and New Moon. Their environment is then free of seawater, creating ideal conditions for an air-breathing, flying insect. The time of pupation must therefore occur at half-monthly intervals co-ordinated with the spring tides, so that the adult animals can hatch in these days, harmonized with the daily rhythm, always at the time of the afternoon low tide. Seasonal, semi-lunar and day-night rhythms are key determinants for these animals.

If we observe these *Clunio rules* and their variants, which we will describe below, we can see that these animals neither adhere opportunistically to any favourable time nor do they stick rigidly to the rules.

After all, they do not hatch at the morning low tide even if the water falls just as low as in the afternoon. However, neither is there any hatching in the evening in the days after the syzygies of the summer half of the year, following serious north-westerly storms, which caused the wind to push the water over the pupa's environment at low tide. In fact it would be quite possible for the midges to hatch even in a flooded environment, as aquarium tests have shown. The mature pupa then develop air at the appropriate time which lifts them to the surface where the insects can hatch. However, hatching takes place quite independently of the weather, and depends on whether it is light or dark. Thus in September in murky weather, swarming can still take place after an early dusk when evening spring low tides fall into the period immediately after New or Full Moon. Swarming behaviour is therefore not determined solely by environmental factors but is also hereditary. However, the inherited factors do not determine behaviour to such an extent that external factors cannot also exercise a certain influence. The synchronization of the animal's own rhythm with that of the environment clearly takes place already at the larval stage, through the influence of the daily light-dark cycle of the Sun, as well as the tide cycle. This was found in laboratory experiments with populations of the marine midge in Heligoland, Normandy, England and Ireland, during which a semi-lunar hatching rhythm was provoked through a 24-hour change of light and dark (twelve hours of light alternating with twelve hours of darkness = LD 12:12), in parallel with a 12.4 hour cycle of simulated tides. Effective factors of the artificial tides turned out to be the components of mechanical water movement: turbulence, underwater noise and floor vibration. With the populations from Normandy, a semi-lunar hatching rhythm was also provoked through LD 12:12 and artificial moonlight (0.4 *lux* during four consecutive nights per month), something which did not work with the populations from further north. With populations from further south, it was only the artificial moonlight, which proved effective in the experiment [26, 27, 28]. Clearly the Moon with its changing phases becomes less important for temporal orientation the further north we go. If in temperate latitudes the Full Moon in summer moves only slightly above the horizon on its daly arc, it does not appear at all during this time of year in high northern latitudes, and at the time of the New Moon it does not grow dark at all at night, or only briefly owing to the position of the Sun.

It is of interest in this connection that the synchronization between own rhythm and external rhythm has also fixed other inherited swarming rules. In places such as Normandy, southern Brittany or Spain (near Santander), where the tide rises as high as in Heligoland — approximately two metres — swarming takes place at half-monthly intervals. Near Tromsø in northern Norway, the local Clunio midges only swarm in June, during the short Arctic summer. This occurs at every low tide, that is twice in the course of a day [29]. Slightly more to the south, in Bergen (Norway), the semi-lunar periodicity still occurs, but in such a way that midges hatch on all days of the synodic month, and mostly at the time of the neap tides around midnight [30, 31].

In the Baltic, where there are practically no tides, the Clunio populations living near Schleimünde Holstein display absolutely no lunar periodicity [32]. The whole process takes place between five and thirty metres underwater. After the midges have hatched about the time of sunset, mating and egg-laying always take place at nightfall on the surface of the water. Irrespective of the annual rhythm, there is thus a daily rhythm, controlled by the alternation of light and dark. Near Varna on the Bulgarian Black Sea coast which is non-tidal, no lunar rhythm can be determined in the Clunio, only a daily and annual rhythm. Here swarming takes place from May to October in the morning between 5 a.m. and 9 a.m. when the mostly seaward breeze dries out their environment [33]. In contrast, the marine midges in the Adriatic display a tidal and semi-lunar hatching periodicity, which is clearly connected with the fact that there is a tidal rhythm, albeit a weak one.

The noted geographical differences in temporal behaviour are, as the laboratory tests show, fixed to a considerable degree so that one can speak of geographical strains of the Clunio midge. The Japanese species (*Clunio pacificus*) even displays a lunar rhythm and only swarms every four weeks in the spring. Even if the individual animal is determined by heredity and environment, a comparison of the different populations of this midge nevertheless show a highly differentiated evolutionary flexibility. Life is never rigid.

CHAPTER 4

A Worm, Fish, Squid and Corals from the Pacific Ocean

The palolo worm

When Christopher Columbus discovered the New World, the modern age began. On October 11, 1492, the night before the discovery of America, a sailor saw the luminescence of glowing worms from his ship off the Bahamas. It was a long time before the origin of this luminescence in the sea was discovered. Only in 1935 did the Englishman, Crawshay, discovered that it was a result of the swarming behaviour during mating of the *Odontosyllis* worm, synchronized with lunar periodicity [4].

In 1705, in *The Ambonese Curiosity Cabinet* the Dutch traveller, Rumphius, described a worm which swarmed shortly after Full Moon off the coast of Ambon, an island in the Moluccas in eastern Indonesia, and which the locals called Wawo (probably *Lysidice oele*). For ten years Rumphius observed its reliable lunar periodicity, but his observations remained initially without consequence, or at best a curiosity [34].

In 1847, a shipment of greenish worms from the south Pacific, the thickness of knitting needles and 20 to 40 centimetres long, reached London, of which the zoologist Gray noted with a shake of his head that all of them were missing their heads. The worms had been sent from the Samoan islands by a missionary by the name of Stair who had included the following report:

> *Palolo* (a contraction of *pa'a-lolo*) is the Samoan name of a remarkable species of sea-worm found under peculiar circumstances in some parts of Samoa. They appear with the greatest regularity and certainty on portions of two days in each of the months of October and November — namely, the day before and the day on which the Moon is in her last quarter. They appear in much greater numbers on the second than on the

first day of their rising, and are only seen for two or three hours in the early part of each morning on the days of their appearance. ... On the second day they appear ... in such countless myriads that the surface of the ocean, near the reef, is covered with them for a considerable extent. ... They are only found in certain parts of the islands, generally near the openings of the reefs or portions of the coast on which much freshwater is found, but this is not always the case. ... The natives are very fond of them and calculate with great exactness the time of their appearance which they always look forward to with much interest.

The worms are caught in small funnel-shaped baskets ... When on shore the worms are tied up in leaves in small bundles and baked. Large quantities are eaten uncooked, but, either cooked or uncooked, they are universally esteemed a great luxury. Such is the strong desire to eat *palolo* shown by all classes, that, immediately the fishing parties return to shore, messengers are despatched in all directions bearing large quantities to parts of the islands on which none are found. [5]

This was an extract from the first report of an European about the palolo worm. In the last century the scepticism adopted by zoologists towards the influence of the Moon, infused with a great deal of superstition, led to increasingly specific studies; first of all on the lifestyle of the worm, then the questionable reliability of the palolo rules, and finally the initial factor analysis of its periodicity. What do we know about it today?

The palolo (*Eunice viridis*) is a marine annelid which is distantly related to our earthworm (*Lumbricus terrestris*). Like the latter, the palolo is made up of many individual ring-like segments. Unlike the earthworm, it is not hermaphroditic but has separate sexes. The male has an orange-yellow colour and the female is blue-green. A series of black dots run along the abdomen, which are pigmented groups of visual cells, but this appearance only applies to that half of the worm, which appears on the surface of the sea. The whole worm is up to sixty centimetres long and lives just below the tidal zone in holes and other cavities of the Porites coral reefs in the South Pacific on the coasts of Samoa, Fiji, Tonga, the New Hebrides, stretching as far as the Gilbert Islands on the equator. The worm consists of a head equipped with hard maxillary points, a zone without gills and numerous gill-bearing

Plate 1. Tropical tree in Tasmania (Erythrina abyssinica, *Fabaceae*). PHOTO: SCHAD

Plate 2. South Asian water buffalo (Bubalus arnee). PHOTO: SUCHANTKE)

Plate 3. Sandbank in the North Sea with seals (Phoca vitulina).

Plate 4. Tidal flat, habitat of the ribbed cockle (Cardium edule). PHOTO: SCHAD

Plate 5. Snails and bivalves of the North Sea: top row from left: mud dog whelk (Nassarius obsoletus), *flat periwinkle* (Littorina obtusata), *rock periwinkle* (Littorina saxatilis), *great periwinkle* (Littorina littorea). *Second row from left: small specimen of the blue mussel* (Mytilus edulis), *the Baltic macoma* (Macoma balthica) *from outside and inside. Third row from left: large specimen of the blue mussel* (Mytilus edulis), *edible and spiny ribbed cockle* (Cardium edule *and* echinatum). *Bottom: soft-shelled clam* (Mya arenaria).

PHOTO: SCHAD

Plate 6. Snails and bivalves of the Atlantic. Top left: green ormer (Haliotis tuberculata). *Top right: limpet* (Patella vulgata). *Middle left: knotted cockle* (Cardium tuberculatum). *Middle right: large scallop* (Pecten maximus). *Bottom left: edible oyster* (Ostrea edulis). *Bottom right: small scallop* (Pecten opercularis).

PHOTO: SCHAD

Plate 7. Shrimp (Palaemon elegans) PHOTO: *PLANTS AND ANIMALS OF THE LAND OF ISRAEL,* JERUSALEM 1983

Plate 8. Mud shrimp (Corophium volutator).
PHOTO: JANKE

Plate 9. Lug worm (Arenicola marina).
PHOTO: JANKE

Plate 10. Fiddler crab (Uca tangeri) PHOTO: ENDRES

Plate 11. Green crab (Carcinus maenas) *with barnacles* (Balanus spec.) *on its carapace; above it in the photo* Portunus depurator *a swimming crab.* PHOTO: SCHAD

Plate 12. Holacanthus ciliaris. PHOTO: MAYLAND

Plate 13. Diadem sea urchin (Diadema setosum). PHOTO: KUHN

Plates 14 and 15. Grunion (Leuresthes tenuis). PHOTOS: STEINHAGE

Plate 16. Asian toad (Bufo melanostictus). PHOTO: SCHAD

Plate 17. Spiny anteater (Tachyglossus aculeatus), *the male develops a pouch each month.*
PHOTO: STORCH

4. A WORM, FISH, SQUID AND CORALS

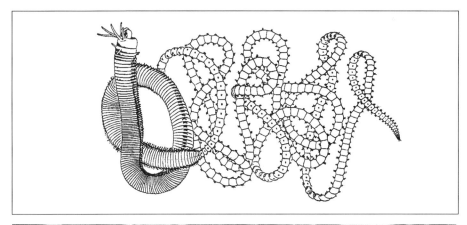

Figure 19. Top: The palolo worm (Eunice viridis), *living among the coral of the South Pacific Islands and atolls, can grow to forty centimetres long. The posterior part of the worm contains the gametes and is broken off at the right lunar phase. (From [533]).*

Bottom: The free ends of the worm are fished as a delicacy [112].

rings (Figure 19). It does not bore its living tube itself but inhabits the bore holes of peanut worms (Sipunculids). Towards October, spring in the southern hemisphere, the worms grow additional posterior segments with eyes on the abdomen, which fill with gametes. In the female these are the blue-green to blue egg cells, and in the male the yellow to yellowy-orange sperm cells. This section, which comprises about half the worm, is called the *epitoke* of the worm in contrast to the *atoke,* the front part of the worm which does not contain any gametes (Greek *a* = without, *tokos* = offspring).

In complete synchronization with the annual, monthly and daily rhythms, the headless epitoke (approximately forty centimetres long) detaches itself from the atoke and swarms to the surface with active wriggling movements following the light, while the atoke, with the head, remains in the coral. The abdominal eyes, which are particularly developed in the epitoke, are clearly of service. On the surface of the water most of them are caught up in the swell from the reefs near the coast. Like with many other annelids, palolo worms also lack an opening for ejecting gametes. Reproduction occurs so that the swell directly smashes the parts of the worm against the reefs to break them apart. This releases the eggs and sperm, fertilization takes place and the fertilized egg cells sink in the ebbing flow of the tide. They develop into larvae with bristles all around, which swim freely for a time and subsequently become attached, inhabiting the coral. The atoke of the worm continues to live with all its functions after the loss of the epitoke and can regenerate the epitoke each year. Not all fully grown worms, however, form sexually mature epitokes each year; about twenty-five per cent fail to do so.

It was difficult to uncover the biology of this animal, because the actual worm, the atoke, was only discovered when larger chunks of coral were removed from the water and chiselled open with great effort. Long before the Europeans, certainly for centuries, the Polynesians were aware of the reproductive method of the palolo, from which its name is derived: *lolo* means 'rich content' in their language, and *bla* or *pa* 'to burst.' Even today, for the Polynesians, the 'richly bursting content' represents caviar gifted by nature once a year in great quantities.

What, then, are the palolo rules? In their spring time the Fijians watch, first of all, for the scarlet blossoms of a bush-like coral tree (*Erythrina indica*) to give them advance warning of when the palolo is approaching. Then the blossoms of a myrtle plant (*Eugenia*) are watched

4. A WORM, FISH, SQUID AND CORALS

Year	Moon's third quarter		Palolo swarming		Other mth
	Oct	Nov	Oct	Nov	
1843	16	14	15/16		
1862	15	14	15/16	14/15	
1864			23	21	
1865	11	9	12/13		
1866	30	28	31	1	
1867	19	18	21		
1868	8	7	8/9	8	
1872	23	22	24		
1873	12	11		11/12	
1874	31	30	31	1	
1881	14	13			March 21
1893	31	29	31	1	
1894	21	19	21/22		
1895	11	9	10/11	8/9	
1896	29	27	28/29		
1897	18	17	16/17/18	15/16	
1898	7	6		4/5	
1925	9	8		7	
1926	27	26	28		
1927	17	15	17		
1928	5	4		4	
1929	24	23	25		
1930	14	13	14/15		
1931	4	2		2	
1932	22	20	21/22	19/20/21	
1933	11	10	10/11		
1934	29	28	28/29		
1935	18	17	18/19		
1936	7	5		5/6	Dec 5/6
1938	15	14	16	13/14/15	
1940	23	22	23/24		
1942	31	30	31		
1943	20	19	20		
1944	8	7	(8/9)	7/9	
1945	27	26	28		
1946	17	15	16/17/18	14/15/16	
1947	6	5	(6/8)	4/5/6	
1948	25	23	24/25		
1949	14	13		12/13	
1950	3	2		3	
1951	22	21	21/22	21	
1952	10	9		9	
1953	29	27	28/29		
1954	18	16	17/18/19		
1955	8	6	8/9	7	
1956	26	24	25/26/27		
1957	16	14		14/15	
1958	5	4		2/3/4	
1959	24	23	24/25/26	22/23	
1960	12	11	11/12	10/11	

Table 1: List of all known data about palolo swarming off the Samoa from 1843–1960 [36].

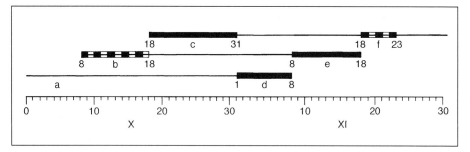

Figure 20. The distribution of Palolo swarming in October (= X) and November (= XI) on Samoa in accordance with the 'Palolo rules'. Solid fat lines: certain swarming period. Broken fat lines: possible swarming periods [36].

for and then it is time to keep an eye on the Moon. At daybreak, when the Moon is positioned very low on the western horizon, that is directly after Full Moon, there are still seven days to go before the palolo feast which starts shortly before sunrise with the catch of the worms.

The dates of the swarming days have been observed and recorded in Samoa over many decades, partly by missionaries living on the islands and partly by scientists. In the period from 1843 to 1960, data is available for fifty years (with interruptions), which is collated in Table 1. As can be seen from the table, the palolo only swarms in October and November (in very rare exceptions in March and December) with a strict link to the third quarter of the Moon. The swarming period is restricted to a maximum of three days and, in this context, the day before and the day after the third lunar quarter are swarming days. The second day is the main swarming day. Sometimes the swarming period can also comprise just one day — or one night to be precise, which in Samoa for example may be the hours after midnight, when the Moon is just rising in the eastern sky as a waning Half Moon and rises to the northern sky (we are in the southern hemisphere). The time difference between the synodic month and the calendar month means that the corresponding last lunar quarter is delayed for the time of swarming from year to year by ten to eleven days. It thus takes place earlier and earlier in October until there is suddenly a jump of twenty-nine to thirty days to the following month. Based on the table both the critical day in October, before which no swarming takes place, and various other regular patterns can be determined. We will present the 'palolo rules' here; they also permit an approximate prediction of future swarming events [36, compare also 35]. A graphical representation of the palolo rules is given in Figure 20.

a) Swarming does not take place before October 8. If the third lunar quarter occurs on October 7 or earlier, the palolo does not swarm until the next third quarter, that is November 6 or in the days preceding that date. The table shows a number of dates before October 8 in brackets. This relates to information about a very minor occurrence of palolo on Savaii. Comparable dates were never reported for other islands such as Upolu or Tutuila.
b) Swarming in October is doubtful if the third lunar quarter occurs between October 8 and 18. Often it fails to occur at all or is only very weak.
c) Swarming is certain in October if the third lunar quarter occurs after October 18.
d) Swarming in November is certain if the third lunar quarter occurs before November 7. There is no second swarming period after this one (compare a).
e) If the third lunar quarter occurs in the period between November 8 and 17, there is a particularly large swarming of the palolo, which can follow an earlier one between October 8 and 18. (compare b).
f) Swarming in November can occur until November 23 as a weak repetition of a main swarming in October (compare c).

Palolo swarming data reveal a nineteen-year period as can be seen in the table, for example, by a comparison of the years 1874 and 1893. This is the Metonic cycle of almost exactly nineteen years after which the Sun, Moon and Earth reach the same position again on the same date.

Experiments have shown that the gametes only mature if the head of the associated atoke of the worm is present. Furthermore, animals examined one month before the swarming event displayed a tenfold concentration of neurosecretory substances in the brain, than animals examined before this time. Clearly sexual maturity is transmitted by neurohormones of the brain. Since the worms only stick their heads out of their tubes, a few metres under very clear water and at night for feeding (algae, etc.), we might also speculate that the change of light in the lunar phases some months before the swarming period, might trigger maturity through the probably light-sensitive brain. On the other hand, both the duration and intensity of the visible moonlight is influenced by the weather, but the weather conditions do not affect the swarming date in any way.

Geographically, it is noticeable that the appearance of the palolo during the day shifts from 8 p.m. to 4 a.m. from the eastern to the western

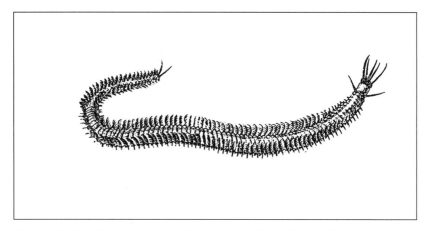

Figure 21. The European Palolo (Eunice harassi) *is a brownish-red colour and can grow up to thirty centimetres in length [1089].*

Samoan islands although there are no tidal differences between the islands and the lunar phase is the same for all of them. So this cannot be a result of lundian sensitivity since the meridian transit for the various islands takes place much more rapidly.

There are also two species related to the Pacific palolo in the Atlantic. Both of them are also synchronized with the Moon with regard to their egg laying rhythms. The Atlantic, or Florida, palolo (*Eunice fucata*) lives off the Antilles, Bermuda and on the neighbouring Florida coast.

Like its Pacific relative, it lays its eggs at night shortly before sunrise in the time of the last lunar quarter or, to be precise, in the three days before the last lunar quarter. In terms of the seasons, its swarming period falls at the height of summer — twenty years of observation allow us to specify the period as lasting from the end of June to the end of July. With more than fifty segments, the Atlantic palolo reaches a total length of up to seventy centimetres, making it longer than the South Pacific palolo. The second palolo occurring in the Atlantic region (European palolo = *Eunice harassi*) (Figure 21) lives in the English Channel. It swarms in the period from May to June, and also in the last lunar quarter. Thus this group of animals also demonstrates what applies to many others: that in particular the reproductive behaviour of the animals is closely related to the movement of the Moon.

The grunion

Let us now look in greater detail at another impressive example of the synchronization of spawning rhythms with the Moon This is the grunion (*Leuresthes tenius*) which occurs in the north-eastern Pacific off the coast of California. It belongs to the Atherinidae family which lives in swarms near coasts, or which migrates to brackish water and occasionally also freshwater. Their unobtrusive shimmering silvery look in the water has led to them also being called *silversides*. One species of Atherinidae, the sand smelt (*Atherina presbyter*) also lives in Europe and off the coasts of Ireland, Britain, the North Sea and France down as far as North Africa, but it does not particularly stand out in terms of its lunar periodicity. In contrast, the sardine Grunion Atherinidae (*Hubbsiella sardina*) from the northern Gulf of California is noticeably tied to the Moon in terms of its periodicity like the grunion.

But back to the grunion. It is a typical, slim Atherinidae with silver stripes and a blue-green back, and a maximum fifteen centimetres in length. It lives in the shallow water a few miles off the sandy coast of southern California at depths of 45 to 120 metres. We do not know a great deal about its life there except that its main distribution is from about Punta Abreojos in Baja California to Point Conception. Occasionally grunions also appear outside this region, sometimes as far as the Oregon coast. In the summer half of the year, thousands of these little fish appear near the coast in dense swarms at regular intervals, but do not, strangely, spawn in the water but on the beach at night during the spring tide. Most of the other species of Atherinidae spawn in the water on water plants. The surprising element being the precise temporal synchronization.

The grunion arrives over three to four nights after New and Full Moon, in the first or second hour after high tide. If the lunar phases are right, this can be as early as the end of February, otherwise early March, and is repeated in late August or early September. If all the circumstances relating to annual, lunar, tidal and daily rhythms are given, then soon after high tide individual males being to appear and allow themselves be washed on to the beach by the surf, swimming against the receding waves until they are stranded. After one or one and a half hours the climax of spawning is reached (Plates 14 and 15). Thousands of fish now lie temporarily on the beach and turn it into a single silvery area. The slightly larger females now also swim to the beach, but only if they are accompanied by one or several males. Otherwise they wait in the water. If males are present, the female allows herself to be carried on to the beach by a large wave and, using her tail and fins digs herself up to ten centimetres into the sand — surrounded by the fertilizing males. After shedding her eggs, the female wriggles out of the sand again leaving the eggs at a depth of about five to eight centimetres. The fish let themselves be carried back into the water by the next wave. The whole act of spawning generally takes about thirty seconds; individual fish may remain for several minutes outside their normal environment without suffering any harm. Thus in the one to three hours of spawning great fluctuations can occur between masses of stranded fish and not a single fish. One wave of spawning fish follows the next until the numbers begin to dwindle. It stops as suddenly as it started. Not a single fish is to be seen for the rest of the night until the next night or until the next spring tide two weeks later. This nightly spectacle has become a tourist attraction in some places on the Californian coast. Observers must behave in a very disciplined manner, as the fish are extremely sensitive. If there is too much movement on the beach or if torches or lights are used they do not come. It is only possible to shine a light and observe the spawning animals between two incoming waves.

How, then, do the young fish develop? On the Californian coast the level of the two daily high tides varies. In spring and summer, the higher of the two high tides, during spring tides, occurs at night. The grunion spawn during these higher tides, that is at night, and during these tides after the water has reached its highest point so that the

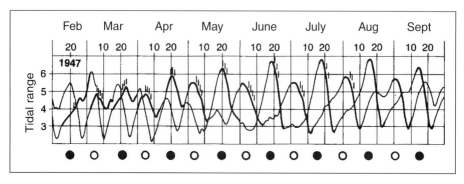

Figure 22. Spawning dates of the grunion on the beach of La Jolla/California in synchronization with the spring tides, here in 1947. The daily high water is shown in two separate lines. The spawning times of the grunion are indicated by vertical lines [91].

spawned eggs are not swept away by subsequent waves but covered up. In contrast to the breakers of the springtides, the waves of the falling tide deposit layer upon layer of sand on the beach. This means that the fertilized eggs end up under sand which is 20 to 40 centimetres thick. The highest waves of the next high tide of the following night no longer reach these eggs, and in a matter of days the spring tides have turned to neap tides (Figure 22).

Under the Californian sun, the surface sand becomes rather hot during the day, but the eggs remain cool and damp in the depths of the sand and the young fish are able to develop. After fourteen days, the corresponding level on the beach is once again reached by the rising tides. Then the next sequence of waves at high spring tide washes the sand away. The eggs are exposed by the last breakers shortly before high water. Two to three minutes later the little grunions hatch from their eggs and are washed into the water by the receding waves. They do not hatch while they are still under the sand. If they are not washed free after fourteen days — for example, because of a strong seaward breeze — the grunions are able to remain in their eggs in the sand for a further fourteen days and only hatch after that, although in smaller numbers.

In fact, the sensory organs of this fish are even more delicately developed. Dependent on the anomalistic month, the two spring tides in the course of a synodic cycle do not reach the same height. One spring tide will have higher water than the other depending on whether the time of greatest proximity to the Earth occurs more closely in time to the Full or New Moon. If the grunion were to spawn in the higher of the two spring tides, it would take a month until they were washed free again. It is evident, however, that the grunion tends to avoid the higher of the two monthly spring tides, just as it does not spawn if the perigee falls exactly on Full or New Moon [24]. In such a case the eggs would only be uncovered again six or seven months later. This perigee would then fall in the opposite phase of the synodic cycle in comparison to the previous one, that is on Full instead of New Moon and vice versa.

Once the young fish have hatched, they grow rapidly. After one year, they are approximately thirteen centimetres long and mature enough to spawn. After spawning for the first time, they continue to grow slowly and not at all during the spawning period, but only in the autumn and winter. The periods of interrupted growth can be seen in the growth layers of the scales. Thus the age of a fish can be read from its scales, just like the age of a tree can be read from its rings. Here it can be observed that the old females spawn first in February and March; in April and May, fish of every age spawn. A female may spawn four to eight times during a season in consecutive 14-to-15-day intervals. Larger females lay about three thousand eggs every two weeks, smaller ones about one thousand. This is how strong populations of these fish are able to maintain themselves, for their lifespan is very short. These animals live for a maximum of three or four years.

The grunion is a particularly fine example among the currently known cases of lunar swarming rhythms. It displays an extremely sensitive synchronization of its reproductive rhythms and embryonic period with the conditions and rhythms of its environment, and has been successful for many thousands of years, but for how much longer? We must ensure that it can continue to do so for the next thousands of years by protecting the fish and its environment.

The pearly nautilus

Let us now acquaint ourselves with another marine animal from the Pacific. It is the most primitive relative of the squid still in existence. Unlike the squid it does not have ink, but instead has the most beautiful shell which the world of the invertebrates can offer the human eye. The shell is a perfect logarithmic spiral with chambers which increase proportionally in size, as can be seen from its longitudinal cross-section (Figure 23). The numerous small chambers are filled with gas in the living animal so that their buoyancy counteracts the weight of the shell [37, 38, 39]. The largest chamber is where the animal lives. From the opening it extends its pair of eyes, together with two olfactory lobes and up to ninety short tentacles, positioned around the mouth like all cephalopods, but which in this case can be hidden under a large hood when not in use. The animal propels itself along at various speeds and with agility at various depths by ejecting jets of water. It lives in the warm latitudes of the Indian and Western Pacific Oceans. Five rare species inhabit the waters off Australia. The sixth species, *Nautilus pompilius* or the 'Pearly Nautilus,' which occurs in abundance off the Philippines, can also be found quite frequently in an area from the Fijian Islands to the coast of East Africa. It is rarely seen alive on the surface of the water but after storms the shells of dead animals are washed ashore, collected by the coastal inhabitants and offered for sale. It is also baited with pieces of meat, which it only accepts at night.

4. A WORM, FISH, SQUID AND CORALS

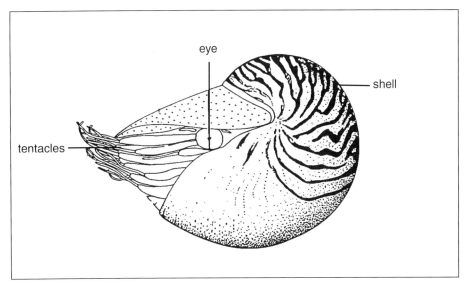

Figure 23. The pearly nautilus (Nautilus pomilius) *of the Indian and Pacific Oceans is a shelled relative of the squid. Top: the living animal [294]. Bottom: the opened shell shows the rhythmically-divided air chambers. (Photo: W. Schad)*

The *Nautilus* represents nothing short of the mathematics of natural logarithms made visible, of which we will introduce a few comprehensible concepts here. Subject and object only become separate through consciousness: if a person planned to undertake a task he, the subject, would need tools and materials to create the object. At a time when the physiological processes were still thought of in terms which were too anthropomorphic, the living substance of each cell was divided into 'building material' and 'active material.' It has long since been realized that there is no sharp delineation between the two, and that each substance is simultaneously building and active material, the one time seen as material and the other time seen through its function. What follows that? All living substance which grows can continue to grow— and more so than before — for the assimilated material can in turn contribute to further growth. It is, roughly, the same situation as with financial capital which grows as interest is added and, in the following year, grows even more than the previous year with interest on the interest (compound interest).

We can look at it this way: if capital of one pound bore interest of one hundred per cent at the end of the year we would have two pounds. But if interest is levied at fifty per cent every six months, the sum credited at the end of the year is even bigger, namely 2.25:

Start of the year: 1.00
End of first six months 1.00 + 0.50 = 1.50
End of second six months 1.50 + 0.75 = 2.25

If interest is levied on a quarterly basis at twenty-five per cent, the final sum rises to 2.44:

Start of year 1.00
First quarter 1.00 + 0.25 = 1.25
Second quarter 1.25 + 0.3125 = 1.5625
Third quarter 1.5625 + 0.3906 = 1.9531
Fourth quarter 1.9531 + 0.4883 = 2.4414

With interest levied on a monthly basis we would get 2.61. How large would the sum grow if interest were levied on a daily, hourly, minute, second basis?

It is an inherent characteristic of life that growth proceeds in each tiny moment for as long as no limit is put on it. Expanding life thus grows through the 'interest' levied in each moment according to 'natural exponential functions' based on the Figure 2.7182818..., described as *Euler's Number, e,* after the mathematician Leonhard Euler (1707–83). This number and its laws applies to the shells of almost all invertebrates. Each snail shell illustrates the steadily swelling build of the growing animal in geometrical spirals. The same is found in many molluscs in the curvature of their shells. The shell of the pearly nautilus has grown just as perfectly in the form of a natural spiral. Each consecutive gas-filled chamber (air chamber) is a little broader and wider than the preceding one — and thus of astonishing regularity in its aesthetics.

In contrast to the snail shells, which spiral into space along one axis, the pearly nautilus is rolled up in a plane (planar spiral). Almost all ammonites, those fossilized shell remains of extinct cephalopods, to which the pearly nautilus is closely related, possess a similarly rolled shell to the *Nautilus*. We know of hundreds of classes and species from the family of animals, like the pearly nautilus, which reach back into even earlier times of the Earth's history (the Ordovician period) than the ammonites (from the Silurian). The pearly nautilus is the only surviving species of these earliest of cephalopods.

All other cephalopods living today — cuttlefish, squid and octopus — have the best eyes of all invertebrates with lens, pupil and eyelids. It is likely that they can perceive shapes. The nautilus, on the other hand, only has a pair of simple eyes without lenses which probably only transmit differences of dark and light, and the direction of the light. The eyes have thus remained conspicuously primitive, and it does not swim to the sunlit, more oxygenated upper ocean zones. During the day it remains in the deeper layers (up to 650 metres deep) of the oceans. Only at night does the *Nautilus* swim upwards to search the substrate for crabs and dead animals.

If we look more closely at the outside of the pearly nautilus we notice the fine scale-like stripes of the whole shell alongside the conspicuous flame-shaped red pattern in parallel with the edge of the opening. These are clearly growth lines. What was particularly baffling for researchers, was that on the inside of the shell there were an average of thirty such growth lines for each air chamber. This immediately raised the question as to whether such a growth line was not formed each day, suggesting that thirty made an air chamber formed on the basis of a lunar rhythm. Such a hypothesis was then also voiced by Kahn and Pompea [40]. In support of this they quoted other molluscs that also showed tidal growth lines in a daily rhythm. Equally, Kahn and Pompea argued, the activity cycle of the *Nautilus,* and the rhythmical rise and fall of the ocean, caused a physiological rhythm which showed itself in the daily periodicity of calcium excretion. In addition, there were indications that the air chamber formation in the *Nautilus* always lasted several weeks. The two researchers then extended their research to the fossilized shells of earlier nautilids. Here it was discovered that the growth lines per air chamber continuously reduced in number the further back in the Earth's history one went; that is, the deeper the layers of undisturbed sediment were located in which the fossils were found. Assuming that the excretion of the inner walls took place in the same way among earlier relatives of *Nautilus*, the orbit of the Moon around the Earth and the distance between the Earth and the Moon can be calculated for earlier periods. These calculations, of course, assume that the masses of the Earth and the Moon have remained constant during the period under consideration. This reveals that the duration of the Moon's orbit has not always been the same, but was shorter in the past. This means that the Moon travelled faster around the Earth, had a smaller orbit and was thus still closer to the Earth. At the time that the Alps were being pushed up in the Tertiary period, the synodic orbit of the Moon lasted about twenty-five days, according to Kahn and Pompea, when the Jurassic system was being formed it was only eighteen days and in the Carboniferous fifteen to sixteen days. In the Devonian period, the distance between the Moon and the Earth was half as much as it is today. The two researchers followed the lunar orbit back to the Upper Ordovician period using fossilized nautilids and recorded the duration of the synodic 'month' as being only nine days on average.

Kahn and Pompea's work was challenged on several counts [41, 42, 43, 44, 45, compare also 46]. On one hand, studies on animals, both in the wild and in the aquarium, suggested that the formation of the air chambers in *Nautilus* required a period of one to four months or longer [compare 47, 48, 49; shell development in aquatic organisms 50, 51, 52, 53], whereby the decisive element for the formation of a new chamber was said not to be the lunar rhythm but the changing requirements of the *Nautilus'* buoyancy. Equally, the extrapolation of present-day growth curves to fossil predecessors appears questionable because studies of closely related ammonites produce different results in this respect.

Even if these results contradict the above work in detail, there are, on the other hand, further studies which point in the same direction, namely that both the length of the days, and the length of the lunar month, have increased in the course of the Earth's history and that the Moon was once positioned closer to the Earth. Indications that this is so were provided by the comparison of fossil and recent shells of corals, molluscs, brachiopods and stromatolites. Calculations based purely on astronomical or (geo-) physical foundations (e.g. observation of eclipses, radar distance measurements of the Moon, analysis of tidal friction and sediment) reach the same conclusions [17, 40, 54, 55, 56, 58, 59, 1078]. The day lengthens by about 0.002 seconds in a century and the distance from the Earth to the Moon grows by about 3.5 centimetres [17, 60]. Over a large part of the Earth's history, the rhythms of the Moon (synodic month) and of the Earth (daily rotation) have been inscribed in the shell patterns of the fossils like in natural memory; only in recent years have we been in a position to begin deciphering it.

The origin of the Moon

From our study of *Nautilus,* we were led to the question of the origin of the Moon. At Full Moon, we see how the illuminated disc reveals large, somewhat darker patches, the 'man' or 'hare' in the Moon. In earlier times, people thought that these darker patches must be oceans and called them *mare*, which they have been called ever since, though there is no water on the Moon. The *mare* are in fact large, flat plains without any noticeable contours. A look through a telescope is enough to show us that. We can also see that the lighter parts are chains of mountains and volcanoes. These highlands were described as *terrae* (Latin *terra* = Earth) in contrast to the *mare*. The *mare* cover about one third of the visible side of the Moon. The other side of the Moon, invisible to us, but in the meantime familiar through space flights, has almost no *mare* but many craters. The whole of the surface of the Moon is like a rocky desert. Since the first manned flight to the Moon in 1969, numerous rock samples have been brought from the Moon to the Earth. The main minerals on the Moon are familiar rock formations on Earth, but new minerals, unknown on Earth, were also found, representing compounds of titanium, magnesium, iron, aluminium and others. No trace of life was found, that is no organic compounds and no water.

There are various hypotheses as to the origin of the Moon [61, 62].

The *Capture Hypothesis,* frequently quoted before the advent of manned flights to the Moon has become less likely because lunar and Earth minerals are too similar. According to this, the Moon, originating in the far reaches of the solar system, was captured from the cosmos by the Earth and forced into its orbit through its gravitational field. Furthermore, the relationship of the oxygen isotopes oxygen–16 and oxygen–18 are almost identical in both, which also suggests a common origin of these two celestial bodies.

According to another hypothesis, the Moon originated by splitting off from the outer layers of the Earth *(Fission Hypothesis).* This was articulated for the first time by George H. Darwin, the second son of the evolutionary biologist Charles Darwin. According to his thinking, the Earth rotated so fast after having formed its core, that a large lump was created at the equator which later separated like a large drop to become the Moon as the Earth and Moon solidified. Apart from the external similarity of the mineral samples, this is also supported by the average density of the Earth's crust and the Moon which at 3.3 g/cm^3 is less than the average density of the Earth as a whole at 5.5 g/cm^3. Since the Moon does not possess a general magnetic field, its core cannot contain any significant amounts of iron. The reason for the lower content of volatile elements such as sodium, potassium, lead or gold in moonstone, as opposed to the surface rock of the Earth, could be that these elements were lost due to high temperatures when the Moon split from the Earth. Just as land and water are irregularly distributed over the Earth, large areas of *mare* are only found on the side of the Moon which faces Earth. Could it be that these large-scale asymmetries complement one another in a way which suggests that they may in the past once have been united? If the Moon had separated at one time from the Earth, is there anything on its surface that might reveal that to us? Particularly noticeable in this respect is a large *mare* surrounded by craters in the southern hemisphere of the Moon, directly behind the visible side of the Full Moon, on the left: the *mare orientale* or *eastern sea.* This *mare* is surrounded by seven concentric rings of mountains. It looks like a scarred navel: the place where the Earth and Moon once separated? But then the Earth would also have to show such traces. The permanent dynamics of continental drift and weathering has largely made such traces disappear on Earth. The largest ocean, the Pacific, has been considered repeatedly to be the place where the Moon separated from the Earth, but the rock on the ocean floor is too young (no older than the Jurassic) for any definitive statements. Above all, this theory raises questions with regard to celestial mechanics. Why does the Moon not orbit the Earth in the plane of the equator rather than in an orbit which is tilted at 5° to the ecliptic? Furthermore, it is doubtful whether the centrifugal force of the Earth was ever large enough to eject the Moon.

A third theory says that the Moon solidified as an independent celestial body from dust and gas clouds of the original solar nebula, at the same time as the Earth *(Co-Accretion Hypothesis)*. The Moon would thus have arisen as the concentration of a ring system which once surrounded the Earth. But this poses the question why the Moon has such a small metallic core in comparison with the Earth.

According to the much-quoted fourth theory, the *Impact Hypothesis,* the impact of a huge celestial body, about the size of Mars, with the young, still-forming Earth threw debris into orbit from which our present neighbour was eventually formed. Here the iron core of the colliding celestial body would largely have been taken up by the Earth and the Moon would have been created from the silicate parts of both collision partners. But all in all, the question of the Moon's origin remains an open one, although the capture hypothesis is the least likely.

Corals and the Moon

Let us return to the greatest, most incomparable treasure of the Earth: to the organisms representing the world of evolution, and the development of ever new things. Let us consider a further outstanding, indeed, unique example of lunar rhythms in the living world: the Great Barrier Reef in the Coral Sea of the Pacific Ocean. As the largest existing reef on Earth, it follows the north-eastern coast of Australia for two thousand kilometres. Coral structures are not known in the North Sea or Mediterranean. They require well ventilated water which is 20°C at minimum, rich in nutrients and clear — conditions which only exist in tropical or sub-tropical regions. Thus we particularly find numerous reef and island-building corals near the coasts between the tropics of the Indian and Pacific Oceans. Here there are still enough nutrients (the sea zones close to the poles are even richer in nutrients) and the surf ensures sufficient oxygen supply. The coral animals can only live up to about 40 metres depth for they all form a symbiotic relationships with single-celled plant organisms which require light for photosynthesis.

The northernmost reef-forming corals in the northern hemisphere are found in the Gulf of Aqaba, the north-eastern branch of the Red

Sea. The multitude of forms stretching up through the Red Sea and the enchantingly shimmering array of colours of the tropical marine animals contrast dramatically here with the surrounding desert landscape. Parallel with, and just off, the coast, a fascinating and beautiful barrier or coastal reef extends from here to the southern tip of the Sinai. This can be enjoyed even just snorkelling. Fish of all different colours (compare Plate 12) and fantastic shapes inhabit the richly structured underwater landscapes of coral and sponge. The silver flashing bodies of passing schools of fish emphasize that, in contrast, the actual coral fishes with the magnificent colour patterns tend to be solitary or in pairs, with defined territories. If insufficient care is taken when diving, the multi-coloured coral can give divers bloody scrapes. Hard material is deposited under a thin softer skin; this is the calcium skeleton of these animals, capable of building reefs as mighty as the Great Barrier Reef.

The majority of the atoll, reef or island forming animals belong to the group of stony corals (Madreporaria or Scleractinia). The soft-structured Actiniaria sea anemones are closely related to them, and are found also in the North Sea and Baltic. Anyone who has not noticed them on rocky shores at low water will probably have discovered them firmly attached to rocks or wood in aquaria. The brightly coloured cylindrical shape is crowned by a ring of tentacles, which surrounds the mouth opening underwater. The first impression is of a plant, which is reflected in their name. Actiniaria, or Madreporaria, are included in the group of Anthozoa *(flower* animals), including coral animals, which comprises about 6500 species. This is the group with the most species within cnidarians (9000 species).

Cnidarians are built rather simply in their fundamental morphology and are thus classed as lower animals in terms of their taxonomy. Their body is only made up of two layers of cells: an outer skin which covers the body and an inner one to line the inside of the body.

The body cavity only has one opening which fulfils both the task of taking in food and excreting. A characteristic feature of many cnidarians is that they can occur in two forms: as polyps and as medusa. Both forms can produce the other. The mostly attached polyp produces the free-moving medusa (commonly called jellyfish) asexually through separation. The latter is the sexual generation with gonads (ovaries or testicles). The medusa produces egg or sperm cells which produce free-floating larvae when they combine, which soon attach themselves and which in turn produce polyps. As cnidarians, corals possess many stinging capsules or nematocysts; these highly specialized cells in the outer skin of the animal can give people a hefty sting if touched.

Corals also belong to Anthozoa. These animals, which occur exclusively in a marine environment, possess the special characteristic that they no longer form a medusa. The polyp either reproduces by sprouting asexually, which keeps all polyps linked to one another, or it can reproduce sexually with sexual organs (gonads) in special folds of the body cavity. In this case the individual polyps can either be of separate sexes or hermaphroditic. New coral polyps are produced by the fertilized egg cells.

Actiniaria are solitary-living larger polyps. The polyps of stony corals (Madreporaria), on the other hand, remain smaller (1 to 30 millimetres) as mostly colony-forming species. However, they are differentiated from the Actiniaria, above all, because the foot plate of the cylindrical body excretes an external skeleton of lime which produces the typical coral structures. The basal plate excreted by a young polyp is strengthened by radial strips. The continued deposit of calcium thus produces an ever higher ridge which pushes the foot plate upwards ahead of itself. Greater and greater numbers of polyps are created through sprouts which excrete lime in the same way, and are many metres high.

In the case of reef-forming polyps, the cytosymbiotic algae, the zooxanthellae [for the present-day taxonomy of this group compare 63], which live in their inner layer of cells, are also of importance. The assimilation activity of these algae removes some of the carbon dioxide from the soluble calcium bicarbonate in the sea water, which means that calcium carbonate is excreted for hard lime formation. Corals inhabited by symbiotic organisms thus produce considerably more lime than corals not inhabited by symbiotic organisms. If the snakelocks anemone *Anemonia viridis* is kept in the dark for several months without food, it loses approximately the same amount in weight as is made up by the photosynthesizing production of its cytosymbiotic organisms in the normal dark-light rhythm [63]. Conversely the algae benefit from the degenerative products of the animal metabolism such as carbon dioxide, phosphate and nitrogen compounds.

The individual little polyps of a reef are not very apparent during the day. They withdraw back into their firmly built fortress. Only at night, when the many plankton drifting in the water come to the surface from the deeper layers of water do the polyps unfold themselves and their tentacles in the hunt for food. The minute plankton, paralysed by the nematocysts are either eaten or smaller particles are guided to the mouth by the tentacles.

The description of the reproductive behaviour of the grunion and the palolo made clear that their orientation towards environmental rhythms affects the behaviour of the individual in relation to the others in a species, in such a way that synchronized reproduction can take place. The special characteristic of the Great Barrier Reef consists of the fact that not only the individuals of a coral species but many different species all release their sex cells at the same time. In order to investigate this phenomenon further, 107 species from 36 genera and 11 families of Scleractinia were studied at five different locations on the Great Barrier Reef up to five hundred kilometres apart [64]. This makes up about one third of all the stony coral species which occur there. Of the species investigated, 105 released their sex cells through their mouth opening in the week following Full Moon. In the location closest to the coast, this occurred in October, and in November in the other four locations. A particularly high level of activity occurred in the third to sixth night after Full Moon. At all five places a total of 87 of the single sex or hermaphroditic species released their female and male sex cells in this time, mostly in the four hours after sunset. In 17 of the 33 species studied for this purpose, the release took place not only at the same time within a population in a location, but within an hour of other populations of the same species in other more distant locations. In other words, whole populations of many species in locations far distant from one another all release their reproductive cells together on the same night of the year within a few hours of one another.

Continuous studies of fifteen species over the course of a year established that spawning only took place once a year. It is known from other studies that this applies also to other species. The same can be assumed to apply to many other species since many samples taken throughout the year show that sex cells were immature at times other than in the southern spring, when the mass event which occurred. On the basis of such studies it is safe to assume that a further twenty species of stony coral from thirteen genera and six families spawn in the three to six nights after Full Moon. A further seven species from four genera belonging to a family other than the one described above release their gametes at this time. Among the stony coral there is another group in which the calcified skeleton is of less importance than an inner skeleton, which provides the main support of these often equally large colonies. This inner skeleton consists of small lime needles or horn-like substances, and includes, for example, the precious red coral which is popular as jewellery.

The male and female sex cells are often released stuck together in round bundles and drift to the surface in that form. About 2.5 hours after spawning the cells unite and fertilize. The developing larval forms can start propelling themselves about thirty-six hours later and start to swim about until they attach themselves and form new growths of coral.

In the northern Gulf of Aqaba, near Eilat, the thirteen ecologically most important species have been studied to see whether they spawn on a mass basis. Although twelve of the thirteen species spawned in a lunar rhythm, they did not do so at the same time but in various seasons, months or lunar phases within a month [65]. This emphasizes the special character of such mass swarming in the Great Barrier Reef, and is thus specifically related to the South Pacific.

The swarming of the palolo only takes place in October and/or November. The grunion, in contrast, swarms during all six of the summer months. Various rhythms of different length are thus superimposed here: the (semi-) lunar and annual rhythm. This is frequently supplemented by the daily or 24-hour rhythm with its alternation of dark and light. Thus the release of gametes in the coral of the Great Barrier Reef takes place in the four hours after sunset. It is precisely events such as this kind of mass swarming which once again raise the issue of their causal forces. How can such precise and comprehensive synchronization take place between all these different individuals? What is the decisive factor here? Do these rhythms still appear when these animals have been taken out of their natural environment and kept under constant conditions? These are questions which we will investigate in the next chapter.

CHAPTER 5

The Causal Question

We have already seen, in many instances, that the rhythms lived by organisms are closely co-ordinated with their environment. The tides, the alternation of day and night, the waxing and waning of the Moon, the course of the year all of these things — are regularly occurring events in the biosphere, which also affect these organisms. The question thus arises whether these biological rhythms are spontaneously produced by the organisms themselves, that is whether they are *endogenous,* or whether they are reacting to periodically changing environmental factors which externally affect them as *exogenous* organisms.

In order to be able to describe biological rhythms adequately, we start with a simple cycle. Figure 24 shows a sine-curve. A given dimension x fluctuates over the course of time t around a mean, or constant, value (straight line). The greatest swing of x in relation to the mean value is called the *amplitude.* The *period* of the oscillation is the time required to pass through the whole cycle. The *frequency* of the oscillation on the other hand specifies how many cycles are passed through in a specific time. Period and frequency are reciprocal values. Two oscillations of equal period may reach their amplitude at different times. These two oscillations are then out of phase.

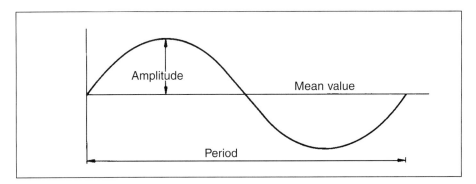

Figure 24. Basic concept of a harmonic oscillation (sine-curve) [238].

One of the key results of research into rhythms has been that many biological rhythms continue even under constant conditions. What was initially found for daily rhythms, that is solar fluctuations of life processes, was also confirmed for the lunar periodicity of organisms: their tidal, semi-lunar and lunar periodicity continues even if the effect of external periodic timers ceases. If we recall, the tidal vertical migration of the turbullarians, *Convoluta,* for example, continued under laboratory conditions, that is without the changes resulting from the tides [66]. The amphipod, *Synchelidium,* only leaves the sandy substrate of his beach habitat when it is covered in water. Animals freshly caught on the Californian coasts were put in seawater in vessels with five to ten millimetres of fine sand at the bottom. With constant permanent light, the temperature of the water in the aquarium was kept at a constant 20°C and the water was continually and carefully stirred. Even under these constant conditions a tidal rhythm could be observed in these crabs. As can be seen from Figure 25, the rhythm of their swimming activity copies the tidal course of their original biotope in detail. Corresponding to the different levels of the two tides occurring on the Californian coast in a day, there is a greater and a lesser maximum of swimming activity. Also the slight temporal variation between the tides, such as occurs in the biotope, is reflected in the maximum values for swimming activity under laboratory conditions [67].

These results were achieved not only among invertebrates but also with vertebrates, such as, for example, the small North American fish, *Fundulus grandis*, which belongs to the killifish. It spawns in Barataria Bay in Louisiana at an interval of thirteen to fourteen days. If the fish are brought into aquaria at a temperature of 23°C and are subjected to light-dark illumination of alternately twelve hours, the previous rhythm is maintained. As shown by one study, there was a clear increase in the number of eggs deposited every fourteen days in five of a total of seven aquaria, over a period of four months. Since all fish in a single aquarium (21 female animals each) as well as the fishes in various aquaria to some extent deposited their eggs at the same time, it may be assumed that there was synchronization through messenger substances across the various aquaria. All the aquaria were connected by the water circulating between them.

A further point worth noting was that the spawning activity for the whole period of the experiment was co-ordinated with the tides occurring simultaneously on the Gulf coast. Spawning always took place at the spring and neap tides. As already mentioned in the chapter on the

5. THE CASUAL QUESTION

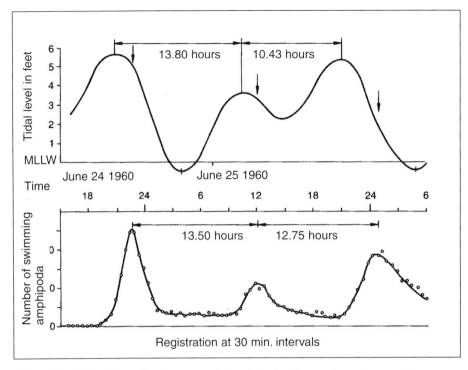

Figure 25. Tidal rhythm of swimming activity of the freshly caught marine amphipod, Synchelidium, *under constant laboratory conditions. For comparison, the tidal rhythm in the natural habitat;* MLLW: *mean lower low water of the two unequal tides on the west coast of North America (see Chapter 2 for San Francisco)* [1056].

tides, a one-day tidal form occurs in the Gulf of Mexico with spring and neap tides in the tropical rhythm of the Moon, that is alternating between 13.66 days and 27.32 days. This rhythm is displayed by both the *Fundulus* fish living in the wild and those in the aquaria [68, 69].

A sea snail, lunar periods and the Earth's magnetic field

In view of such parallels, the question arises whether there were not continuing exogenous effects in the studies which have been described, and which have so far been left out of account. The factors which are kept constant are mostly light conditions, temperature, current, salt content etc, but just like the water mantle of the Earth, the atmosphere and the solid Earth also display lunar tidal periodicity. Thus certain layers of the ionosphere at the equator fluctuate up and down several kilometres with the orbit of the Moon. Something similar applies to the magnetic field of the Earth, air pressure, polarization of celestial light, rainfall etc. [1093-1101]. Some of these factors can only be excluded from experiments with great difficulty and at great expense. Could some of these influences not continue to be effective in laboratory experiments to enable organisms to maintain their rhythms?

In some organisms, any continuation of a tidal rhythm was absent when they were transferred from their natural habitat into an unchanging artificial one. In others the tidal rhythm waned within a few days and thereafter could no longer be distinguished. In others again, only irregular patterns of activity could be determined. The question must be asked whether the disappearance, or irregular appearance, of the rhythms in these cases is due to unsuitable and too artificial conditions in the laboratory. A registration box under artificial light can hardly replace the familiar habitat.

Behaviour in various magnetic fields was investigated using the sea slug, *Tritonia diomedia*. To begin with, the orientation of the longitudinal bodily axis was observed in the Earth's magnetic field which had not been interfered with. The majority of the animals turned from a northerly direction towards the east with a deviation of approximately 88°. Then the direction was studied in a second magnetic field in which the horizontal component of the Earth's magnetic field was neutralized, in that an artificial field of the same intensity, but in the opposite direction, was generated. The distribution of the animals according to the points of the compass was now random (Figure 26). This means

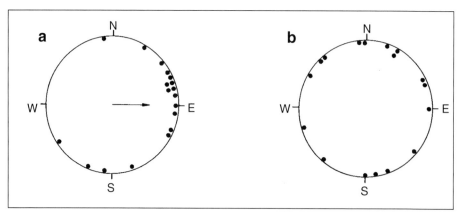

Figure 26. The orientation of the the Pacific sea slug, Tritonia diomedia, *in two different magnetic fileds. a) in the Earth's magnetic field, b) in a magnetic field in which the horizontal component of the Earth's magnetic field was neutralized by an artificial field of the same intensity.*

that under normal circumstances the animals can orientate themselves according to the geomagnetic field; if the horizontal component of the same is neutralized, they can no longer do so.

The direction of the bodily axis in the Earth's magnetic field was checked regularly over a period of four months. It turned out that the preferred orientation towards the east changed for no apparent reason. Only when someone hit on the idea of relating the direction of the animals on various days to the respective phases of the Moon did it emerged that the east was only preferred with a waxing Half Moon. In the course of the continuing synodic cycle the direction then changed from north (Full Moon) via west (waning Half Moon) to south (New Moon) and finally back to east.

If these experiments were initially concerned with the longitudinal axis of the body (an animal situated towards the west could display an easterly orientation), an attempt was subsequently made to resolve the question whether any orientation was displayed in the movement behaviour. For this purpose the animals were placed in the base of a Y-shaped tube with the bodily axis oriented initially towards the south-east (135°). When the animal moved forwards it had the opportunity, at the fork, to move to the left (corresponding to east in the Earth's magnetic field), or to the right (corresponding to south in the Earth's magnetic field). In the preceding experiments, east to north of the Earth's magnetic field was the preferred orientation in the week before Full Moon. In the experiments with the tube during the corresponding period it was to be expected that the movement at the junction would tend to be towards the left (towards the east), rather than to the right (towards the south). This was indeed the case with seventeen of the twenty-one animals [70].

These experiments show that the movement behaviour of the small sea slug *Tritonia* is influenced by magnetic fields whose dimensions are of the order of the Earth's magnetic field. It was also shown that the preference for a specific orientation of the bodily axis, or a specific direction of movement in the Earth's magnetic field, progressed in accordance with the lunar rhythm. It remains an open question, of course, whether these discovered characteristics are of any significance in the natural environment. It could be that after about a month the animals arrive approximately back where they started; this could be significant in terms of reproduction. The animals would then have crept in a counter-clockwise circle. It could also be the case, however, that the animals react under laboratory conditions, but that the Earth's magnetic field is irrelevant under natural conditions. This can only be shown by field experiments, that is, observations directly in the sea.

We know that other organisms (earthworms [71], snails [72, 73, 74], bees [75, 76, 77, 78], birds [79], mammals — including human beings [80, 81, 82, 83, 84]) are also sensitive to magnetic and electromagnetic fields [compare also 85, 86]. Thus it is known that some birds orient themselves by the Earth's magnetic field on their long-distance flights. This knowledge indicates that, for example, changes in the Earth's magnetic field, connected with lunar periods, can still be effective as an exogenous timer when the latter function can be excluded for other factors. The same applies to other rhythmically changing environmental factors which can be experimentally excluded only with great difficulty.

Endogenous rhythms?

Experiments relating to the continuation of biological rhythms under laboratory conditions led to a particularly important discovery. As an example, we will quote the marine isopod, *Excirolana chiltoni,* from the surf zone of the Californian sand coast. In experiments under constant laboratory conditions with animals from La Jolla in southern California, this isopod maintained its tidal rhythm of swimming activity for a period of two months. At the beginning, the two maximums for swimming activity still coincided with the high tide on the coast, but over time the phases of the two cycles drifted apart so that after about two months maximum activity had fallen about five to six hours behind the corresponding phase of the tidal rhythm. Because of the mixed, mainly semi-diurnal tides, it appears appropriate to calculate

Figure 27. Activity rhythm of an individual animal of Macrophthalamus hirtipes, *a fiddler crab from the western Pacific, under constant conditions. Top: the hourly activity of a single crab over a day, or, since the period is about 24 hours 50 minutes, slightly more than a day. The mean value calculated from all the hourly values for a day is represented by the straight line. The dashed line below, shows an 'e' for each hour in which the activity was equal to or larger than the mean for the first peak, and 'o' for the second peak. An 'x' shows the mean value for the activity has been exceeded without reaching a full peak.*

Several days taken together produce the middle graph. Here the hours are counted beyond the 24 of the first day; the 25th hour is the same as 1 o'clock. The angle of the straight line drawn through all the points 'e' and 'o', show activity periods. They are greater than 24 as they do not run vertically. Under laboratory conditions, the period of activity lengthens from 24.8 to 25.4 hours. The mean data of many experiments are graphically represented at the bottom [88].

5. THE CASUAL QUESTION

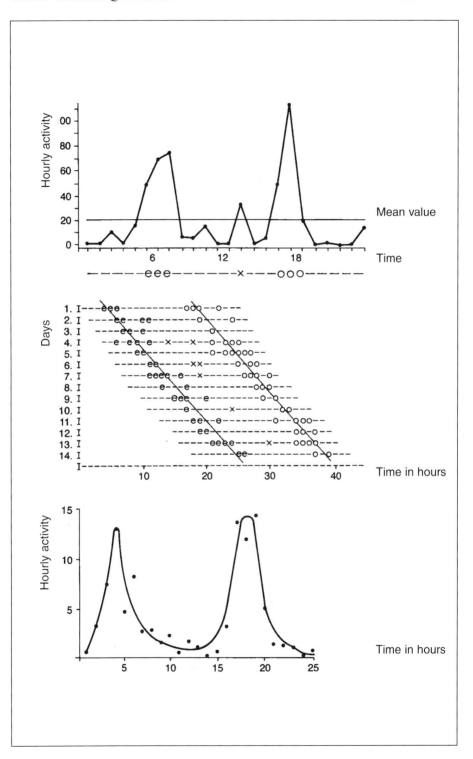

the period from one higher high water, through the lower high water, to the next higher high water, as a total cycle. From this perspective, we have the result that the average period of the laboratory rhythm at approximately 24 hours 55 minutes is about five to six minutes longer than the average period of the tides at 25 hours 50 minutes [50]. The tidal rhythm of the isopod still has a lunar rhythm superimposed on it, whose periodicity produced values of 26 to 33 days under laboratory conditions.

The same phenomena of *Excirolana chiltoni* can be found in many other organisms, such as in individually studied crabs of the species *Helice crassa*, *Macrophthalamus hirtipes*, *Uca pugnax* and *Uca pugilator* (for Uca compare Plate 10). With these the period of their (lunar diurnal) movement activity also changed within a range of 21 to 28 hours. The oscillation period of the rhythms can thus shorten or lengthen if the natural timers are no longer given under constant maintained conditions [88]. Figure 27 shows the results for *Macrophthalamus hirtipes*.

The simplest interpretation of these observations would be to assume an endogenous, self-generated rhythm which continues to oscillate under constant conditions with its own periodicity or spontaneous frequency (free-running rhythms), and which continues to be synchronized with events in the environment through periodically effective external factors working as timers. The periods of biological rhythms, which also continue under constant conditions, and which change in relation to environmental rhythms, are one of the best arguments in favour of their endogenous character. Specifically, if we assume that other external factors to the ones already considered are at work, the period should not change in relation to the environment other than if we try to explain the diverging periodic length of the biological rhythms through frequency transformation in the organism. But then we would have to ask why this happened in relation to more subtle factors, and not in relation to such obvious ones as light and temperature.

Since the freely running rhythms no longer correspond exactly to the period of the environmental rhythms — that is, are no longer exactly tidal or semi-lunar, but are only approximate (Latin = *circa*) — we refer to the endogenous rhythms as circa-rhythms. The rhythm of *Macrophthalamus hirtipes* described above is thus a circa-tidal one. Similarly, we refer to *circa-semi-lunar* or *circalunar* rhythms (compare above for *Excirolana chiltoni*). What has been described here in terms of lunar rhythms, was shown first and can still be shown today in much larger scope for daily rhythms [89, 90, 91, 92]. But equally we know today that annual rhythms can be endogenous in nature. In other words, we know of free-running rhythms with circannual periods [93, 94, 95].

Exogenous rhythms?

The occurrence of endogenous rhythms does not, however, mean that there are no rhythms which are exclusively exogenously determined. Tulips, for example, open and close their blossoms daily. If they are kept at a constant temperature, this movement does not occur. Getting colder or warmer therefore acts as an exogenous factor for the opening and closing of the blossom. If the sea anemone (*Actinia equina*) is kept underwater in the laboratory, we can control whether it opens or closes its tentacles by light. If there is light for twelve hours and then the same period of darkness, the animals extend their tentacles for twelve hours and contract in the darkness. If the change from light to darkness is made every two hours, the polyp also changes every two hours between extension and contraction [96]. On the other hand, we know biological rhythms which exist as exclusively free-running or spontaneous rhythms, without any interaction with the environment (e.g. in nerve action). From the perspective of the interaction between organism and environment, we can thus divide biological rhythms into three groups:

— EXOGENOUS or EXO-RHYTHMS are exclusively controlled by the environmental periodicity;
— EXOGENOUS-ENDOGENOUS or exo-endo-rhythms are rhythms produced by the organism itself, which are maintained under constant environmental conditions. the periodicity of which can be synchronized through timers with environmental rhythms of similar duration;
— ENDOGENOUS or ENDO-RHYTHMS are produced spontaneously by the organism itself, and do not display any immediate relationship to environmental periodicity.

Thus — as we have seen — lunar rhythms are only examples in all three modes of general biological rhythms as a whole.

The American, Frank A. Brown Jnr., is the main representative among those researchers who have repeatedly raised the question whether the exo-endo-rhythms, which continue under constant conditions, are not controlled by rhythmic systems from the environment [97, 98, 99, 100, 101, 102, 103, compare also 104]. One of his arguments is based on studies of the fiddler crab (*Uca*). It continued to display its daily rhythmical changes of skin colour under light and temperature conditions, which were kept artificially constant: during the day it had a dark colour which lightened during the night. The degree of colour change can easily be determined by the pigmented cells of the skin covering, the so-called *chromatophores*. If this rhythmical change of skin colour is produced by these crabs themselves, and the organism is constituted by nothing other than material and chemical processes — thus Brown's thinking — then this chemical clock should be changeable by means of temperature changes. In chemistry *van't Hoff's rule* states that chemical processes occur twice as fast when the temperature rises by 10°C, and at half the speed when the temperature falls by 10°C. Biological processes as a rule also increase by a factor of two to three with a temperature increase of 10°C. Brown, then, on one occasion kept his study animals at 25°C instead of 16°C, and then at 5°C. Although the colouring at the higher temperature was more intense, the period of colour change in the *Uca* crabs remained the same in all three temperatures of the experiment. This still came to precisely — not approximately — 24 hours in the fiddler crabs, even after several weeks under constant conditions [86, 105, 106, 107, 108, 109, 110, 111 p.111ff, 112 Ch.15; on the question of the temperature-dependence of biorhythms compare also 113, 114, 115 p.442ff]. These and other studies on the daily and lunar periodic movement rhythms, and on oxygen consumption in fiddler crabs, potatoes, carrots respectively etc., under so-called constant conditions, made Brown and his fellow researchers increasingly doubt that endogenous timers alone could provide a full explanation of the observed phenomena. The only explanation they could find for the behaviour of the *Uca* crabs was to assume that the Moon exercised a direct influence on them, through its strongest physical effects on the Earth such as gravity or magnetism. They gave much greater weight to the exogenous component in all life cycles running in parallel with the terrestrial or cosmic environment.

However, Brown's interpretation of an exogenous cosmic link can hardly be maintained today. As early as 1960, Hammer tested the theory in an expedition to the Antarctic when several organisms, whose daily periodicity had been well researched, were put on a rotating disc near the South Pole which was turned counter to the Earth's rotation in a 24-hour rhythm. The plants and animals concerned (a mushroom, beans, cockroaches, flies and a golden hamster) were thus exposed to the stellar cosmos almost completely uniformly. Their position in relation to the cosmos did not change over 24 hours, yet their night-day rhythm was fully maintained [116, compare 117 p.17ff]. In addition, there are indications today of a continuation of the daily rhythm with constant parameters under conditions of weightlessness in space [118, 119, 120, 121, 122]. Such experiments clearly show the endogenous nature of the daily and lunar periodicity of organisms.

However, the endogenous character of these rhythms must not be overestimated. When the botanist, Erwin Bünning (whose name is particularly linked with chronobiological research in Germany), germinated and grew seeds of the scarlet runner bean under constant conditions — either under constant light or constant darkness — their typical rhythmical leaf movement only appeared when he gave the young plants a short light or dark stimulus. This stimulus triggered the inherited endogenous rhythm of the plant which was then maintained for a long period of time under constant conditions [90 Ch.3, 112 Ch.8]. In this case the endogenous rhythm was dependent on an environmental stimulus which was needed to trigger the endogenous rhythm, or to couple the various individual physiological functions to it.

Life indivisible

On closer inspection, the alternative of endogenous or exogenous rhythms is thus in many cases a false one. Endogenous rhythms are dependent on exogenous influences *(triggers, timers)*, and the supposed exogenous rhythmicity has an endogenous base. This is also expressed in the exo-endo-rhythms.

We can consider these facts from yet another perspective. If we speak of inherited endogenous rhythms, modern genetics knows that no traits, capabilities or other characteristic of an organism are inherited but, strictly speaking, only flexible norms for reacting to the effect of suitable environmental influences by developing the relevant characteristics or skills. What are inherited are merely *reaction norms*. Some examples will clarify what that means. Every green plant has its chlorophyll due to heredity because its parent plants were also green. After all, 'heredity' means nothing more than the similarity of a living organism with its ancestors.

However, heredity as the basis for the production of chlorophyll is not in itself sufficient if, for example, the potato shoots have developed in a dark cellar. They remain white, and only turn green when exposed to light. Thus the only thing which is passed on is the reaction norm of producing chlorophyll on being exposed to light. But neither is it just the light which produced such an effect on its own. The human organism, for example, reacts to light by forming vitamin D in the skin; here heredity produces a different effect. Heredity cannot do without environmental influences, and environmental influences cannot do without heredity. Fundamentally, to divide them disregards their mutual interdependence. A much more relevant question is whether the reaction norm is broader (more flexible) or narrower (less flexible). For heredity is present even in broad reaction norms, even if the transformation of characteristics under environmental influences is referred to as a modification which, by definition, is not based on heredity. The band width of modification is specified in terms of heredity by the width of the reaction norm. The dispute about nature or nurture is largely redundant. The only point at issue is how broadly or narrowly the inherited reaction norms are.

In this sense the endogenous rhythms are not inherited, only the readiness of the organism to develop them at the appropriate time (sensitive phase), under suitable environmental influences. However, in this context it also becomes clear — on closer examination — that the influence of the environment only exercises a trigger function, because the living organism is in a position to accept environmental stimuli such that its own physiology is preserved. Otherwise we would have to consider it to be wholly part of its environment and would ignore its physiological autonomy regarding heredity for example. The concept of organism would then be just as superfluous as that of the environment.

We will use a characteristic example to clarify this linkage between the organism and its environment. The fertilized eggs of bladderwrack (*Fucus vesiculosus*) undertake the division of the cell as their first incisive development process, in which the polarity of the future organs of attachment (rhizoids) and the plant body (thallus) is established between the two subsidiary cells. Normally this differentiation is triggered by the reduction of light intensity in the cell. The illuminated side of the fertilized egg cells turns into the thallus cell after first division, the dark side into the rhizoid cell. At first sight, light appears to be the cause of this polarization, but closer investigation reveals that the light does not play a specific role at all. The light has no effect immediately after fertilization. If illumination is delayed beyond the appropriate time — that is, beyond the sensitive phase — the *Fucus* egg accepts any other spatial trigger, be it the gravitational field, a one-sided touch trigger, a magnetic or electrical field, other surrounding and closely attached zygotes, or, as the case may be, the light-dark difference in the egg. Only if there is a complete absence of external direction-giving factors, something which can only be achieved in an artificial experiment, do the rhizoids in *Fucus* zygotes arise in arbitrary spots [122a]. The important thing, then, is that any environmental polarization of whatever kind is provided as the trigger for the own physiological polarization, the possibility of which increases with time. The one will not succeed without the other.

We have repeatedly referred to *timers*, those environmental stimuli by which the endogenous rhythms of the organisms are synchronized with the rhythms of the environment, and by which they are linked with the corresponding order of the environment. For coastal organisms with tidal rhythms of movement activity (e.g. surfacing from the sand, feeding, migration on the substrate) the following timers could be discovered, for example: fluctuations in temperature triggered by the changing tides, turbulence connected with water movement, being flushed with water, fluctuations in standing water pressure or salt content, light changes correlating with the tidal changes (increased darkness because of churned-up mud or sand, for example) [28, 123, 124]. In the study of semi-lunar and lunar rhythms, moonlight could be identified as the timer in many instances. We need only recall the various populations of the marine midge *Clunio*. A lunar swarming rhythm could be triggered in the marine annelid, *Platynereis dumerilii*, through weak nocturnal illumination (0.02 to 0.1 lux) for six days in each month [125, 126, compare also *Typosyllis prolifera*]. Success was also achieved with artificial moonlight as the trigger in the tropical crab, *Sesarma haematocheir*. This land-living crab releases its larvae in a freshwater river just above where it flows into the sea. This happens every fourteen days after sunset at the time of the Full and New Moon. In addition to a dark-light alternation of 14 hours to 10 hours, artificial moonlight was given in the dark phase, the light cycle of which was delayed by a week in relation to natural moonlight. The release of the larvae could still be observed every two weeks, but now in parallel to the artificial 'New and Full Moon' [127, 128].

A link between the potential time order of the living organism and

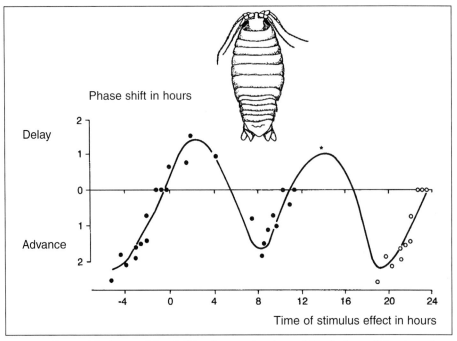

Figure 28. Above: The Californian marine isopod Excirolana chiltoni *at three times its natural size (slightly changed from [629]).*

Below: The curve shows a tidally changing readiness to change in response to a two-hour vibration. The following rhythm of its own activity can be delayed or brought forward depending on the phase of the free-running previous rhythm at which the vibration took place. The maximum of the first movement activity corresponding to the first of two tides in a day was chosen as the reference point (= 0^h on the horizontal).

Black circles: experiments with many animals.
Star: with single animals.
White circles: continuation of black circles. A similar type of thing happens in nature during storms [123].

the periodic influences of the environment can thus always be shown. Indeed, the environment is not something separate from the organism but is part of its existence, just as the organism exists as part of the world to which it belongs. That, clearly, is the nature of life.

Uniformity or multiplicity in temporal behaviour

As we saw in the isopod *Excirolana chiltoni* (p. 92) the tidal rhythm of swimming activity also continued under constant laboratory conditions. A further study was designed to show the effect on the free-running rhythm of a single stimulus — vibration for a period of two hours

5. THE CASUAL QUESTION

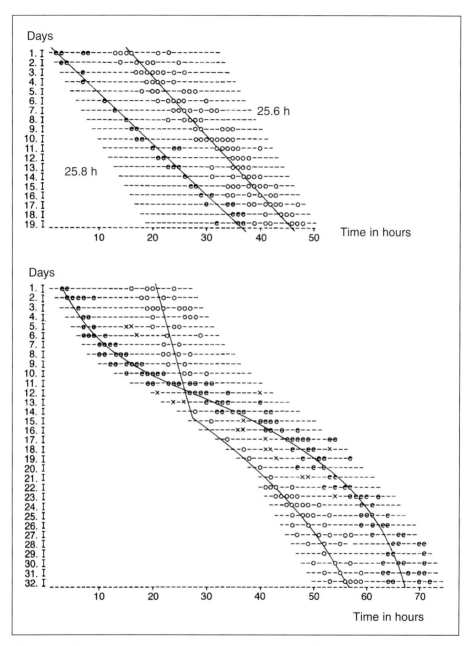

Figure 29. The free-running rhythms of movement activity of individually studied crabs. Top: Helice crasse, *below:* Macrophthalamus hirtipes. *For explanation see Fig. 27 [88].*

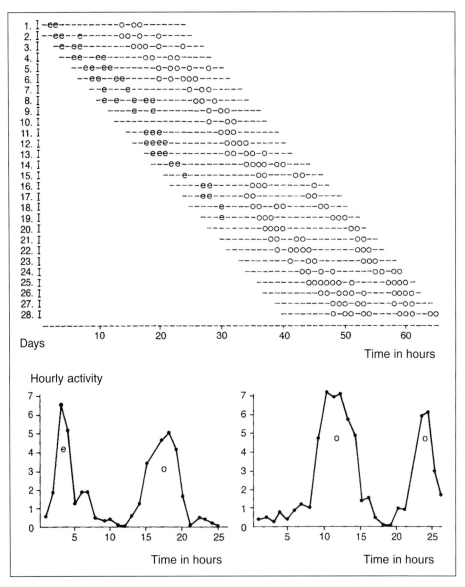

Figure 30. Free-running rhythms of movement activity of an individual fiddler crab (Macrophthalamus hirtipes).
Top: the period of both activity phases of this fiddler crab initially comes to 25.6 hours. On the fourteenth or fifteenth day one of the two activity phases (o-peak) divides into two; both develop an interval of approximately twelve hours and retain this until the twenty-eighth day. The other activity phase (e-peak) is extinguished after about two-and-a-half weeks.
Bottom: for creating the left curve the data from the first two weeks and for the right curve the data from the last ten days above were included.
For explanation compare Figure 27, p.92f [88].

as simulated water movement — which was given at different phases of the rhythm. Figure 28 shows the effect for a series of measurement points producing the phase-response curve. From this curve we can see that the same stimulus leads to a delayed phase in the sense that the phase of the rhythm, after the single stimulus, is delayed or brought forward to a certain extent, in relation to the free-running rhythm before the stimulus. This phase shift means, in the first instance, that the artificial stimulus can clearly act as timer for the endogenous movement activity of this little crab, but it is also evident that this depends on the time at which the timer occurs. If the stimulus was given in advance of the expected maximum movement activity, the phase shift was brought forward. In contrast, application of the stimulus during, or shortly after, the maximum produced a delay. The reaction of this animal to one and the same specific stimulus can thus be quite different depending on its physiological state at the time that the stimulus takes effect [129, compare also 28 p.360f 123, 130]. The maximum value for a delay or advance of the movement rhythm was no greater than two hours. As far as the situation in its known environment is concerned, this indicates on the one hand an openness for entirely new situations and, on the other, that the organisms can counter temporary irregular conditions with a certain inertia, such as storms, which can greatly change the normal tidal rhythm.

The amphipods *Synchelidium* and *Excirolana* on the Californian coast are similar in this sense, in that their rhythms in their natural environment display two different activity phases within one lunar day. It seems obvious in this case to consider the lunar day as a period since the pattern of activity is repeated with this periodicity, but the question remains — to what extent are the two activity phases within one lunar day related to one another? Revealing results were found in studies on other species of Californian crabs. Figure 29 (top) shows the free-running movement activity rhythms of *Helice crassa*. It can be clearly seen that the two activity phases falling in a lunar day have different free-running phases in the laboratory, namely 25.8 and 25.6 hours. This tendency is magnified in *Macrophthalamus hirtipes* (Figure 29 bottom). In the laboratory, different periods sometimes occurred in the laboratory in individual animals of this species, so that it appeared as if one activity phase crossed the other. Other variations again occurred with fiddler crabs (*Uca*). Thus one of the activity phases gradually disappeared while the other divided into two activity phases at almost the same time (Figure 30).

The results of these laboratory experiments can be understood if we assume that these two activity phases, which under natural conditions are coupled and run synchronously with the tides, are based on inner periods which are independent of one another. The periods of both can change under certain conditions. One can divide without the same occurring in the other one. One can disappear while the other remains. Results of this kind were found for five species from three genera and two families of coastal crabs [88, compare 111, 131]. The two activity phases of the specified species falling in a lunar day are thus based on two independent rhythms with the common period of the lunar day, whose maximum phase shift can be about 180° (= half a period). According to these experiments, then, the biological rhythms in the tidal rhythm should be described as *circalundian* rather than *circa-semi-lundian*.

The period of the lunar day only differs by about fifty minutes from that of the solar daily rhythm and there is an overlap between the two under laboratory conditions. Thus it is also possible to ask whether the two rhythms are not based on one and the same endogenous, flexible foundation. Various timers (for example, factors connected with the tides for lundian, and light for the light-dark rhythm) would then lead to the appearance of the basic general, though sensitive, rhythm, as solar diurnal rhythm on one occasion and as lunar diurnal rhythm on the other. This is supported by the fact that we also know that organisms with circadian rhythms have periods that can spontaneously change [132, 133, 134] or divide as in Figure 30 [124 p.296, 135, 136]. In addition, both rhythms show a similar sensitivity towards certain substances (e.g. deuterium oxide or azadirachtin) [88, 111 p.120ff, 124 p.292ff, 137, 138, 139, 140; on the interaction of solar and lunar diurnal rhythms compare also 141]. There are, nevertheless, also observations which support another model; namely that both rhythms are based on two different biological clocks. The beach crab, *Carcinus maenas* (Plate 11), which only display the solar diurnal rhythm, displayed a tidal rhythm with a period of 12.4 hours after being cooled down to a temperature of 4°C for 15 hours, although the animals had never been exposed to a rhythm with this periodicity in terms of their environmental conditions. The possibility of a 12.4 hour rhythm clearly appears to be inherited, and an expression of an independent internal clock independent of the solar diurnal rhythm [142].

Interestingly, after hyposmotic shock treatment (the animals were placed in diluted salt water) the same species displayed a two-peak

5. THE CASUAL QUESTION

rhythm which is clearly semi-diurnal. Since the phase position of the two activity peaks in relation to one another did not remain stable, it may be assumed that the two activity phases are based on two independent rhythms with circatidal periodicity, whose phases have shifted in relation to one another [143].

With these results, then, no specific categories can be established and the reality might simply be that both, or several, variations are always possible. Here we encounter one of the basic experiences in chronobiology: the close interrelationship of the rich complexity of factors which we consider in isolation. Life is interconnected in many different ways not only in its spatial shape, but also in its respective species-specific *Zeitgestalt,** or temporal form.

As a whole, detailed analysis of these factor demonstrates that the chronobiological behaviour of these coastal crabs is clearly capable of compensating in two directions. If *Excirolana* is mistreated with two hours of artificially stormy water, it responds — depending on the stage of its inner phases — with an advance or a delay, to synchronize its own rhythm with the disturbance. If exactly the opposite is done, and all the environmental factors are provided with the greatest possible uniformity, the animal's own otherwise synchronous, uniform rhythm divides. With non-physiological change (= disturbance), the organism unifies its temporal behaviour; with non-physiological uniformity, it provides multiplicity. In respect of its rhythms life is always intent on softening the extremes without excluding the physiological element.

With several organisms of the tidal zone, we noted that they follow both the solar diurnal rhythm and the lunar semi-diurnal or diurnal tidal rhythm at the same time. As we have seen, these rhythms, like all characteristics, are inherited through reaction norms, that is endogenously. That applies equally to semi-lunar and lunar rhythms. The superimposition of the two former rhythms produces a cycle leading to an exceptional value every fourteen to fifteen days, when the maximums or minimums of these two rhythms coincide. Does the endogenous semi-lunar or endogenous lunar rhythm therefore result from the interaction of the circadian and the circatidal or circalundian rhythm, that is from the superimposition of daily rhythm and tidal rhythm?

Experiments to discover whether this is the case were carried out with the seaweed *Dictyota*. Changes in circadian solar periodicity,

---

* *Zeitgestalt* is German for 'a shape in time, showing a wholeness.'

produced through the appropriate changes in the timer periods, did produce a shortening or lengthening respectively of lunar rhythm, but these changes did not correspond to what was anticipated on the basis of the calculations [91 p.124, 144 p.171ff, 145]. This might be due to the circatidal or circalundian rhythms also being affected alongside the circadian one by the changed timer periodicity. Results from other experiments have not yet produced a uniform picture either; some results [91 p.123ff] support the view that the semilunar or lunar rhythm is the result of the superimposition of the two shorter rhythms; others do not [90, 109, 124 p.298, 125, 146, compare 130 p.32ff, 144 p.171f, 145, 147 p.79f, 148, 149, 150 p.121f, 151, 152, 153, 154].

CHAPTER 6

Moon Rhythms in Non-Marine Organisms

Freshwater animals

In European freshwaters, mayfly larvae can be easily recognized by their two or three long feathered tail fans, which they use to propel themselves forwards. When the larval development, which can take several years, comes to an end, the larvae leave the water and transform themselves through their last but one moulting into a winged pseudo insect, the 'sub-imago' which moults one last time to become the finished insect, the imago (Figure 31). Their life in the air often lasts only a few hours, a few days at best, and is devoted to reproduction. Around

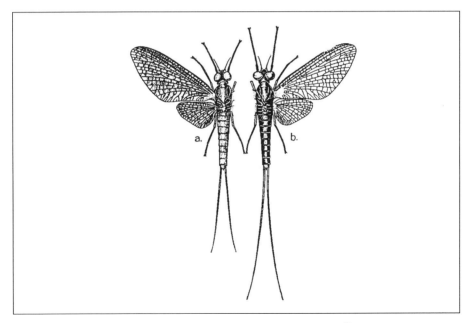

Figure 31. Sub-imago (a) and imago (b) of Isonychia siccus Walsh. ♀, *a European mayfly with a body length of twelve millimetres (from [1090]).*

sunset the males gather in dense clusters for mating displays. Females flying into the swarm are immediately grasped by the males in the air and mating takes place. The females then deposit their eggs in water without delay.

An African mayfly, *Povilla adusta* on Lake Victoria, was discovered to hatch and swarm only in the five to six days before and after Full Moon, and specifically on the second day after Full Moon. In the climatic uniformity of the tropics, and particularly the tropical lakes, it is the lunar cycle which in this case ensures reproduction. The *Povilla* imagines only live for about an hour and a half which means that the hatching of the larvae must be well co-ordinated. This happens when light from the rising Full Moon lengthens the short tropical periods of twilight, in which the mayflies undertake their mating display. Since the lunar rhythm of hatching continued in mayflies brought into the laboratory, we can assume an endogenous component. In the following examples, however, it must be said that few studies exist which have examined the extent to which the observed rhythms also continue under constant conditions.

If in the tropics the Moon is important as a timer, the relative constancy of the environmental conditions means that reproduction can take place throughout the year. In 37 species of insects from four orders occurring on Lake Victoria, just four were identified whose frequency fluctuated rhythmically with the phases of the Moon: *Athripsodes stigma* (maximum in the fourth lunar quarter); *Athripsodes ugandanus* (maximum shortly after the first lunar quarter); *Clinotanypus claripennis* (maximum at first lunar quarter); and *Tanytarsus balteatus* (maximum two to five days after New Moon).

The studies were carried out by means of a light trap in which artificial light was used as bait for the insects. The above results contradict the objection that the insect catch rate is due to the apparent variation in brightness of the light trap against the changing moonlight in the background. It is, in fact, evident that it is a rhythm connected with reproduction, namely of hatching and swarming, from the fact that the maximum appearances of the four species fall at quite different times of the lunar cycle, and not all at the time of the New Moon (that is at the time when the light trap is at its relative brightest). Furthermore, no cycles could be observed in the other species [see also 155, but compare 156].

Thus the many other results which displayed semi-lunar or lunar fluctuations in the frequency of the various insects, using light traps, have to be partly interpreted as corresponding reproductive rhythms. This also includes species which are not tied to water for their development. Similar observations were made also in India, North America and Australia, and in other latitudes than the tropics (see Catalogue of Species).

Remaining in Africa, the population density of various free-swimming crabs (*Crustacea*) fluctuated in a lunar cycle in Lake Cahora Bassa in Mozambique. They occur most frequently just before New Moon, but this periodicity is not the result of varying reproductive activity but is caused by the feeding behaviour of a fish. The predatory *Limnothrissa miodon* from the herring family prefers to hunt for the crabs on nights when there is bright moonlight, otherwise hunting for other prey. This gives the populations of the various freshwater crabs the opportunity to recover their numbers.

Land animals

In comparison to life by and in the sea, reproductive rhythms with lunar periodicity has been found in relatively few land animals. The South Asian water buffalo (*Bubalus arnee*) is said to mate mainly at New Moon and occasionally at Full Moon (Plate 2). Mating in the African brindled gnu (eastern white-bearded wildebeest, *Connochaetes taurinus albojubatus*) and rutting in the local Impala are also said to be influenced by the lunar cycle. In one Australian kangaroo, the Tammar Wallaby (*Macropus eugenii*), embryonic development is regularly arrested for a certain period so that the further maturing and birth of the young animals falls in the most favourable time of the year. It was observed in caged animals that the continuation of embryonic development is linked with the Full Moon of the summer solstice.

Several field and laboratory studies show that the activity of nocturnal animals is influenced by moonlight. The porcupines of the Negev Desert in Israel, for example, during the winter (October to March) only leave their lairs briefly for minimum exposure to the moonlight. Many other small mammals, such as rodents and bats, also display reduced activity during nights when there is a bright Moon. The maximum night activity of the kangaroo rat (*Dipodomys spectabilis*), for example, occurs in the time after sunset, and then reduces for the rest of the night. If the Moon was high in the sky, night activity was reduced by a third, and the animal moved from more open into bush-covered terrain. With the South American leaf-nosed bat (*Artibeus lituratus)* moonlight also inhibits activity. Accordingly the activity of the bat is reduced during the first half of the night with a waxing Moon, during the second half of the night with a waning Moon, and for the whole night during Full Moon. The Egyptian fruit bat *(Rousettus aegyptiacus)* has its maximum activity with a waxing Moon, while another bat, the spear-nosed bat *(Phyllostomus hastatus)* does so with a waning Moon. Both species display minimum activity when the Moon is full. The activity curve of the South American nocturnal monkey *(Aotus trivirgatus)* runs exactly contrary to this. Kept in a cage with natural light, it displayed a strong dependence on lunar phases during the night. During New Moon, its activity is restricted largely to dusk and dawn, while it is active for the whole of the night during Full Moon. Mice and bats tend to use their hearing for orientation, monkeys their sight.

Laboratory experiments make it likely that the activity periods are often determined by the circadian rhythm, but that sun and moonlight can directly vary behaviour as exogenous timers. The extent to which a general endogenous lunar periodic component is involved in animals which are active at night, remains an open question.

Lunar periodic activity rhythms are also found in other terrestrial vertebrates. Birds in particular reveal themselves to be creatures of the eye. In Britain, it can be observed in lapwings *(Vanellus vanellus)* that their normal behaviour (the search for food largely during the day, rest at night) is reversed during the winter months at the time of Full Moon. This is not the case when the light of the Full Moon is obscured by thick cloud cover. Cranes *(Grus grus)* seek their sleeping grounds much later after sunset the stronger the light of the Moon is.

The lettered terrapin or red-eared turtle *(Pseudemys scripta elegans)* from North America has been studied under constant light and temperature conditions, and indications of a lundian rhythm of movement activity was found. A slight maximum occurred at the lower culmination of the Moon, and a slight minimum at the upper culmination. Experiments with Faraday cages indicated that the rhythm, and thus the organism, could be influenced by electrostatic fields.

Beyond the rhythms of movement activity, periodicities of inner physiology were also found. In the North American Leopard frog *(Rana pipiens)*, for example, the two daily maximums of calcium transport activity of the duodenum coincide at Full Moon, at the first and third quarter and at New Moon (only these four days were investigated), with the upper and lower culmination of the Moon. The calcium ion concentration in the blood plasma is also at its maximum at these times. In the North American Eastern Red-Spotted Newt (*Triturus* = *Diemictylus viridescens*) oxygen consumption oscillates in a lunar rhythm with the maximum value three to four days before Full Moon.

Land plants

As for ocean regions, most studies on land have been carried out on animal organisms, but as current studies of plants show, detailed investigation reveals processes with lunar periodicity here too. (A number of examples were cited in the introductory chapter.) As early as 1923, E. S. Semmens referred to an increased germination rate in mustard seeds under the influence of moonlight. She linked this with the polarization of the reflected light at certain periods of the lunar cycle, and with the effect this has on the hydrolysing, starch-splitting enzyme diastase [157]. Also in the 1920s, Lily Kolisko was able to observe experimentally a link between the synodic lunar rhythm and plant growth [158]. Seeds of various vegetable, garden and seed plants (lettuce, white cabbage, leeks, tomatoes, peas, beans, lovage, yarrow, lemon balm, larkspur, maize, wheat, oats, barley) sown two days before Full Moon germinated better, grew larger, formed more numerous blossoms and produced a larger harvest, than seeds sown two days before New Moon. The lunar phase at the time of germination thus determines the whole growth cycle, blossoming and fruit production [158, 159 p.13ff, compare 160 p.174, 161, 162; on the link between the Moon and plant growth compare also 163]. More recently, Rounds was able to show a semi-lunar rhythm in certain substances in quite different species of blossoming plants (*Coleus blumei*, common bean = *Phaseolus vulgaris*, *Philodendron sagittifolium*, *Forsythia spec.*, *Lilium tigrinum*, *Ulmus americana*, *Geranium spec.*). Extracts were gathered from the leaves of the plants, which produced an increase in the heart rate of the American cockroach (*Periplaneta americana*). This stimulating effect was minimal in extracts taken shortly after Full or New Moon. The results of such experiments make it seem likely that the time when medicinal plants are harvested, for example, is not irrelevant. Depending on the purpose, the pharmaceutical effects could be optimized by the choice of the appropriate harvesting date. This provides us with new insight into ancient traditions and folklore about the connection between the Moon and the living world.

The evolutionary aspect of biological rhythms

On the basis of the progress of current research, we can expect it to be only a matter of time and effort before lunar rhythms are revealed in just as many organisms on Earth as has already been observed in innumerable living creatures for the biological solar rhythms (e.g. circadian rhythms, photo periodicity, annual cycles, etc.). We were able to give no more than an introductory, albeit typical, look at this field of chronobiology. If we look back on the examples collected here, a first evolutionary line can be discerned.

As regards primitive organisms without a cell nucleus — cyanobacteria (blue algae), eubacteria and archaebacteria, jointly called procaryonts [compare 164] — it was a matter of dispute for a long time whether they could produce solar or lunar periodicity spontaneously on their own. After all, their lifetime in an optimum environment lies in the range of hours if not minutes. How can they display a circadian rhythm, for example, if they do not even live for a day like mayflies, but may only exist for ten minutes, for instance, until their next division?

Now it is quite possible to observe rhythmically expanding and contracting physiological processes with longer-term periods in procaryonts, but the question remains whether such rhythms are also produced endogenously in an environment in which the conditions are kept constant. It is only a few years ago that such a rhythm was found for the first time, namely the rhythm for nitrogen fixation in the marine blue algae *Synechococcus* with diurnal periodicity [165, 166, 167, 168, 169, 170, compare also 171]. Apart from this, there is as yet no other evidence for a lunar diurnal rhythm or longer-term lunar periods in procaryonts, but no specific studies have yet been carried out with this purpose.

However, if no other lunar periodic rhythms under constant conditions can be shown to exist, this does not mean that cyanobacteria and eubacteria do not display a multitude of lunar cycles, not endogenously but exogenously determined by the environment. That is understandable on the basis of their organization. Procaryonts belong to the smallest organisms. They possess a great deal of surface area in relation to their volume and thus have optimum contact with their environment. The consequences of this are considerable. Owing to their ability to multiply in myriads over a short period of time, the procaryonts form a mighty biomass everywhere in the Earth's biosphere, which actively provides for a great flux of substances in the environment. The largest part of the oxygen and carbon dioxide cycle in the Earth's atmosphere is not produced by higher plants and animals but by the innumerable microscopic algae, protozoa and procaryonts without a nucleus. Alongside the tiny *picoplankton* [172, 173], free chlorophyll grain organelles [174, 175, 176] must also be considered. These are called prochlorophytes [177, 178, 179, 180] and live in immense quantities in the sea. They have only been known for a few years [compare also 181, 182]. Finally, viruses which are not visible under a normal microscope, and which inhabit the oceans of the world in myriad numbers as femtoplankton [183, 184], are also likely to be very significant for the geochemical cycles of our oceans since they attack and dissolve bacteria, leading to the release of carbon and nitrogen [185]. If we also include the assimilating and dissimilating unicellular organisms, (that is, the well-known producers and decomposers), the microbial realm by far exceeds all multi-cellular plants and animals on Earth in terms of its ecological effect. Such significant integration into the environment means that they hardly possess any internalized rhythms of their own, but are dependent on the rhythms provided by the environment. The same applies to the many multi-cellular organisms which they inhabit symbiotically. Lichen, for example, are fungi which often live in a symbiotic relationship with blue algae. They only grow under certain weather conditions. Crustose lichens in particular exercise a considerable ecological effect in this respect since they cause about half of all rock erosion on Earth [186].

Accordingly, the tides also have a direct and immediate effect on the life cycles of bacterial and blue algae cover on the floor of the tidal flats. In this sense these prokaryonts live purely exogenously in terms of lunar periodicity. They are organisms which are so highly adapted to their environment that they are largely determined by any of its changes at any moment [for instance compare 187 on the temperature changes of the floor of the tidal flats].

Blue algae and bacteria also have nucleic substance (DNA, RNS) but without the delimitation from cell plasma. This is only achieved in the protista, true unicellular organisms with a nucleus, which are collectively described as eukaryonts (Greek *eu* = good, *karyon* = nucleus) with all multi-cellular organisms. Physiological periodicity begins to develop clearly independent, endogenous components, in parallel with this additional protection of those cell parts which are of importance for heredity, and through which each living organism is linked to a greater extent in genetic continuity to its generational series, than to the environment. Parts without a cellular nucleus of the unicellular algae *Acetabularia* display diurnal rhythmical photosynthesis activity over many weeks [188 p.192], but if a nucleus from another cell is inserted, which displays the opposite rhythm to the first cell, this fragment assumes the phase of the implanted nucleus [189]. In crossing related strains with different period lengths for circadian rhythms, both intermediary and additional hereditary processes could be observed [90 p.15ff, 190, 191]. The nucleus thus contains constitutive elements of what, as the *inner clock*, enables a first level of autonomy of the physiological conditions related to time [compare 192].

Our collection of all known findings (see Catalogue of Species) illustrates the large degree of lunar periodicity in the realm of multicellular organisms, with seaweeds, coelenterates, worms, echinoderms and tidal zone molluscs of all coasts especially represented. The insects, determined by air space and daylight, show less of a lunar rhythmicity; and when they do, it is often their larvae, living in the sea or in freshwater, which accompany the external lunar periodicity with their endogenous activity. In all these invertebrates, the nervous system is decentrally organized in the form of a diffuse ring-like or rope ladder-like network of nerves positioned directly in the soft body particularly in the vicinity of the stomach (ventral nerve cord).

In vertebrates, by contrast, the nervous system, fully centralized from the beginning, is intensely screened from the biological environment: in the development of the spinal cord (in clear contrast to the ventral nerve cord, which is close to the ground in all *Articulata*: annelids and arthropods such as insects) and above all the brain. These organs are encapsulated by the bony mantles of the vertebrae and the skull, and are also mechanically protected from the environment. A physiological blood/brain barrier is even erected with regard to the organism's own blood supply which blocks some substances. Among vertebrates we also find endogenously secured lunar periodicities, noticeably in species, which live in water or are nocturnal. In a statistical analysis of the pregnancy periods of 213 terrestrial higher mammals (placental mammals) — including human beings — the period of thirty days and its whole-number multipliers turned out to be values which stood out [193, 194]. These periods are almost identical with the synodic lunar orbit, even if the pregnancy periods are not correlated with the Moon. The lunar periodicity is most strongly internalized precisely in the fact that it is produced spontaneously and no longer requires the link to the Moon. This applies to a large extent to female menstruation in humans and apes. This chronobiological emancipation of higher living organisms contrasts the overwhelmingly exogenous character of the lowest organisms, the prokaryonts. Human beings have gradually become estranged from surrounding nature and its rhythms, not only physiologically but also psychologically and thus culturally, and in terms of their civilization. Indeed, the fact that they are in the process of losing all connection with nature, can also be seen in that humans are freeing themselves from these evolutionary reminiscences: female menstruation is becoming increasingly irregular. In other fields, however, such as human sensory perceptions, we continue to be strictly linked to the external Moon — even when we do not see it. Thus the rhythmical spectrum of human beings still combines the whole range of externally and internally generated life processes. We are not just biologically internalized, living organisms but continuously carry all stages of evolution in us. We will discuss that further in the next chapter.

CHAPTER 7

Moon Rhythms in Human Beings

Sensitivity of the eyes to Sun and Moon rhythms

Imagine yourself in a blossoming summer garden as the Sun is setting. You are surrounded by many different flowers and experience the onset of night with open eyes. There is no artificial light. You notice that the colours begin to fade as the twilight increases, but the eye gradually becomes used to the growing darkness so that the contours of plants and other objects remain recognisable for a considerable time. After half an hour or a whole hour in the nocturnal darkness your eyes have become so sensitive that you can walk through the woods without a torch even on moonless nights. You simply no longer see any colours, just delicate variations of dark and grey.

If we observe this transition from daytime vision to night-time vision attentively, we discover that the restriction of our colour vision first affects red and orange, then yellow and green, and finally blue and violet colour tones. The bright red blossoms of the garden poppy are the first to turn black, while the blue flowers of the bluebell can still be recognized as such for a long time. We observe the same sequence when the increasingly weak evening light shines through the bright stained glass of a cathedral window. The first to go dark are the red panes; the blue panes also grow paler, but appear comparatively brighter. At dawn, the process is reversed: first blue colours brighten up, then yellows, oranges and reds. This phenomenon is today described in the physiology of the eye as the *Purkinje phenomenon*. Jan Evangelista Purkinje, the Czech naturalist and contemporary of Goethe, who lived at the turn of the eighteenth to the nineteenth century, was born in 1787, in Libochovice, near Litoměřice, in Bohemia. Destined to be a monk, he left the priesthood at 21 years of age shortly before his ordination, studied philosophy and medicine and began to read Schiller, Novalis and Fichte. However, his study of Goethe's theory of colour was of decisive importance to him, and it led him to many new discoveries of his own *(Seeing from a Subjective Perspective* [195]). In 1822, he discussed his ideas with Goethe, and in 1823 received his first professorship in Breslau (now Wrocław) through the latter's good offices. He subsequently taught in Prague where he laid the foundations for the classic physiology of the senses and histology. He died there in 1869. Even today certain brain cells are named *Purkinje cells* after him. Purkinje's fibres is the name for muscle fibres in the heart, which have been transformed into impulse-conducting fibres, and ophthalmologists refer to *Purkinje's vascular tree,* this being the blood vessels in the retina which can be seen when light shines on it from the side. When he was still in Breslau, he set up the first physiological laboratory, and in 1825 discovered the *Purkinje phenomenon* described above as the result of his exact investigations. All that is required to repeat it is a slide frame half covered with a red and a blue film of the same brightness. The colour fields are projected and, once the eye has become accustomed to them, they are gradually covered with a grey film until they are almost dark. The blue appears brighter than the red for a long time.

This discovery led to the hypothesis that we have two different types of visual cells in our eyes, which on one hand are better suited to daylight and can see the red side of the colour spectrum better, and on the other, enable vision in subdued light and are more sensitive to the blue side. This theory was later confirmed through the discovery by microscope of the smaller *cones* (daytime vision), and the longer *rods* (night time vision), as the two effective types of visual cells in the retina. During the day the cones see all colours, while at night the rods perceive different levels of brightness without colour (which is why we see everything in shades of grey), but when there are low light levels (twilight) the cones are most sensitive in the yellow range (560 nanometres), which is more on the red side, while the rods are more sensitive in the green range (510 nm), that is tending towards the blue side. When the visual effectiveness of the cones begins to fail at twilight, the rods can still see greens and blues as bright. Nocturnal animals such as owls and cats only have rods in their eyes.

However, things became even more interesting in the early 1940s [196, 197, 198] when it was discovered that this sensitivity threshold does not remain constant, but fluctuates rhythmically independently of our control. In exact measurements during laboratory experiments to determine the colour efficiency of test persons with completely healthy vision, it emerged that the determined values could not be repeated a few weeks later. At first, researchers suspected faults in the lamp apparatus until measurements over a whole year showed that the colour

7. MOON RHYTHMS IN HUMAN BEINGS

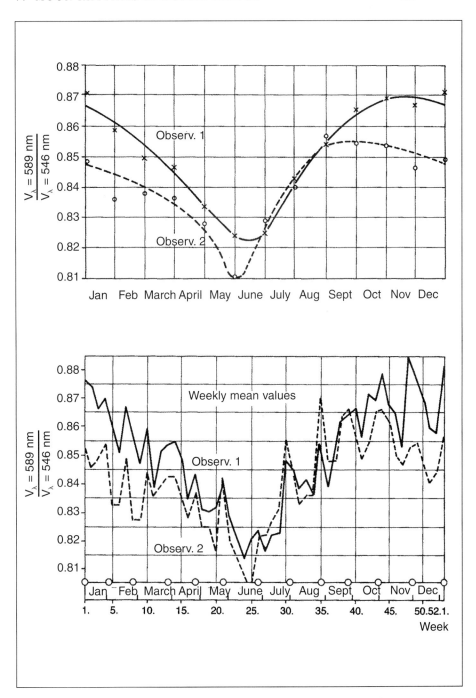

Figure 32. The dependence of spectral brightness sensitivity on the time of year in human beings for two test persons (observers). Top: monthly mean values for $V_\lambda = 589$ nm/$V_\lambda = 546$ nm. Bottom: weekly mean values. [197].

sensitivity of the human eye when seeing at twilight fluctuates in three superimposed rhythms: a daily rhythm; an annual rhythm; and a synodic lunar rhythm. Thus each person is more sensitive towards blue than red at the height of summer, and additionally throughout the year, is more sensitive to red than blue in the days around Full Moon, irrespective of whether we have observed the changes in the lunar phases or not. Figure 32 top shows the light sensitivity in two test persons measured over the course of a whole year for the colours orange (589 nm) and yellow-green (546 nm). The eye test was carried out each week and then the monthly mean was entered. If sensitivity to long waves (orange appears brighter) is predominant, the value of the fraction 589 nm/546 nm on the vertical axis rises. If the short wave sensitivity of the eye increases, the value drops — as happens in June. Figure 32 bottom shows the same arrangement. Here the weekly mean values are entered, as well as the Full Moon dates for the various months. We can see immediately that the value of the fraction rises at Full Moon, that is that the two test persons see the long wave colours more brightly.

Through the work of Lang [199–206] we know that the lunar periodicity of colour sensitivity also exists in fish. The Guppy *(Lebistes reticulates)*, a fish frequently kept in aquaria, comes from pools, ponds and other small still waters in Venezuela and some Caribbean islands. The normal position of the fish, with its back uppermost, is not only dependent on the gravitational field perceived by the gravity sensitive organs (static organ in the ear labyrinth) of the animal, but also on the light in the water which mostly shines from above. If the fish is artificially illuminated from the side, it moves its back into a sloping position which increases the more intensely it perceives the light. The angle of slope is a measure for the light sensitivity of the fish. This changes in the Guppy over the synodic month in such a way that yellow is perceived most strongly at Full Moon, and least at New Moon. For the boundary colours of the spectrum, red and violet, the maximums and minimums were precisely opposite.

That human beings and a fish fluctuate in their colour perception in parallel with the phases of the Moon, leads to the assumption that all vertebrates may have this ability, and not just their highest and one of their lowest representatives. Can something similar be shown for other sensory fields if the measurement were precise enough? From 1963 measurements of light sensitivity for the same colours were carried out by Lang on several occasions. It frequently happened that the extreme

7. MOON RHYTHMS IN HUMAN BEINGS

monthly values not only coincided with the days of the syzygies, but even with the hour of Full or New Moon. Divergences from this pattern were the same for all animals studied. Lunar rhythms could also be observed with the same phases and amplitudes in animals, which had been kept under constant illumination, or under so-called constant conditions, since birth. These are indications of an exogenous cause for the rhythms in question. A factor oscillating with the same period must thus have been effective in the laboratory despite the *constant conditions*, but so far no physical factor has yet been discovered in real terms.

There is another human rhythm which is connected, among other things, with sight. In 1977, Californian doctors became aware of a young man who had come for treatment because of sleep problems [207, 208]. The strange thing was, that at regular intervals of about two to three weeks he suffered from nocturnal sleeplessness and a high degree of tiredness during the day, strongly affecting both his work and his leisure activities. He was psychologically a completely normal twenty-eight-year-old man, with a degree in biostatistics from one of the larger universities, and was now pursuing his career under normal circumstances. The fact that every two weeks or so he lost his normal day-night rhythm indicated that his sleeping-waking rhythm was not oriented to the solar day but to the lunar day of almost 25 hours. When, as part of his medical examination, he was allowed to sleep and wake as he wanted over a period of almost four weeks, his need for sleep coincided almost precisely with the local low water on the Californian coast.

What was unusual about this young man was that he had been blind from birth, and could not perceive the daily light-dark alternation. With an alarm clock and sleeping pills he forced himself into the day-night rhythm of his fellow human beings during critical periods. Figure 33 shows the growing trend during Phase A (at home) to sleep longer in the day, and to go to bed very late at night. Phase B, during the stay in hospital, shows his own free rhythm of sleeping and waking. The sleeping phases clearly slip out of synchronization each day by just under an hour with the light-dark rhythm dictated by the Sun — just as the tides do. Phase C, at home again, represents the hopeless attempt to synchronize his own rhythm with that of the solar day.

If sighted people are studied over a longer period of time, shielded from any outer periodicity, it is frequently found that their sleeping-waking rhythm has a period of about twenty-five hour hours [see, for

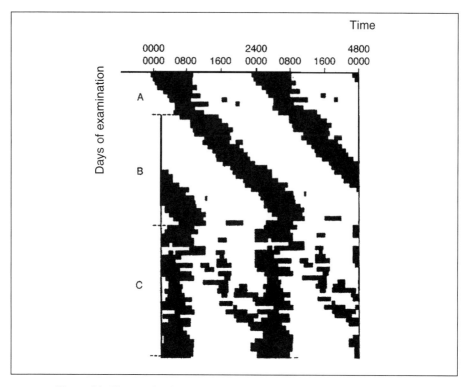

Figure 33. Sleeping/waking rhythm of a blind man. The dark thick lines represent sleeping phases. In the clinic he was allowed to sleep when he wanted (Phase B), in his professional life the sleeping rhythm is disturbed by the enforced rhythm of the solar day (Phase A and C) [208].

instance 209, 210, and compare 211]. When left to its own devices the daily rhythm corresponds to the lunar day. The above example thus shows how in sighted people the rhythmical stimulus of light links our basic underlying rhythm of 24.8 hours to a true day-night rhythm [compare 212, 213 Ch.26, 214, 215, 216]. In the blind man this link was clearly absent which meant that the course of his day coincided with the changing tides [but see 217]. Since the eye is a sensory organ which is particularly pronounced in humans — the majority of mammals orient themselves primarily by their sense of smell — we achieve our normal daily rhythm, of waking and sleeping, largely in harmony with the daily course of the Sun and the own rotation of the Earth. In addition, social factors can determine the daily rhythm in humans as timers.

Kidneys, lungs and Moon rhythms

A further phenomenon in human beings, which is dependent on the orbit of the Moon around the Earth, was described for the first time by Schweig [218] in 1843. In studies on himself, he discovered what was later confirmed by other researchers in further studies [219, 220, compare also 222] — namely that the rhythm of uric acid excretion is governed by the phases of the Moon. In all analyses there was reduced excretion at New and Full Moon, and on the following day. A particularly low uric acid value was also always observed on the fourth day after New Moon. The agreement between the various studies is particularly noteworthy since the uric acid value of the excretions is dependent on many factors.

Alongside uric acid, urea is another substance which the kidney must excrete, and is the greater as regards quantity. It is the result of protein decomposition, that is of the metabolism of the cell plasma of all body cells, is easily soluble in water and can thus be easily washed out by the kidney. Uric acid, in contrast, is only present in urine in small quantities, is the result of the decomposition of nucleic acid (especially the purines) in all body cells, is difficult to dissolve and can only be transported by special means (liquid protective colloids) in lymph, blood and urine. If uric acid is deposited in the lymph of the joints, it results in gout; if it is deposited in the renal pelvis or the bladder uric acid calculi are created. Uric acid mineralizes much more easily than urea, which in turn is formed in the liver by carbon dioxide and ammonia. The urea and uric acid values thus reflect basic physical metabolic processes. Every muscle activity resulting from physical movement affects the uric acid value. It is also dependent on the type and quantity of food. In order to achieve controlled values, studies were undertaken on subjects with stomach illnesses who were being fed with a standard feeding tube. In these cases, too, the above rhythm was shown to exist.

Illnesses associated with the uric acid value have rapidly increased in the past decades. Gout generally used to be a disease of old age, but is appearing increasingly among young people today. Medical practitioners have noticed that this illness tends to affect people with high performance levels or high achievers. A research project therefore investigated whether gout-causing uric acid in the blood was connected with success itself. Initial studies of various social groups have shown that the average uric acid value rises in line with the rise in social group [222].

Does the physiology of uric acid then have a psychosomatic side? The body-soul problem has a long history. The current hypothesis is that phenomena of consciousness are the result of highly complex peak achievements of biological processes, but too many facts contradict this. Konrad Lorenz, for example, drew attention to the fact that the nerve structures, which are available to our waking consciousness, are less complex than those which transmit unconscious organic and psychological actions, the co-ordination of which is far more complex [223]. We would be kept rather busy if we had to consciously regulate just the highly complex chronorhythms of our body! We are mostly relieved of that through more effective unconscious procedures.

Car Fortlage, an unjustly forgotten nineteenth-century psychologist was probably the first to draw attention to the relationship between physical and psychological processes which, far from running in parallel, are the reverse of one another. When we sleep, when we shed our waking consciousness, the biological growth processes are undertaken much more effectively than when we are awake. In childhood, when the body needs to perform special feats of growth, we slept a great deal; during the foetal period, at the time of our greatest biological performance, we *only* slept. If, however, we are concentrating our attention and intellect, the substances of decomposition accumulate particularly in the nerve cells, substances which after further processing into urea and uric acid, for example, are excreted via the blood circulation, the kidneys and the bladder.

Mental presence does not, then, mean an increase in biological performance but the decomposition of the latter, and the organs which keep us alive during the day, such as the digestive and nutritional organs of the abdominal cavity, still remain outside the arbitrary control of consciousness. Fortlage wrote:

> Only when we are asleep are we alive; as soon as we are awake we begin to die in that we expend more vitality than we acquire. And yet we consider expending life in this way as our real life and mere unconscious sleep [is regarded] as nothing and miserable. We despise pure life which is nothing other than itself. For the start of our dying process, and nothing else means true life to us. As paradoxical as this may sound, it is no more than a simple fact of physiology. [224]

In contrast to liver functions, which in healthy human beings continue day and night without stopping, urine production in the kidneys ceases for a few hours during deep sleep at night. Only as we get close to waking up and during the day is there an increase in the excreted substances which have been decomposed. The formation and excretion of uric acid thus appears to be an unconscious physiological reflex to the actions of waking consciousness. As we mentioned earlier gout used to be a disease of old age and signalled increasing decomposition in contrast to vital regeneration. If it occurred in middle age, it was seen as a 'disease of academics.' Nowadays it is beginning to appear in the early decades of life, this may well be connected with one-sided over-intellectualization, particularly in schools.

As we have seen, the uric acid values at the time of the Full and New Moon are lower than at other lunar phases. Does this mean that we might be more strongly inclined towards daydreaming at the syzygies than at other times of the synodic cycle? Full Moon and New Moon are often named as characteristic days when the irrational layers in the depths of the psyche surface. Thus in Florida murder and grievous bodily harm are inflicted more frequently at the time of Full Moon [225, 226]. In Baltimore, USA, 22,079 telephone calls to report poisoning cases were related to the lunar phases over a period of thirteen lunar cycles. It turned out that a higher proportion of all calls, as well as of calls relating to unintentional poisoning, occurred during the Full Moon period (= Full Moon ± two days). Suicide attempts and cases of drug abuse were relatively more frequent during the New Moon period (= New Moon ± two days). [227]. In those periods of the synodic lunar cycle, then, in which — based on the values for uric acid excretion — human beings tend towards a reduction in consciousness, they are clearly in particular danger of succumbing to their subconscious forces.

Nevertheless, the studies with a positive result in the above-named areas are clearly in the minority when taken as a whole. A review undertaken in 1991 of twenty studies from the previous twenty-eight years on actual or attempted suicides, and their relationship to lunar phases, showed that the few studies with a positive result contradicted one another, or could not be replicated [228, compare also 229]. A more recent review evaluating the literature of the last twenty years on the connection between emergency calls to the police, other emergency services, advice facilities etc., and lunar phases also reached the conclusions: firstly, that the majority of studies did not produce any evidence of a connection; and, secondly, that the positive findings contradicted one another in respect of lunar phases [230, see also 229, 231, 232, 233, 234].

The clinical increase in frequency of eclampsia on the fifth day after New Moon is connected with the lunar periodicity of the uric acid level. Eclampsia results in muscular convulsions in heavily pregnant women when the kidneys can no longer cope with the additional uric substances from the growing child, and they accumulate in the blood, leading to convulsive muscular contractions. Such dangerous states can be controlled today by means of rapid diagnosis, and measures to relieve the kidneys. The studies evaluated findings from the 1920s and 1930s [235].

Another illness in humans indicating lunar periodicity is pneumonia, and in particular the form of pneumonia, which occurs independently and not as the consequence of other illnesses. Almost two thousand instances were evaluated by Brunner as early as 1898 and 1899 [236, 237], when it was shown statistically that the start of water accumulation in the lung (invasion of the oedema) occurred more frequently during the waxing than the waning Moon.

Birth and death

The preceding examples give an impression of the effect of the Moon on human health. The question presents itself whether the largest incisions in life — birth and death — are also affected by lunar rhythms. Are there periodicities for mortality frequency? For the solar year they were easy to find, particularly as in the northern hemisphere they lie in opposition to the southern hemisphere, in the same way that the seasons do. The least deaths occur in the respective summers [238 p.94]. Finding evidence within a lunar rhythm is more difficult. Schweig found a slight increase in deaths when the Moon was waxing, Hecker when the Moon was waning [221 p.65ff]. Hecker also pointed to correspondences between birth and mortality against the background of the synodic lunar month. For both aspects he found a fourfold periodicity in corresponding observational series (Figure 34). In both curves the maximums at New Moon are greater than those at Full Moon. At the quadratures a contrasting tendency in the maximum values becomes evident: more births in the first than the third quarter, more deaths in the third than the first quarter [221 p.73]. Overall, the respectively four-peaked curves of birth and death rates appear in opposite phases.

7. MOON RHYTHMS IN HUMAN BEINGS

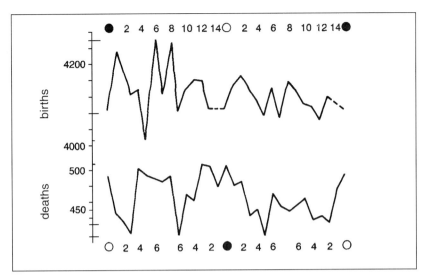

Figure 34. The lunar rhythms of births and deaths run in opposition. Deaths in the city of Hamburg from 1934. Births according to various authors [221].

With regard to births, we encounter a variety of results. In New York, for example, when measured in the total number of 370,000 births in the 1950s, the birth rate was significantly higher in the three days around Full Moon than in the three days around New Moon [239]. In the early 1970s, in the same city, 500,000 live births were correlated with lunar phases over a period of three years. Close to the limit of statistical significance, it became evident once again that an above average number of birth dates lie before Full Moon, and a below average number lie after Full Moon [240].

In comparison, between 1968 and 1974 a statistical study of almost six million (5,927,978) births in France produced the result that more births took place between the third quarter and New Moon, and less births in the course of the first quarter [241]. Nevertheless, a study from 1988 which reported on twenty-one studies from the last fifty years and reanalysed them as far as possible, reached the conclusion that by far the largest number of results does not support a connection between birth rate and lunar phases. The few positive studies contradicted one another in their results [242, compare also 229, 243]. This generally makes clear that a single result cannot simply be generalised but that it must be seen in its geographical and social context.

As we have already seen, the lunar periodicity in both plants and animals appears mainly as a reproductive rhythm, which also applies to human beings. Every doctor learns that the length of a human pregnancy is on average 280 days, calculated from the last period. The doctors refer to ten 'lunar' months of twenty-eight days each [244], but we can also speak of ten sidereal months or nine calendar months. The female menstruation cycle itself corresponds to approximately the period of a synodic lunar rhythm. These parallel characteristics are variously described as purely coincidental, but if we take into account the tendency towards an increasing internalization of biological rhythms in the course of evolution, we reach the conclusion that the menstruation cycle in human beings has simply been individualized to a much greater extent, and has emancipated itself from the external Moon in comparison to related groups. Much research effort has been invested in finding a time-based dependence, but even if we accept that there is a higher birth rate before Full Moon, for example, this effect could also be explained by social factors such as a generally higher level of sleeplessness at Full Moon leading to a higher level of conceptions nine months before birth [compare 245 in this respect]. This claim could no longer be made if the menstruation cycle itself were tied to lunar phases, but studies with this in mind produced differing results here too. Researchers at the University of Pennsylvania found that in 312 women, whose menstrual cycle lasted 29.5 days, ovulation tended to occur with statistical significance in the dark phase of the Moon, that is in the period from the last quarter over New Moon to the first quarter. Women with irregular menstrual cycles also tended to ovulate during this period [246]. Another study with 826 women aged from 16 to 25 found that 28.3 per cent of them menstruated at New Moon, while at other times of the lunar cycle this percentage ranged from 8.5 to 12.6 per cent [247]. Other studies found no evidence of a connection between the external orbit of the Moon and monthly menstruation [248, 249, 250, 251].

There are also indications that the sex of human beings may be connected with the effect of the changing lunar phases. In human beings, as in many animals, sex is determined biologically at the moment when the egg is fertilized by the penetrating sperm (to be precise, fertilization is the actual combination of the two cell nuclei). The male semen contains millions of two sorts of sperm cells which respectively determine the male or female sex. The male-determining sperm cells possess a Y chromosome and the female ones an X chromosome. The mature egg cell always contains an X chromosome. The combination of the Y and X chromosome constitutes the male set of chromosomes on fertilization for all future cell nuclei of the developing individual, whereas two X chromosomes determine the female set. Since both sorts of sperm are formed in equal quantities, on average the same number of male and female births should occur, given the equal probability of fertilization; sex would thus be determined by statistical chance.

However, the reality is different. In general, slightly more boys than girls are born in all countries. For every 100 girls that are born, there are 105 [252] to 106 [253, 254 p.80] boys. This relationship is strikingly constant. It has been postulated that more female than male embryos die during the embryonic period. Now the number of naturally aborted embryos in the first months of embryonic development is so high — it is estimated that about one third of all fertilized egg cells die within the first four weeks after conception — that it is hardly possible to count them by sex in this early phase, a process which would only be possible with the help of genetics. The earlier the embryos die, the more frequently they display chromosomal anomalies, of which the faulty distribution of sex chromosomes make up about one third [255, compare also 256]. However, if the premature births and miscarriages of later phases are taken into account, we get a ratio between the sexes of 125 to 150 male to 100 female foetuses. According to these figures, then, a far greater number of male embryos is created than is to be expected statistically. Then there is the factor that the average ratio between the sexes of 106 to 100 clearly rises after wars [253, 254 p.80]. The decimation of the male population is curiously partly countered in this way — a process which is entirely unconscious. These are indications that the unification of the gametes during conception is not just subject to pure statistical chance. Correspondingly, the biological processes are also much more complex than was previously assumed. We know today that the first sperm which reaches the egg does not fertilize it at all. A series of sperms must first locally soften the jelly-like protective membrane covering the egg (zone pellucida) until it admits a single sperm. Clearly we have here an active, not to say selective, participation of the egg — after all, it has an extraordinary number of sperms to choose from. So, we must be careful in saying that the sperm cell alone is sex-determining.

This is supplemented by a further discovery made in 1940. In a compilation of 33,000 births from the registry office of Freiburg, Germany, between 1925 and 1938, Walter Bühler noted fluctuations in the boy/girl ratio related to the lunar periods. The frequency of births of both sexes fluctuated in a 29.5-day rhythm, that is in a synodic rhythm. Girls are born slightly more frequently shortly before New Moon, boys before Full Moon [257; on the question of the statistical basis of these results compare 221 p.35ff, 258, 259].

The precondition for fertilization is always the release of egg and sperm cells. In mammals, there are two different modes of release of the egg from the surrounding tissue (ovulation): on the one hand it occurs at periodic intervals (spontaneous ovulation in the rutting period) independently of the mating process; on the other hand it can be triggered by the mating process itself (provoked ovulation). The former mostly occurs in the large mammals, hoofed animals for example; the latter in small mammals such as rodents. Human beings are not fixed in this respect either, but both modes are possible and individually given according to the constitution of the woman. Thus conception is possible over almost the whole female monthly cycle; but it is most probable during the time of spontaneous ovulation which lies almost half-way between one menstruation and the next, that is fourteen days after the last menstruation.

The egg only remains capable of fertilization for about 24 hours; otherwise it is aborted at the next menstruation. Since birth follows on average 266 days after conception, conception and birth fall into the same lunar phase from a statistical perspective, for 266 days correspond more or less exactly to nine synodic lunar months. In girls, this will be New Moon again, in boys, Full Moon. If we include the fact that birth within fourteen days before or after the due date is considered to be perfectly normal, and that provoked ovulation is also possible (in other words that both dates can be spread over the whole month) we are justified in assuming that the certain determination of the date of conception would bring out the connection between the determination of sex and lunar phases even more concisely than with the birth dates. Indeed, in 1953 Gunda Bühler was partly able to confirm her husband's results with 55,000 births from Mannheim and 91,000 births from Stuttgart for the years from 1925 to 1938. If the birth rates from Mannheim, Stuttgart and Freiburg are taken together, the cyclical change in birth rates over the synodic month can be confirmed for boys, but not for girls [260, see also 259]. Is this due to the different types of population? The Freiburg registry office covers a large catchment area in its hospitals and university clinics which not only includes an urban population but also large parts of the rural southern Black Forest. The Mannheim area (including Ludwigshafen) is largely industrial like Stuttgart and the area of the mid-Neckar river. The stronger links with nature of the Black Forest population at that time perhaps led to the positive results of the study, while urbanization has probably caused the link with the natural rhythms to be broken. Thus today the fluctuation of the female cycle today varies to a degree, which, a few decades ago, would have been seen as pathological but which is now considered normal because it is so common.

Human beings pay for their growing freedom with the loss of their natural connections. To lament their passing or to try and restore them would only subject human beings once again to the dictates of their unconscious nature. However, as beings completely alienated from the world and only relating to themselves, they would, on the other hand, destroy themselves spiritually and physically. The question for the future can therefore only be whether human beings are prepared to place their lives consciously in the increasingly well-studied, and known, context of the natural environment. To do so, we must know the rhythms of the cosmos and their connection with our own life processes. That is, lastly, also the contribution which this book wishes to make.

CHAPTER 8

The Rhythmical Spectrum of Human Beings

In a rhythm, a particular event is repeated after a specific time in a similar way. In machine production, the important thing is to produce standard products, and so the machine must repeat exactly the same action in all parts. Repetition of the same action here takes place in identically replicable phases with the constancy of a metronome. Basically, all machines operate on a regular beat.

This does not happen in a living context. A beech forms each leaf individually. None is exactly the same as another yet they are all similar. Thus the periods in a living rhythm are never exactly the same, but always similar. Rhythm is thus a repeat of what is similar, not of what is the same. A repetition in time and space can therefore take place either rhythmically or mechanically [261, compare also 262].

Cosmic repetitions are also rhythms. The constellations are constantly changing. Each day repeats the previous day in its own way. In rhythms old and new are interlinked. In life the old is repeated in a new way. The change in the length of the day is most noticeable in higher latitudes during the course of the year, and even the day-night rhythm is not constant if it is measured exactly. Earth and Moon travel in ellipses around their joint centre of gravity, so there are times when they travel faster, and times when they travel more slowly, which are in turn influenced by the changing distance of the Earth from the Sun in the course of the year. No day is the same as another as far as its duration is concerned, no month like any other month, no year like any other year. Rhythm is always also constant renewal.

In addition, rhythms in life are also flexible. A repetition can be absent on occasion without this automatically implying that the rhythm no longer exists. In other instances, a repetition can stand out in comparison to others, yet it still remains integrated into the rhythmical proceeding. The rhythms of organisms possess the ability to be flexibly adaptable. Adaptation, from a physiological point of view, is always an active ability and action of the organism [263].

The moment of polarity in which the rhythm takes place is equally characteristic: action and rest, tension and relaxation, activity and passivity, inhalation and exhalation etc. alternate over time with one another. If one extreme manifests itself, the rhythm always already contains the other potential. Actuality is transformed into potentiality and vice versa. Rising and fading are constituted in the rhythm; and are entirely different without excluding one another. Rhythm is development which is constantly renewed and adapted in polar tension.

The quality of rhythm is also the character of life. There is no living metabolism without cyclical processes, the results of which form the beginning of new conditions. Every organism — be it a unicellular, plant, animal or human being — maintains its rhythm despite irregular repetition, despite beat and chaos, in a movable, always flexible, active and reactive repetition of what is similar.

On closer inspection, we find in each living organism a rich sequence of many hierarchically interlinked rhythms, and just as each organism is integrated in space with its surroundings, so in time rhythms are connected with those of its environment.

Biological rhythms

The human rhythm spectrum has been thoroughly researched by chronobiology in more recent times so that a first review is possible [compare 238, 264, 265, 266 p.30ff] (Figure 35). The fastest rhythms (that is those with the highest frequency) in human beings are fluctuations of tiny electrical currents along our nerves and in the nerve centres. Depending on the type of nerve and its stimuli, rhythms ranging from one second to one eight-hundredth of a second can be recorded here. There is an experiment we can do without measuring equipment to make these *shortest wave* rhythms audible. We place a finger in each of our ears and tense our arm muscles. The gently rustling *barrage* of dull sound allows us to hear the frequency developed by the arm muscles together with the nerves which supply them. This is referred to as the *muscle tone,* the level of which is different in each person and can easily be determined at the piano. A similarly rapid rhythm can be seen in the sequence of beats of the ciliated epithelium in the wind pipe. If we have had to inhale a lot of dust and clean our nose after that, our nasal mucus on the next day is still coloured by dust: innumerable fine

8. THE RHYTHMICAL SPECTRUM OF HUMAN BEINGS

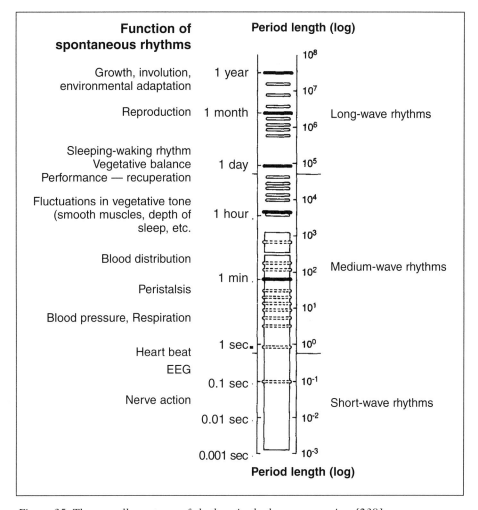

Figure 35. The overall spectrum of rhythms in the human organism [238].

microscopically small 'hairs' transport the dust with a strong upwards beat out of the lung back into the nasal cavity — an important self-cleaning mechanism.

Our cardiac rhythm changes directly with each physical or emotional exertion or respite. Seventy-two pulse beats per minute is considered to be the value for a healthy adult at rest. Thus this period is slightly less than a second. Our respiration is approximately four times slower, but the ratio between pulse and respiration is very different in different people under the same strain. Thus values ranging from 7:1 to 2.5:1 are wholly within the range of a normal healthy constitution in a waking daytime state. At night, the ratio is regulated to the classic 4:1

ratio in every healthy sleeper, particularly between midnight and 6 a.m., with the best synchronization at about 3 a.m. We can see from this that nightly recuperation is not essentially about organs simply resting, but about the resynchronization of their activity in relation to our arrhythmic daytime existence, in order to achieve the greatest degree of co-ordination through such synchronization [on the age-relation of such synchronization compare 267]. Here the faster high frequency rhythms are integrated into the respectively higher-level, long-wave rhythms in whole number multiples, thus synchronizing the great variety of various different organic activities into a common whole. We experience such successful regeneration of time-related processes in the morning as the fresh feeling after a good night's sleep. The activity of the organs, which lie outside our waking consciousness is also ordered into the rhythm spectrum of the whole organism. Since it is precisely these organs which determine the long-period rhythms (see Figure 35), they ensure the co-ordination of the short-period rhythms among themselves. The contractions of the intestinal muscles (peristalsis) still lie in the second to minute range, as do various rhythms of the circulation. The tone of many organs with smooth muscles lies in the minute to hourly range, such as the rhythmical contractions of the ureter, the contractions of the uterus during labour, or of the vegetative nervous system.

Among all these complex regulations governing the organs, the one-and-a-half hour rhythm is an important one. At night, we become restless every ninety minutes, turn over, breath more rapidly with rapid eye movement. These restless phases of sleep are also called REM (Rapid Eye Movement) sleep. During the day this rhythm also continues to oscillate noticeably. Thus in each concentrated activity we need a rest after one-and-a-half hours at the latest. We find it hard to listen to a lecture for longer than that without a break, for example.

The circadian rhythm of human organic functions has been the subject of particularly intensive research. Various hormone values fluctuate as regularly in a day-night rhythm as do body temperature and the immune condition. The dates when operations take place and the dispensing of medicines are increasingly fixed in accordance with our understanding of this rhythm, as different times can greatly influence effectiveness [268, 269].

The weekly rhythm, which for a long time was considered to be nothing more than a conventional cultural rhythm, is also just as much of a biological rhythm. Its special feature is that it is not produced spontaneously by the organism like the other rhythms mentioned so far, but is triggered in reaction to external, interfering stimuli [266 p.39ff, 270, 271 p.19f, 272, 273]. The weekly rhythm applies in particular to the healing of injuries. Sprains and swellings, for example, gradually reduce in a weekly rhythm. The red blood corpuscles are replenished from the bone marrow in a weekly rhythm after a person has donated blood or suffered an accident with major blood loss. The fever curve in the uninterrupted course of scarlet fever also displays weekly dynamics. Such regeneration, which ensures that we stay alive, takes place far removed from the sphere of arbitrary decision-making [compare for example also 274, 275, 276, 277].

Even if chronobiological research in humans took a long time to discover the circaseptan rhythm, it has been known to physicians for more than two thousand years. In the description of illnesses of Hippocrates, Galenus and Avicenna, the critical days of illnesses are clustered around whole-number multiples of the week (Figure 36).

The threshold of arbitrary action is the reaction time, that short period of time which we need to react consciously and specifically to a perceptual stimulus. The reaction time can be reduced by training oneself, for example, to press a key in response to a light signal with an indicator then showing the reaction time — vital training for racing drivers. However, the test person cannot reduce this measured value indefinitely, only to a minimum which cannot be reduced any further, no matter how much training we do. This threshold value once again fluctuates in a seven-day rhythm [266 pp.118f & 189].

Biographical rhythms

We have already devoted a chapter to the monthly or lunar rhythm in human beings; it is even less subject to arbitrary influence. The same applies to annual rhythm. We only notice our dependence on that when the characteristic weather conditions do not happen in a given season. Only the anomaly lets us see that we are still unconsciously embedded in the annual rhythm. We experience this long-term rhythm even more strongly as our own physically based periodicity when people from temperate latitudes for example live in the tropics for several years. The absence of the six-monthly lengthening of the days and their subsequent shortening is experienced particularly painfully with feelings of indisposition and fatigue. People native to the tropics do not

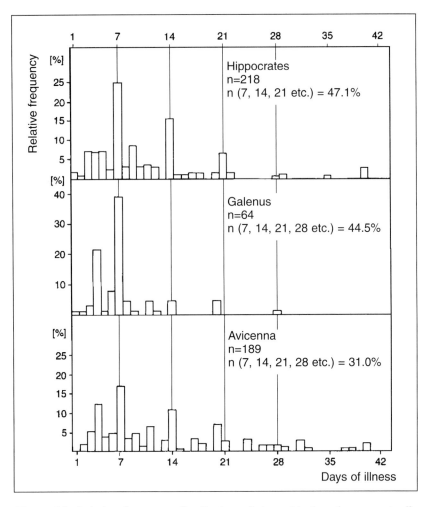

Figure 36. Relative frequency distribution of the critical and prognostically important days of illness from the Corpus Hippocraticum, *the collection of the existing writings of the Greek physician Hippocrates (460–370 BC), the Roman physician Galenus (AD 129–199) and the Persian physician Avicenna (AD 980–1037) [273].*

experience such vegetative reactions, but react to the cycle of the dry and wet seasons which occur twice a year.

The development of the physical body is also connected with annual rhythm. The average period of ten sidereal lunar months from the last menstruation, or nine synodic months from conception to birth, is needed for building up the body. The following three months, the so-called *trimenon* are also of importance for subsequent development. At the moment when the umbilical cord is cut the child becomes spatially,

8. THE RHYTHMICAL SPECTRUM OF HUMAN BEINGS

but not yet physiologically, independent of its mother. Only once the trimenon has passed without incident can the transition from life in the womb to life in the external world be considered to have succeeded. Furthermore, at the beginning the infant still carries immunological substances (antibodies) from its mother in its blood which were taken from the mother's blood before birth. Initially, these largely protect it from infection, but after about three months this natural immunological protection is broken down as it consists of foreign protein [1076]. It is only now — with the start of the second annual cycle as calculated from conception — that the child's organism begins to come to terms independently with the environment. Again exactly one year later, that is one year and three months after birth, the milk teeth with their hardest form of the enamel crown are finished, even if they are still partially hidden under the gums. This represents another important step in the development of the physical body [compare 278].

The annual rhythm, in turn, is embedded in even longer biographical rhythms such as the seven-year rhythm. The visible changing of the teeth commences during the seventh year, and after thirteen to fourteen years sexual maturity with its polarization begins. In the first years of the second decade of life, the last epiphysial line of the large limb bones close: height growth stops, the person has 'grown up' together with the maturing of the skeleton.

There is even one organic area in the human body which is involved in the seven-year rhythm several times: the back molars in the sequence of their appearance. The milk teeth have no precursors of these back molars. The back molars are thus not replacement teeth but additional teeth. Three such molars break through in sequence on each side of the jaw. The first molar —the sixth tooth counting from the middle to the sides — appears in the seventh year (at six years and three months as a statistical average), and often indicates the beginning of the visible changing of the teeth. The second molar appears at age twelve to thirteen, moving into the final bite level at fourteen. From seventeen to forty, the third molar — the *wisdom tooth* — can become visible as the last one, statistically most often at age twenty-one.

In the rest of life, the seven-year rhythm makes itself biographically felt. It no longer displays any great effects in the biological field but characteristically tends to become evident in the psychological and spiritual transformation, which is life-long personality development. The art historian, Wilhelm Pinder, interpreted the seven-year rhythms in the life of Rembrandt using more than one hundred self-portraits of the artist [279]. Human beings, interesting subjects throughout their life and constantly developing, are able to mature not only in the first three seven-year periods, which are occupied with the development of the physical body, but also continue to develop themselves — whether they know it or not.

This brings us to the question whether in the final instance the average lifespan of human beings is also a measure which is integrated into the planetary rhythms, like day, month and year. To discuss this question, we need to digress a little. Days and years are given through the rotation of the Earth and its orbit around the Sun in planetary space. The projection of the Earth's equator onto the celestial sphere is called the celestial equator, which divides the northern and southern starry sky. The projection of the Earth's orbital plane around the Sun is the ecliptic, known since ancient times and lined by the band of constellations of the zodiac. The annual orbital plane of the Sun — as we see it — passes through the latter. The ecliptic stands at an angle of 23 degrees in relation to the celestial equator, corresponding to the angle of the Earth's axis to its orbital plane around the Sun. Thus the Sun — as we see it — is north of the celestial equator on its ecliptic for half a year. During that time the northern hemisphere has its summer. In the next half year the Sun passes through the southern part of the ecliptic lying south of the celestial equator. It is summer in the southern hemisphere, while it is winter in the north. In spring and autumn the Sun is positioned once at each intersection of ecliptic and celestial equator: this is the equinox. The spring and autumn equinoxes are today positioned in the constellations of Pisces and Virgo, but this is not rigid either. During the course of centuries and millennia these intersections travel slowly through the zodiac. The vernal equinox, for example, travels through one sign of the zodiac on average every 2160 years, so that it *advances* each year by a tiny amount in the heavens in contrast to the annual motion of the Sun. This advance is described in Latin literally as *praecession*. The precession around the whole of the celestial sphere is the Platonic or cosmic year and lasts $12 \times 2160 = 25{,}920$ — or approximately 26,000 years — an immense period of geological dimensions.

As a reference, there are two figures which are better known: the last ice age finished about 10,000 years ago, or not even half a Platonic year ago. The half-life of the element plutonium, produced by humans for the first time last century, is about 22,000 years — that is about eighty-five per cent of a Platonic year. After this period half of the plutonium now produced will have died, so approximately six such Platonic years are required for it to reduce to less than one per cent.

The rhythm spectrum of a human life is intertwined on a miniature scale with the macrocosmic rhythm of the precession. If we take the average duration of a fulfilled human life as a good seventy years, such a life contains 26,000 days. Hence we live on average as many days as the Platonic year contains Earth years. We see as many sunrises in a lifetime as springtimes occur one part of the Earth during the great cosmic year. The physician Gotthilf Heinrich Schubert (1780–1860) was probably the first to describe this connection [280, 281].

One of the long-term lunar rhythms is also of interest in this respect. In Chapter 1 we referred to the period when the lunar eclipse returns to the same position again as a good eighteen years. This period contains as many Earth years, that is orbits around the Sun, as the breaths people take on average in a minute, namely eighteen. If we count our breaths for a day and a night, the result is eighteen breaths per minute; in twenty-four hours this is then 60 x 24 as many breaths: 25,920 breaths per day and night. What, in humans, is the number of breaths in a day, is the number of orbits of the Earth around the Sun in a Platonic year, which could by comparison be called a *cosmic day*. A good *minute* in this 'cosmic day' is represented by the Saros cycle in which the Sun, Earth and Moon return to the same position such as in an eclipse. Thus the large, long-term rhythms of human life are not strictly identical with those of the cosmos, but linked with them in a smaller microcosmic order of time.

Conclusion

After this brief characteristic of the human spectrum of rhythms, let us summarize once more the most important aspects. The whole of the human spectrum of rhythms is shown in graphical form in Figure 35 (see also [238, 264] in this respect). The short waves with their high frequencies are opposed by the long waves with low frequencies. The short waves — by which we mean the nerve rhythms mediating the alertness of our reactions — are oriented towards a great variety of performance requirements (ergotropism), and react to individual, arbitrarily chosen intentions or modes of behaviour required by chance by the environment. If the nerve action of a nerve is measured which, for example, serves for heat perception, and if heat is applied to the associated skin area, the number of fluctuations in electrical current per unit of time, that is the frequency, increases drastically; the amplitude remains unchanged, however; not the amplitude of the rhythm oscillation but its frequency can change continuously. This is referred to as *frequency modulation with amplitude stability*. It also means that here we have the least synchronization with other oscillations because frequencies can change constantly at random. The short-term rhythms are least tied to other rhythms and in this sense are most free. This physiological freedom is needed for the arbitrary use at will of our sensory perception and the formation of thoughts in the sensory and nervous system. This means, however, that they can also produce the greatest arrhythmic effect.

Biological long-waves, by contrast, are not only synchronized among themselves to a great extent but also with the great life rhythms of the environment (day, month, year, etc.). This multiplicity, which is also a *harmony*, integrates the individual organism into the whole of the cosmos, thus harmonizing and preserving it (trophotropism). The synchronization of the rhythms with the environment always has a physiologically restorative effect if the rhythms harmonize with one another, that is represent whole-number multiples of the next slowest rhythm; the strength of the function is less important here than its regularity. The long-term rhythms are extraordinarily stable as regards their frequency, but their amplitude is modulated. We have amplitude modulation with frequency stability. Thus the daily production of hormones takes place with great regularity but the quantity of the hormones produced may vary. The degree of freedom no longer relates to

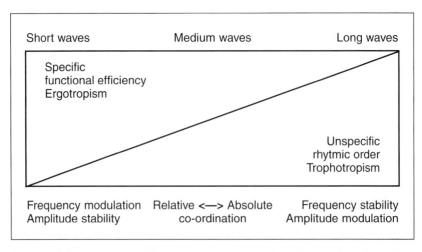

Figure 37. The structure of the spectrum of human biological rhythms [281a].

the progression of time. In this respect, then, human beings remain subject to nature to a greater degree.

Finally, the medium-waves are flexibly changeable without losing their synchronization. No breath, nor heart beat is exactly the same as the next. Each rhythmic wave is determined by physical or mental stress or relaxation, and yet they do not just have a free run but are subject to overall synchronization. Thus the way that the organs are synchronized in human beings always seeks to ensure, for example, that there is a contraction of the main chambers of the heart (respiratory systole) when inhalation starts. Or another example: at the end of a fast sprint we can relax mentally at the finish, happy to have reached it, but our heart beat and pulse only slow down gradually; thus the effort we have made, momentarily applied and suddenly ceasing, is adapted to more slowly reacting regenerative rhythms in order to compensate for the physiological loss of substance and co-ordination, through subsequent regeneration and the re-establishment of rhythmical synchronization.

For the short-term we are thus always free. In the long-term we are also parts of the preserving order which surrounds us. The interplay between respiration and circulation also contains the secret that it allows both of these opposites, which in principle are mutually exclusive, to supplement one another, allowing us in this sense to live in the alternation of freedom and physical necessity. The *Zeitgestalt* or temporal form of any living being is neither a purely open, nor a purely closed, system but the differentiated capacity to be both at the same time. Medium wave rhythms thus allow us to see the capacity of rhythm to incorporate and combine polar moments with particular clarity.

Lunar rhythms in organisms, which in particular were the subject of this book, thus belong to the sphere of biological long waves. With them we encounter conditions of unconscious states which integrate the processes occurring during a human life-time, ranging from the lunar day via the semi-lunar and lunar months to the half and full lunar nodes cycles (the nutation period), and as far as the astronomical processes of the environment — without losing the flexibility which leaves human beings, in particular, at liberty.

The threefold nature of the human temporal form, the *Zeitgestalt,* revealed by the human rhythm spectrum, also applies to the spatial arrangement within the organism. The human form, too, turns out to be a polarity with a unifying centre. In the head sphere, we encounter a

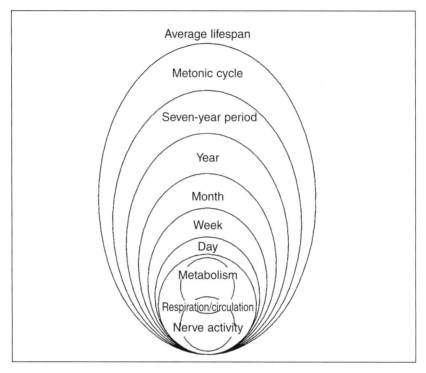

Figure 38. Hierarchical order and interaction of the rhythms in the human being [264].

striking accumulation of telesensory organs. The waking consciousness functions in human beings are based on them and the nerves combining in the brain. The hard, immovable, bony capsule surrounding the brain indicates the predominant tendency at the head pole towards mineralization and hardening. The organs of the soft abdominal cavity, not enclosed by any kind of bone, however, serve the creation of new substance, the assimilation of ingested nutrients into highly specific own substance. The tendency towards the regeneration of substance and movement predominates at this *metabolic pole* of the human organism. Human beings use the sensory system to orient themselves in the environment. In the metabolic system they live their own biological existence to a far greater extent, forming their own biological individuality down to its immunology. Between these two systems we have, in functional and spatial terms, the blood circulation and respiratory system. Its processes are active throughout the whole of the physical body, thus in turn penetrating the other two regions. Indeed, there are no regions of the body where each of the three systems does not

8. THE RHYTHMICAL SPECTRUM OF HUMAN BEINGS

have a direct effect. The upper rib cage with its rounded enclosure has greater similarity to the closed skull, while the lower rib cage opens up to the soft organs of the digestive tract.

The lungs are in pairs like all organs of the head pole. The whole of the rib cage, too, is constructed in pairs like a frozen rhythm in visible spatial form with its alternation of ribs and intercostal space. The heart, positioned lower down in the chest, like most abdominal organs, stands alone, although the structure of its inner chambers is paired. In its spatially mediating position it mediates in functional terms too. It is precisely the blood flow with the heart at the centre which unites all the organs of the body in a closed unity. With respiration, the lung, like the sensory organs of the head, in turn opens itself to the outside world. The lung, immovable in itself, thus turns out to be more closely related to the upper pole of the human organism. The heart, with its own motion, is more closely related to the metabolic and limb systems.

Lungs and heart are positioned side by side in the centre of the body, not only in a spatial respect. As rhythmically pulsing organs, they achieve a balance through their temporal interaction in inhalation and exhalation, systole and diastole. In other words, the actual harmonizing process is not contained in one organ with the *most central* position, but in the way that respiration and heart beat harmonically complement one another in time. The spatial threefold structure of the human organism is a visible expression of the basic time-based organization of its spectrum of rhythms.

The daily rhythm already comprises the constituting sub-rhythms of the organism as a whole: the nerve rhythms, the centrally mediating respiration and circulation rhythms and the primary metabolic rhythms together form the primary rhythm spectrum.

More comprehensive, because it includes the totality of organs, is the biological periodicity of the week, month and year, as well as the seven-year period, the rhythmicity of the lunar nodes and a person's normal life as a whole. The lunar periodicities dealt with in this book comprise the lunar day, month, the lunar nodes cycle, and their sub-divisions. They all are naturally integrated into the seven comprehensive periods of time with which we are placed in the overall context of nature, in a way which includes the natural world around us (Figure 38).

CHAPTER 9

The Quality of Time

Rhythm is not the repetition of the same but of the similar. With this quality, rhythm also gives us an insight into the quality of time. Something similar is no longer something that is the same, and yet it is not completely different either. It is, curiously, both at the same time. Reason allows us to comprehend what being similar means, the intellect does not. The intellect demands that a clear difference be made between what is the *same* and what is *different*: either it is one thing or it is another. However, living rhythm is never exactly the same as itself but always flows into something new as well as into its origins, thus changing them. That is why the intellect has such difficulty in understanding the particular characteristics of time. It demands unchanging laws with unequivocal 'equal' signs. If the intellect had its way, the world would be a non-developing spatial structure whose content could be localized, and for which it would be sufficient to frame its relationships in unchanging concepts. Goethe characterized this in the self-observation: 'The concept of something new developing is completely alien to us; that is why when we see something new developing we think that it has already existed. That is why we have no difficulty in understanding the system of preformation' [282]. In the eighteenth century, it was above all the Swiss naturalist, Albrecht von Haller (1708–77), who gave prominence to the theory of the pre-formation of living beings: the tiny descendant is already present in the egg cell and only needs to enlarge. It took a long time for this theory to be discredited. The female ovaries would already have to contain the next generation in finished form. This means that the ovaries of the mother of humankind, Eve, would have had to contain all future generations in finished form. When, on the basis of the development of chickens in the egg, the physician and anatomist, Caspar Friedrich Wolff (1733–94), first discovered, about a generation later, that organs develop as something new (epigenesis), and empirically refuted the preformation theory, this development was welcomed by Goethe. In life, the continuity of the old (= heredity) must also be supplemented by the new (= development).

The riddle of time

How can we understand the nature of time? In the final instance, we only experience time because we ourselves are constantly changing without losing our identity. Similarly in the course of the world, time always produces something new at every moment, and yet maintains a continuum. Around 500 BC, the Greek thinker, Heraclitus of Ephesus, was probably the first to bring this dual nature of time to consciousness. He recognized that, in time, logical opposites cancel each other out and yet remain unchanged: 'We enter the same river and yet do not, we are and yet we are not [283].' When we enter the river the second time with its new water it is no longer the same but similar. The same is true of human beings: they are never human beings while they remain what they were, but only if they develop and change. In Greece, Heraclitus remained a *lateral thinker*. Intellectual thinking was just beginning to awaken at that time. All other Greek philosophers simply called him *The Obscure*; they did not understand him. Even the greatest thinker among them, Aristotle (384–322 BC), had difficulties with the concept of time: is there such a thing as time? In his work *Physics* he writes:

> The following considerations would make one suspect that it either does not exist at all or barely, and in an obscure way. One part of it has been and is not, while the other is going to be and is not yet. Yet time-both infinite time and any time you like to take-is made up of these. One would naturally suppose that what is made up of things which do not exist could have no share in reality. [284]

If, then, past and future can never be found in the here-and-now, that is, if they can be denied in principle, does the present at least exist as a point in time? Aristotle continues:

> Again, the 'now' which seems to bound the past and the future — does it always remain one and the same or is it always other and other? It is hard to say ... The 'now' is the link of time, as has been said (for it connects past and future time), and it is a limit of time (for it is the beginning of the one and the end of the other) ... It divides potentially, and in so far as it is dividing the 'now' is always different, but in so far as it connects it is always the same. [284]

9. THE QUALITY OF TIME

Aristotle constantly encountered the logical contradictions in the nature of time.

In about AD 400, the early Christian thinker, Augustine, succeeded in taking a step forward in the comprehension of time. At the end of his autobiography, *Confessions*, he theologizes a great deal about the question of the eternal nature of God, but before that he tries to understand the riddle of time. Soon he realizes, like Aristotle, that the past cannot be found in the past and the future cannot be found in the future, since neither can be found to exist, because either they have already been or have yet to be. He recognizes that we do not know the past as such, but only what we can remember of it in the present. Memory is thus the present of the past. Equally we do not know the future but can anticipate the future in the present. We know neither the past nor the future but can know the past in the present and the future in the present as memory and anticipation. The present itself, however, is not an infinitely small moment, nor a nothing, but the content-filled 'presence of the present' in each direct encounter of the world, in the reality-filled activity of experience. Augustine wrote at the time:

> Can we speak of three periods of time? But this is what is clear now: neither the future exists nor the past and we cannot rightly say that there are three periods of time, past, present and future, either. Perhaps it would be more correct to say: there are three periods of time, present of the past, present of what is present, and present of the future. For these three are in the soul and I see them nowhere else. Present of the past is memory, present of what is present is perception, present of the future is anticipation. If we may say that, then I do indeed see these three periods of time and have to admit: there are three. Yet it does not bother me if people say: there are three periods of time, past, present and future. Let people say it as it is commonly misused. See, it does not bother me, I do not contradict and argue as long as people understand what they are saying and do not think that the future is now already or the past is still in existence. For there is little that we say with precision. Mostly we say things imprecisely but people understand what we want to say. [285]

He thus discovered that the segments of time past, present and future which are normally thought of in linear terms, following on from one another, can only be present simultaneously in the soul. Any other conception of time is a hypothetically extrapolating projection beyond the present. In our personal experience time is not simply a one-dimensional sequence but always also appears in our consciousness as the simultaneous interrelationship of all three qualities of time.

In particular, great artists experienced this immediacy of time as the source of their productivity. The creative person is only such, after all, as a developing person who does not remain caught in what has been, but can link what has been practised in the past into an elevated present to its present potential, with what otherwise remains in the future but can already be presaged, thus realizing something new.

Mozart once described with uninhibited joy this experience of creative simultaneity while composing:

> It warms my soul if I am not disturbed; it grows larger and larger and I expand it wider and wider and brighter and brighter and the thing is truly almost finished in my head even if it is long so that I can afterwards see it at a glance in the spirit like a beautiful picture or an attractive person, and I do not hear it in sequence like it will have to happen later but in the imagination everything at once. That is a real feast! [286]

This description of unbroken, filled present is written with Mozartian lightness. Goethe tended to experience active time through the pleasure of listening to music. Thus his friend in old age, the art collector, Sulpiz Boisserée, reported in his diary (September 16, 1815) a conversation with him and the congenial co-poet of the *West-Eastern Divan*, Marianne von Willemer:

> The little women remarked and Goethe concurred that time during music passes infinitely slowly — the greatest compositions are compressed into a short period of time and it appears, with the greatest interest, that a long time has passed. [287]

Everyone knows the way that time can change speed in our subjective, psychological experience. That is why all processes which attempt to be objective seek precisely to eliminate this *unreliability*, and to replace it with a physical time measure — from the water clock of the ancient Egyptians to the caesium atomic clock of today. Only with Heisenberg's discovery of the uncertainty principle was it realized that time is neither an objective nor a subjective state. On one hand, it cannot be found as a real-state existing object, an experience which the classical Sophists already noted with disappointment; on the other hand it is more than a mere subjective, a *priori*, as Kant thought. Instead, it contains the evolutionary events of nature as much as of consciousness. However, it is precisely in this necessary disillusionment, that time is neither objective (I cannot display it in a bottle) nor subjective (it has its own laws in life independent of consciousness), that we get the positive boundary experience: since time is neither the one nor the other; it is the only thing which can mediate between the two modes of existence which a dualistic approach can only separately see as subject and object. It can do so because it can integrate the link to the past (the causal approach of the physical side of the world) with future orientation (all psychological intentionality) in the reality of the present, which is more than both other aspects. 'Time itself is an element' [288] was Goethe's answer to Kant.

If the integration of time is discovered equally in the so-called *external world* as in the *inner world* of consciousness, then this can form the start of a meaningful monism, which allows for the common elements in both. This is proclaimed, in the most concentrated and simplest fashion, in the *West-Eastern Divan*:

> How glorious my 'heritance spread sublime!
> Time my possession, my ploughland — time. [289]

Time is here characterized as an inheritance — it is, firstly, what has been passed on from the past to the present; secondly, it is property which can be used now; and, thirdly, it is farmland, in which the seeds of the future can be planted in the present. True art can express profound thoughts in simple words.

Time integration in life

The integration of the three periods of time into their near-simultaneity does not, however, take place just in human consciousness, but comprises every natural life process in its constant fulfilment. What is contained in human beings, for instance, as a real link to the past as a continuous presence, the line of heredity, is called the *genotype*. At the same time each living being has a restoration potential, which can lie dormant throughout life, but is always present to correct faults, heal injuries or regenerate losses. Appropriately, this ability is described in physiology as the *prospective potential*. It is the sum of the anticipatory possibilities in relation to the uncertainties of the future, but which exists in the here-and-now. Thus, in each living phenomenon both retrospective heredity and prospective potential represent the *de facto* integration of the associated biological past and future, in simultaneous concurrence.

However, this characterization too — taken in absolute terms — is also too much of a generalization and thus too abstract. If time existed in pure simultaneity, it would be more space than time. As it always does life itself once again demonstrates, in its phenomena, the solution to our urgent questions about time: something of what has been is always repeated, thus keeping the past present, and this process always takes place with slight variations each time, thus allowing the potential of the future to break into the present. Life is repetition of what is similar — not of what is the same. Preservation and evolution can interlink in this way and that, precisely, is what happens in each rhythm. All physiological rhythms manage to integrate time without turning time completely into space. The insufficient explanatory power of the linear image of the arrow of time, and of the circular image of the ring of time, means that we had to alternate between the different, partly contradictory, conceptual models of time in our search for an appropriate understanding of time, and will have to continue to do so. Hegel referred to this need as the 'unhappy consciousness' — unhappy only because of the too simplistic expectation of finding a final formula to explain the world. However, this desire is just as much a happy one, if the necessity of a living way of thinking, which remains flexible, is given prominence.

Everything which we have learnt about the rhythms connected with lunar periodicity shows, when we seek to understand it, the diverse way in which each living being can separate itself from its environment in its rhythmicity as much as it can be integrated into it. This can be tightly or loosely, in mutual synchronization up to, and including, identity, through heredity, yet still reacting to habitat. Is the organism autonomous or simply a link in an environment with a much higher level of autonomy? We become aware that this question touches on most intimate issues connected with our very approach; for example, how we pose questions. To what extent do we ask our own questions, or are we prompted to ask them by the natural world? To what extent, then, do we project our own riddles on nature outside the human being in our search for causes? Goethe once said in his *Reflections*: 'Human beings will never understand how anthropomorphic they are.' But Goethe understood it, otherwise he could not have said it. If this thought is thoroughly digested, we can also discover that the human search for knowledge itself is similarly structured to the phenomena of chronobiology.

For the sake of clarity, we press for the conceptual virtue of a self-supporting lack of inconsistency with a minimum of received axioms. However, any approach to the reality of the world always remains asymptotic, that is incomplete. Thus knowledge, as we experience it, is torn between its declared autonomy and its equally declared objectivity. Extreme positions remain the least productive ones: on the one hand, deducing everything autonomously from the logic of ideas, leading to ideology and fundamentalism; on the other, declaring the inability to know anything since the world is not accessible to the insufficiency of human thinking, leaving only the gathering of scientific facts, and thus leading to mere conceptual nihilism, and excusing every laziness in thinking.

How, then, is human understanding still possible? Does the living world itself not give a meaningful answer? Before our eyes it constantly solves the apparent contradiction by combining autonomy and openness to the environment in every organism, the endogenous and exogenous in fruitful productive tension. If we adopt the same approach in the way we think, we can speak precisely of the living thinking, which the organic rhythms of nature constantly invite us to undertake.

Species Catalogue

The *Index of Species* (p. 289) should be used to find a specified species in the species catalogue.

Only a small number of the organisms presented in the species catalogue have been studied under constant laboratory conditions for the continuation of the rhythms connected with lunar periodicity. In the majority of cases, the observations were made in their familiar habitat and their behaviour in the various lunar periods, such as they appear in the form of high and low tide and in the changing light of the Moon, was compared. Whether the biological rhythms recorded in this way can also be considered to be endogenous cannot be evaluated on the basis of such field observations alone. We should only speak of 'circa rhythms' in the strict sense of the word if under constant conditions a continuation of the rhythms in the wild can be observed such that their period is circa = approximately equal to the one which can be observed under natural conditions. We have endeavoured to use the term *circa* in this sense in the following species catalogue. In many species for which experiments under constant conditions have not so far been carried out, we will nevertheless be entitled to assume that a continuation of the rhythms could be shown. The species discussed in the previous chapters can serve as a basis for the evaluation of the different systematic groups in this respect. In order to allow the reader to gain a targeted insight and perspective of this complex subject matter, we have always listed the relevant literature. It should also be remembered that even in those organisms which have been studied in laboratory experiments, the constancy of only a few factors such as temperature, light, water movement, water pressure, air pressure etc. are involved. Electrical and magnetic field fluctuations, for example, are not usually taken into consideration. It has been shown [86, 290, 291] that organisms are sensitive to the earth's magnetic field and that their rhythmical behaviour can be influenced by it.

As far as the taxonomy is concerned, we use well-known and mostly easily accessible works and standard works [primarily 293, 295, as well as 292, 294, 296] as a basis. The single celled species of animals listed at the beginning belong to a group of forms in which it is difficult to make a distinction between the plant and animal kingdom. They are thus treated both in the botanical and zoological system in the literature [compare also 164]. For these species we therefore specify the taxonomic categories of both kingdoms (left: eucaryontic algae, right: protozoa).

A second name — in parenthesis — is sometimes given for the genus or species name in order to make identification in the literature easier. After the species name, the name of the person who first described it, the year of first description as well as the English names are given as relevant. The geographical information in the following lines names the region in which the species under consideration was studied locally or collected for laboratory experiments. If not specified in the literature, the locality of the experimenting institution is given. In order to use the species catalogue, the following summary of the specialist terms for the rhythms should be used as an orientation – if not already seen in the introductory chapter.

Period (duration of cycle)	Nomenclature
12.4 hours	tidal, semi-diurnal (lunar) (sometimes semi-lundian)
24.0 hours	diurnal or daily rhythm, (solar) diurnal (sometimes dian)
24.8 hours	lundian, (lunar) diurnal
14.75 days	semi-lunar or syzygian lunar
29.5 days	lunar, synodic or monthly rhythm (sometimes trigintan)
1 year	annual, yearly rhythm
6939.9 days = 19 solar years	Metonic cycle

Eucaryontic algae Protozoa

Class: Euglenophyceae (bot. system)
Order: Euglenoidina (zool. system)

Euglena limosa (= obtusa) Gard
Avon Estuary near Bristol, England
Vertical migration rhythm: only rises to surface of mud during low tide [297] (compare *Hantzschia virgata, Bacillariophyceae*).

Class: Dinophyceae (bot. system)
Order: Dinoflagellata (zool. system)

Amphidinium herdmaniae Kofoid & Swezy (= *operculatum* Herdman)
Isle of Man [302, 303, 304, 305]

Gymnodinium spec. Stein, 1878
Barnstable Harbour, Cape Cod, Massachusetts, USA [301]

Polykrikos spec. Bütschli, 1873 [301]
Vertical migration rhythm: only rises to the mud surface during the day at low tide (compare *Hantzschia virgata, Bacillariophyceae*).

Gonyaulax excavata (Braarud) Balech
Monhegan Island, Maine, USA
Epidemic blooming in line with lunar rhythms during the May spring tide of each month [299], but compare [300] on issue of connection between phytoplankton blooms and spring neap tide cycle.

Nematodinium spec. Kofoid & Swezy, 1921

Warnowia rubescens (Kofoid & Swezy) Lindem.
Mar del Plata-Region, SW Atlantic
Lunar cycle of appearance. The organisms could be collected during the waning quarter of the Moon [298].

Oxyrrhis spec. Duj.
Tidal pool in the bay of Saint Malo, France
In harmony with the cycle of the tides, an approx. eighteen-hour encysted rest phase alternates with approx. six-hour active swimming during low tide. This rhythm is maintained during constant laboratory conditions for 5–6 days [306] (compare *Strombidium oculatum, Spirotricha*).

Class: Chrysophyceae (bot. system)
Order: Chrysomonadina (zool. system)

Chromulina psammobia (luminous algae)
Brittany, France

Vertical migration rhythm: only rises to the mud surface during the day at low time [307] (compare *Hantzschia virgata, Bacillariophyceae*).

Class: Bacillariophyceae, diatoms

Diatomeae
Monhegan Island, Maine, USA
The diatoms bloom in a lunar cycle. During the main spring tide of each month most of the diatoms can be found on the surface of the water [299].

Hantzschia virgata (= amphioxys)
Ehrenberg var. *intermedia* Grun

Barnstable Harbour, Cape Cod, Massachusetts, USA
Vertical migration rhythm: during the day at low tide, the cells travel to the surface of the mudflats (positive phototropism, photosynthesis is possible) and – hours later – at rising tide deep into the sand or sediment (negative phototropism), where they agglutinate. This migration moves each day in relation to the solar day in parallel to the cycle of the tides. There is no vertical

migration during the night. Every 15 days, when the daytime low tide and migration fall into the evening, migration changes on the following days to the low tide in the early hours of the morning which lies 12 hours earlier [301, 308].

Other species of diatoms showing a vertical migration rhythm are listed in [25] and include: **Biddulphia aurita, Cylindrotheca signata, Navicula ammophila, Nitzschia colsterium, Pleurosigma aestuarii, Scoliopleura latestriata, Stauroneis salina, Surirella gemma, Tropidoneis vitrae.** Information on the literature in [25], where the issue of the continuation of these rhythms under constant conditions is also dealt with.

Class: Phaeophyceae, brown algae
Order: Chordariales

Nemorma tingitana Schonsboe
Moroccan coast
Semi-lunar egg-laying rhythm, at maximum at neap tides [309].

Order: Dictyotales

Dictyota ciliolata
Jamaica
Lunar reproductive rhythm: gametes swarm at Full Moon [310].

Dictyota dentata
Jamaica
Lunar reproductive rhythm: gametes swarm at Full Moon [310].

Dictyota dichotoma (Hudson) Lamour
Heligoland, Germany; Plymouth, England; Bangor, Wales; Naples, Italy; North Carolina, USA
Circa-semi-lunar rhythm of gamete production: gametes swarm on 10th and 26th day after night exposure to light; in North Carolina: circalunar rhythm with swarming of gametes at Full Moon [145, 146, 310, 311, 312, 313, 314, 315].

Order: Laminariales

Ecklonia maxina (Osbeck) Papenf.
Kommetjie, west coast of Cape peninsula, South Africa
Various parts of the algae show lunar periodicity of content of substances with activity similar to cytokines. maximum value for activity was found in the time around Full Moon and the last quarter [316].

Order: Fucales

Fucus spec. Linnaeus
Woods Hole, Massachusetts, USA
Semi-diurnal (lunar) rhythm for oxygen consumption, at maximum when the Moon is setting, at minimum when the Moon is rising [97, 317, 318].

Fucus ceranoides Linnaeus
River mouths of Isle of Man
Semi-lunar rhythm of gamete release, at maximum some days around the syzygy at daytime high tide [319].

Fucus vesiculosus Linnaeus
(bladderwrack)
Baltic Sea
Circadian and semi-lunar rhythm for release of gametes, at maximum 2 full days before Full and New Moon in the period from 6 p.m. to 10 p.m. [320].

Fucus virsoides
Northern Adriatic near Trieste, Italy
Lunar rhythm in content of alginic acid, at maximum at Full Moon [321].

Sargassum enerve (gulf weed)
Misaki, Japan
Semi-lunar rhythm of reproduction [322, 323].

Sargassum muticum (Yendo) Fensholf
Scotland
This algae releases its eggs over several days. The firsts release occurs 2 days

respectively after New or Full Moon, i.e. in a semi-lunar rhythm [324].

Sargassum vestitum *(R. Brown & Turner) C. Agardh*
South Australia
The release of the eggs occurs in the period of 4 days before and after New Moon [325].

Class: Chlorophyceae
Order: Chlorococcales

Pediastrum boryanum *(Turpin) Menegh*
Pediastrum duplex *Meyen*
Scenedesmus quadricauda *Bred*
San Joaquin River near San Francisco, California, USA
Lunar rhythm of reproduction: rise in first quarter, strongest reproduction at Full Moon [326].

Order: Ulotrichales

Enteromorpha intestinalis *Link (gut weed)*
Wales
Semi-lunar rhythm of gamete and zoospore production: 3–5 days before Full and New Moon [327].

Ulva angusta
Ulva linza
Ulva lobata *Kütz*
Ulva stenophylla
Ulva taeniata
Monterey peninsula, California, USA
Semi-lunar rhythm, release of gametes at Full and New Moon. Release of zoospores 2–5 days later [328].

Order: Siphonales

Halicystis (= Derbesia) ovalis *Aresch.*
Heligoland, Germany; California, USA
Semi-lunar rhythm of gamete release: at Full and New Moon [329, 330].

Halineda discordea
Halineda tuna
Halineda hederacea
Great Barrier Reef, Australia
Lunar rhythm of gamete formation probable [331].

Vascular plants

Division: spermatophyta, seed plants

Sub-division: angiospermae (= Magnoliophytina), angiosperms

Class: Dicotyledoneae (= Magnoliatae), dicotyledon angiosperms

Dicotyledon angiosperms

Coleus blumei *(common coleus)*, ***Phaseolus vulgaris*** *(common bean)*, ***Philodendron sagittifolium*** *(arrow-leafed philodendron)*, ***Forsythia spec.*** *(forsythia)*, ***Lilium tigrinum*** *(tiger lily)*, ***Ulmus americana*** *(American elm)*, ***Geranium spec.*** *(wild geranium)*
The leaves of these plants provide extracts with cordial activity (increase of heart rate in cockroaches, comparable to the effect of the neurotransmitters noradrenalin or acetylcholine). This stimulating effect is minimal in extracts which are taken shortly before Full or New Moon (semi-lunar rhythm) [332, see also 333, 334, 335, 336, 337, 338].

Lactuca sativa capitata *(head lettuce)*, ***Brassica oleracea*** *(white cabbage)*, ***Allium porrum*** *(leek or porrum)*, ***Solanum lycopersicum*** *(tomatoes)*, ***Pisum sativum*** *(peas)*, ***Phaseolus spec.*** *(beans)*, ***Daucus carota*** *(carrots)*, ***Levisticum officinale*** *(lovage)*, ***Achillea spec.*** *(yarrow)*, ***Melissa officinalis*** *(lemon balm)*, ***Delphinium***

consolida (larkspur) etc., ***Zea mais*** *(maize),* ***Triticum spec.*** *(wheat),* ***Avena spec.*** *(oats),* ***Hordeum spec.*** *(barley),* also in the African tropical tree species ***Maesopsis eminii***
Seeds sown in the propagator 2 days before Full Moon germinate better, grow larger, form greater numbers of blossoms and produce a greater harvest than those sown 2 days before New Moon [158, 159, 160, 161, 162, 339, 340].

Anemone nemorosa *(wood anemone),* ***Corydalis cava*** *(bulbous corydalis),* ***Aegopodium podagraria*** *(ground elder),* ***Symphytum tuberosum*** *(tuberous comfrey),* ***Campanula rapunculoides*** *(creeping bellflower)*
Padua, Italy
Lunar growth rhythm, Full Moon: regular; New Moon: stronger or weaker depending on temperature [7].

Seeds of ***Lycopersicum esculentum*** *(tomato)* ***Helianthus annuus*** *(sunflower)*
Semi-lunar rhythm of gaseous exchange in permanent darkness under constant temperature: the values determined in the morning (10.30 a.m. to 2.30 p.m.) and afternoon (2.30 to 6.30 p.m.) for oxygen consumption oscillate in relation to one another with greater values for the morning approximately every 2 weeks in correlation with New and Full Moon. [341].

With four long-day plants: ***Hordeum distichon*** *(pearl barley),* ***Triticum vulgare*** *(common wheat),* ***Iberis amara*** *(rocket candytuft),* ***Agrostemma githago*** *(corn cockle),* the period to blossoming is shortened by a few days through the effect of moonlight, while with the short-day plants: ***Soja hispida*** *(soya bean)* and ***Pharbites hispida*** the period until the blossoms appear is delayed by a few days by moonlight [344].

Order: Fabales
Family: Fabaceae

Phaseolus vulgaris Linnaeus *(common bean)*
Seven-day rhythm of water absorption of seeds, at maximum at Full and New Moon and at waxing and waning Half Moon [345].

Circa-seven-day and circannual rhythm of water absorption of seeds, at maximum in the time between the beginning lunar quarters; also at maximum August/September and at minimum February/March respectively [346].

Vicia faba Linnaeus *(broad bean)*
Partly semi-lunar rhythm, sprouts grow in rhythms of 11, 14 and 27 days [7, 347].

Order: Araliales
Family: Apiaceae

Daucus carota Linnaeus *(wild carrot)*
Woods Hole, Massachusetts, USA
Lunar rhythm of oxygen consumption, at maximum in third quarter and at New Moon, at minimum at Full Moon [317, 348].

Order: Rhamnales
Family: Vitaceae

Vitis vinifera Linnaeus *(grape vine)*
Statistically better grape harvest if New Moon occurs in the first half of June at the time of the vine blossom. Analysis over 425 years from start of fifteenth to start of nineteenth century [quoted from 349].

SPECIES CATALOGUE   *161*

Order: Capparales
Family: Brassicaceae

Lepidium sativum Linnaeus
(garden cress)
Ontario, Canada
Lunar growth rhythm, Full Moon: weaker growth, New Moon: stronger growth, in ionised air [350].

Seedling of **Lepidium sativum**
(garden cress)
Lunar rhythm of longitudinal growth of roots in permanent darkness at constant temperatures [341, compare 342, 343].

Order: Scrophulariales
Family: Solanaceae

Solanum tuberosum Linnaeus *(potato)*
Woods Hole, Massachusetts, USA
Lundian and semi-lunar rhythm of oxygen consumption: maximum at setting Moon, minimum at rising Moon and maximum at New and Full Moon respectively [317].

Laboratory experiments over 10 years with potatoe tubers under constant conditions
Lunar diurnal rhythm of oxygen consumption only in the period from September to February and not for remainder of the year [351].

Class: Monocotyledoneae, monocotolydon angiosperms
Order: Helobiae
Family: Hydrocharitaceae

Enhalus acoroides
Banda Island, Indonesia
This water plant only blossoms at low tide during the spring tides. The male blossoms drift to the surface, open and are captured by the female blossoms. With a rising tide the female blossoms close, keeping the male ones inside them. *Enhalus* also sheds pollen directly on the surface of the water [352].

Family: Cymodoceaceae

Halodule pinifolia
Sand beaches of the Fiji Islands
The *Halodule* populations fall almost dry at extreme low water, that is twice per month at the time of the spring tides. Precisely at that point the stamen release their pollen threads which are capable of swimming. These are washed to the tidal pools by the rising water where the equally thread-like stigma of the female plants are drifting [352].

Animals

Sub-kingdom: Protozoa, unicellular organisms

Phylum: Protozoa
Class: Rhizopoda, rhizopods
Order: Foraminifera

Globigerinoides sacculifer Brady
Gulf of Eilat, Red Sea
The maximum reproduction rate for the whole population can be observed about 9 days before Full Moon [353, compare 354].

Gloligerinoides ruber D'Orbigny
Globigerinella siphonifera D'Orbigny
Gulf of Eilat, Red Sea
Semi-lunar rhythm of reproduction in respect of whole population: the frequency peak around New Moon is not quite as pronounced as the one prior to Full Moon [353].

Class: Ciliata, ciliates
Order: Holotricha, uniform covering of cilia

Conchophthirius lamellidens Gosh
Ecto-parasite on the gills of the fresh water mussel Lamellidens marginalis in Calcutta, India
Lunar rhythm of sexual reproduction, conjugation 1 day after New Moon [355].

Order: Spirotricha

Strombidium oculatum Gruber
Brittany, France
Approximately half an hour before the tidal pool inhabited by this ciliate is flooded at high tide, it attaches itself to the substrate and forms a cyst. The encysted rest stage is maintained for a period of 2 high and one low tide. About half an hour after the second high tide has receded, the animal excysts itself and swims around until the next high tide. In synchronisation with the diurnal tidal cycle, an approx. eighteen-hour encysted rest stage alternates with approx. six-hour active swimming during low tide. This rhythm is maintained for 5–6 days during constant laboratory conditions [306, 356, 357].

Sub-kingdom: Metazoa, multi-celled organisms

Phylum: Porifera, sponges
Class: Demospongiae
Order: Poecilosclerida
Family: Esperiopsidae

Neofibularia nolitangere Duchassaing & Michelotti, 1864
Carmabi Buoy Zero, Piscaderabaai, Curaçao (Lesser Antilles)
Lunar spawning rhythm, starting on 3rd day after Full Moon over three days in the afternoon and evening hours until shortly after sunset [358, 359]. Similar in Mycale spec. [360].

Order: Petrosida

Xestospongia testudinaria Lamarck, 1815
Pioneer Bay, Orpheus Island, Great Barrier Reef, Australia
Semi-lunar rhythm of gamete release, in October and November at Full and New Moon in the early afternoon hours [361, compare 362 for mass swarming of this species in the Indonesian archipelago].

Phylum: Cnidaria, cnidarians

Cnidaria, 18 species
West coast of Okinawa, Japan
Most species spawn only in June on the 8th day following Full Moon. Some species also spawn during the following months until September. The time of spawning remains constant in relation to the Full Moon (= lunar rhythm) [363].

Class: Hydrozoa
Order: Hydroida
Family: Campanulariidae

Campanularia flexuosa
Laboratory test
Semi-lunar rhythm for life of hydranth (cup-like upper part of polyp): longevity is at a maximum during New and Full Moon [364].

Laomedea (= Obelia) geniculata
Linnaeus, 1758
Millport, Great Cumbrae, Firth of Clyde, Scotland
Lunar reproductive rhythm: ejection of medusa form in July, August and September. For 10 days after the third lunar quarter [365].

Class: Anthozoa

Corals
Great Barrier Reef, Australia
The spawning of 105 species of coral was observed in five different locations of the Great Barrier Reef; these locations were a maximum of 500 km apart. Spawning took place in October in one locality and exactly one lunar month later in November in the four others. In all locations 87 of the 105 species released their gametes within the third to sixth night after Full Moon, frequently in the 4 hours after sunset [64, 366, compare also 367].

Beacon Islands, Wallaby Group, Houtman Abrolhos Islands, Western Australia
Four species from the order of stony corals (= Madreporaria = Scleractinia) spawned in 1987 in the ninth, 24 species in the tenth and 19 species in the eleventh night after the March Full Moon. Reference to other work makes it appear likely that approx. 60% of 184 species from the group of stony corals on the reefs of the Houtman Abrolhos Islands and the surrounds spawn in late summer within one week of Full Moon [368].

Northern Gulf of Eilat, Red Sea
Twelve to thirteen studied, ecologically important species spawn in lunar rhythm, although not at the same time but in various seasons, various months or various phases within one month [65].

Sub-class: Hexacorallia
Order: Actiniaria, Aktinien, see anemone

Family: Actiniidae

Actinia equina Linnaeus, 1758
(beadlet anemone)
Roscoff, Brittany, France
Tidal rhythm of opening and closing and oxygen consumption, maximum at high tide, minimum at low tide [369, 370, 371, 372, 373, 374, 375]. This rhythm is not continued under constant laboratory conditions [compare 96].

Order: Madreporaria (= Scleractinia), stony coral

Madreporaria, rezent
The banded growth of various growth zones may take place in a monthly rhythm [404].

Family: Acroporidae

Acropora cuneata
Heron Island Reef, Great Barrier Reef, Australia
The gametes mature in the (Australian) autumn, the sperm are released about the time of the first quarter [376].

Acropora palifera
Heron Island Reef, Great Barrier Reef, Australia
The gametes mature in the (Australian) spring, the sperm are released about the time of the last quarter [376].

Family: Faviidae

Favia fragum
Coast of Puerto Rico
Lunar rhythm of oogenesis and spermatogenesis. Ovulation and presumably the swarming of the spermatozoa always takes place on the days after Full Moon. The release of the planula larvae (planulation) also takes place with lunar periodicity with the greatest rate from the eighth to the eleventh day after New Moon [377, 378].

North coast of Panama
Lunar planulation rhythm, at maximum in the days before Full Moon [379].

Goniastrea aspera
Magnetic Island, Queensland, Australia
Annual and lunar spawning rhythm, in the week after Full Moon in October [382].

Montastrea annularis
Montastrea cavernosa
La Parguera, south coast of Puerto Rico
Lunar spawning rhythm, at maximum about one week after Full Moon [383].

Montastrea valenciennesi
NW Pacific, south-west of Japan
Spawning takes place annually in the period from July to August. The majority of spawning occurrences fall in the period from the third quarter to New Moon [384].

Montastrea annularis Ellis & Solanthe (= Ma)
Montastrea cavernosa Linnaeus (= Mc)
Diploria strigosa Dana (= Ds)
Bermuda
In the laboratory: Ma, Mc and Ds spawn in the 4–6 days about the third quarter. Here there is a partial overlap between the three species. With Ma and Mc, a close coincidence of the spawning time was observed in the laboratory and in the field [385].

East Flower Garden Bank, NW Gulf of Mexico
Overlapping lunar swarming rhythm at or in the hours after sunset in the time of the third quarter [386].

Family: Fungiidae

Fungia actiniformis var. palawensis Döderlein (sea anemone-type mushroom coral)
Iwayama Bay, Palao, Japan
Swarming of the planula larvae about New Moon at low tide and high temperatures [380, 381].

Family: Meandrinidae

Manicina areolata Linnaeus, 1758
San Blas Islands, NE Panama
In the months of June and July the planula larvae are released in the observed population synchronously within 2 days of the New Moon (= lunar rhythm) between 2 and 5 a.m. [387].

Family: Pocilloporidae

Pocillopora damicornis Linnaeus, 1758
Costa Rica; Gulf of Chiriqui, Panama
Lunar spawning rhythm, mature gametes appear a few days before and after Full Moon. Limited spawning occasionally also at New Moon [388].

Marshall Islands and Hawaii
Lunar rhythm of larvae release, maximum values vary between geographically distant populations and in different species within a population [389]. As laboratory tests have shown, the lunar rhythm is synchronised through nocturnal illumination by the Moon [390].

Salmon Point, Mary Cove, Rottnest Island, Western Australia
Lunar rhythm of gametogenesis and planulation [391].

Hawaii
The lipid content of the polyps changes cyclically over the period of one lunar month. This fluctuation cannot be explained by lipid loss during monthly planulation [392].

Pocillopora elegans Dana
Costa Rica; Gulf of Chiriqui, Panama
Lunar rhythm of spawning, mature gametes appear a few days before and after Full Moon. Limited spawning also occasionally at New Moon [388].

Pocillopora bulbosa
Great Barrier Reef, Australia
Lunar rhythm of planula ejection at New Moon in summer (December to April), at Full Moon in winter (June to September) [393].

Pocillopora verrucosa Ellis & Solander, 1786
Pocillopora eydouxi
Sesoko Island, Okinawa, Japan
Lunar spawning rhythm, *P. verrucosa*: 1–6 days after New Moon, *P. eydouxi*: 1–5 days after Full Moon [394].

Family: Poritidae

Porites astreoides
Porites furcata
San Blas Islands, Panama
Lunar rhythm of mucous secretion into the environment, *P. furcata* at Full Moon, *P. astreoides* at the first quarter [395, 396, 397].

Porites astreoides
Discovery Bay, Jamaica
Lunar reproductive rhythm, spawning of male gametes above all at Full Moon, larvae release above all at New Moon [398].

Porites spec.
The alternation of thicker and thinner growth bands of the calcium skeleton, 10-14 each year, corresponds approximately to the number of annual synodic lunar months [399, 400, but also compare 401].

Porites lobata
Costa Rica; Panama; Galapagos Islands
Advance stages of gonad development can be widely observed at New or Full Moon, particularly clearly in the case of gametogenesis, less in the case of spermatogenesis [402].

Order: Zoantharia
Family: Zoanthidae

Protopalythoa spec.
Orpheus Island, Gt Barrier Reef, Australia
Annual and lunar spawning cycle: always 4–6 nights after Full Moon in November, simultaneously with mass spawning by other stony corals [403].

Sub-class: Octocorallia
Order: Helioporida
Family: Helioporidae

Heliopora coerulea Pallas, 1766 (blue coral)
Orpheus Island, Great Barrier Reef, Australia
Lunar or semi-lunar rhythm of reproductive behaviour [405].

Order: Alcyonaria, leather coral
Family: Cornulariidae

Clavularia hamra
Coral Nature Reserve, Eilat, Red Sea
The studied population spawned once per year in a single night between 9 p.m. and midnight a few days before New Moon at the height of summer [406].

Phylum: Plathelminthes, flat worms
Class: Turbellaria, turbellarians
Order: Acoela
Family: Convolutidae

Convoluta roscoffensis Graff, 1891
Roscoff, Brittany, France; England
Semi-lunar rhythm: eggs deposited at neap tide; circatidal rhythm: rise to surface of tidal flats at low tide during the day, buried in sand during high tide and at night [66, 370, 374, 407, 408, 409, 410, 411].

Order: Seriata
Family: Planariidae

Dugesia dorotocephala (black planaria)
Evanston, Illinois, USA
Lunar rhythm in the cold part of the year; semi-lunar rhythm in the warm part of the year; the planaria will avoid weak gamma-radiation from the side if

they are crawling in a NS or SN direction. Maximum divergence 4 days before Full or New Moon, maximum convergence 2 days after Full or New Moon [412].

Evanston, Illinois, USA
Lunar rhythm: negative phototaxis is least pronounced during the period of the Full Moon. The spread of statistical variation of this behaviour reaches its maximum in December and its minimum in June [413, compare also 414]. Control groups exposed to deuterium oxide display the same rhythm [415].

Euplanaria gonocephala Duges
(river planaria)
Göttingen, Germany
Lunar periodic rhythm of light sensitivity; greater at Full Moon than at New Moon with 2 maximums: at waxing Half Moon and soon after Full Moon [416].

Phylum: Mollusca, soft-bodied animals
Class: Polyplacophora

Acanthozostera gemmata
Great Barrier Reef, Australia
Lunar spawning rhythm with diurnal rhythm superimposed: at Full Moon, during the night from end of August to April, maximum in December [417].

Order: Ischnochitonida
Family: Chitonidae

Acanthopleura japonica Lischke
Shirahama and Tanabe Bay, Japan
Circa-semi-lunar rhythm of spawning in the days around Full and New Moon, can be induced in the laboratory through artificial daylight and artificial tidal rhythms [418, 419].

Chaetopleura apiculata Say
Woods Hole, Massachusetts, USA
Lunar spawning rhythm with diurnal rhythm superimposed, maximum between Full Moon and last quarter from 19.30 to 22.30 in summer [420, 421].

Chiton tuberculatus Linnaeus, 1758
(West Indian chiton)
Bermuda; Woods Hole, Massachusetts, USA
Lunar spawning rhythm, maximum between Full Moon and last quarter [420].

Class: Gastropoda, snails
Sub-class: Streptoneura
Order: Diotocardia
Family: Haliotidae

Haliotis tuberculata Linnaeus, 1758
(tuberculated sea bear)
Roscoff, Brittany, France
Tidal rhythm of oxygen consumption, maximum at high tide, minimum at low tide [371].

Family: Neritidae

Melanerita atramentosa Reeve
Rocky coast at Arrawarra, NSW Australia
Circatidal activity rhythm with a diurnal rhythm superimposed; maximum at high tide, particularly at night [422, 423].

Retina plicata Linnaeus
Coral bank of Heron Island, Queensland, Australia
Tidal activity rhythm with diurnal rhythm superimposed; maximum at high water, particularly at night [423].

Family: Patellidae, limpets

Cellana exarata Reeve
Hawaii
Lunar reproductive rhythm [424].

Patella vulgata Linnaeus, 1758
(common limpet)
Roscoff, Brittany, France
Tidal rhythm of oxygen consumption, maximum at high tide, minimum at low tide [425, 371].

Anglesey, Wales
Circatidal and circadian rhythm of search for food, maximum at nocturnal low tide [426, 427, compare 428, 429].

Isle of Man
Search for food occurs only in the hours of high water during the day [430].

Family: Trochidae

Austrocochlea obtusa Dillwyn
Evans Head, NSW, Australia
Tidal activity rhythm, at maximum following high tide [423].

Cittarium pica *Linnaeus (West Indian topshell)*
Lee Stocking Island, Exuma Cays, Bahamas
Eight of nine spontaneous spawning events in the period from June to November occurred in the 4 days either of New or Full Moon [431].

Trochus (= Tectus) niloticus Linnaeus, 1758 *(giant topshell)*
Observations on reared cultures
Lunar rhythm of spawning, largely a few days after New Moon [432].

Moorea, Society Islands, Pacific
The observed population behaves asynchronously with regard to reproduction taking place over the whole year; but various sub-groups can be distinguished which spawn in a synchronised manner at the time of the New Moon and the three following days [433].

Order: Monotocardia
Family: Littorinidae, periwinkles

Bembicium nanum Lamarck
Evans Head, NSW, Australia
Tidal activity rhythm, maximum after high tide [423].

Littorina angulifera Lamarck
Biscayne, Virginia Keys, Florida, USA
(Semi-)lunar reproductive rhythm, around Full and New Moon [434].

Littorina littorea Linnaeus, 1767 *(common periwinkle)*
Wimereux, Boulogne, France; Woods Hole, Massachusetts, USA
Diurnal and weak tidal respiratory activity: minimum towards 2.30 a.m. and 1.30 p.m. or one hour before high tide respectively [435].

White Sea; Porth Cwyfan, Anglesey, Wales; Whitstable, Kent, England
Diurnal and lunar spawning rhythm, at the nightly spring tides around the time of New Moon [436, 437, 438].

Littorina neritoides Linnaeus, 1758 *(European rock periwinkle)*
Plymouth, England
Semi-lunar spawning rhythm: maximum at Full and New Moon from September to April, also with specimens which are always under water [439].

Littorina obtusata Linnaeus, 1767 *(flat or smooth periwinkle)*
Wimereux, Boulogne, France
Tidal activity rhythm [407].

Littorina rudis *(rough periwinkle)*
Wimereux, Boulogne, France
Circa-semi-lunar activity rhythm with maximum at time of spring tides; continues over several months at constant laboratory conditions [407, 440, 441].

Littorina saxatilis Olivi, 1792 *(rough periwinkle)*
Firth of Forth, Scotland
The release of the juvenile animals from the pouch took place faster in periods close to the New Moon (2 or more juvenile animals per female animal and day) than close to Full Moon (average 1 or less juvenile animals per female and day) [442].

Littoraria strigata
Malaysia
Littoraria strigata spawns regularly at the spring tides. About 89% of the eggs

appear in the period between 1 day before and 5 days after New or Full Moon. During spawning at Full Moon at the highest spring tides 7 times more eggs are found on average than at the weaker spring tides at New Moon. Under constant conditions spawning takes place at all times of the day and night [443].

Melarhaphe (= Littorinopsis) scabra Linnaeus
Palao, Japan
Eggs deposited at Half Moon [444, 445].

Nodilittorina pyramidalis
Malaysia
Spawning takes place in strict adherence to the spring tides. Approximately 90% of the eggs appear in the period 2 days before to 5 days after New or Full Moon. Under constant conditions: spawning at all times of the day and night [443].

Family: Muricidae, murex

Morula marginalba Blainville
Evans Head, NSW, Australia
Tidal activity rhythm, at maximum after high tide [423].

Urosalpinx cinerea Say, 1821
Woods Hole, Massachusetts, USA
Diurnal and tidal respiratory activity: at maximum between 4.30 and 6.30 a.m. and 7.30 and 9.30 p.m. respectively and 2-3 hours before low tide [435].

Family: Nassariidae

Nassarius (= Kyanassa) obsoletus
(mud dog whelk)
Woods Hole, Massachusetts, USA
Tidal activity rhythm of oxygen consumption, also lundian and lunar rhythm of creeping behaviour; varies with the earth's magnetic field [73, 74, 446, 447, compare 72].

Sub-class: Euthyneura
Order: Basommatophora, Wasserlungenschnecken

Family: Siphonariidae

Siphonaria atra Quey & Gainard
(Athecate hydroids)
Palao Islands; Japan
Eggs deposited at Half Moon [448, 449].

Siphonaria denticulata
Sydney, NSW, Australia
Semi-lunar spawning rhythm. This occurs 3-4 days after Full or New Moon. Animals kept in aquaria do not display this behaviour. Spawning in aquariums only occurs in animals which were collected during a spring tide and then only once soon after collection and no more thereafter [450].

Siphonaria japonica Donovan
(Athecate hydroids)
Asamushi Station in Mutsu Bay, N Japan
Hermaphrodite, mating mostly around the first quarter, eggs deposited at Full Moon, larvae hatch at New Moon from May to July [449, 451].

Siphonaria sipho (Athecate hydroids)
Palao Islands; Japan
Eggs deposited at Half Moon [448, 449].

Siphonaria virgulata
Probably similar to Siphonaria denticulata [450].

Family: Ellobiidae

Melampus bidentatus Say
Salt marsh, Woods Hole & Little Sippewisset Marsh, Massachusetts, USA
Semi-lunar rhythm of egg deposit, within about four days, if the habitat is flooded by water during the spring tides in the months May to July. Animals brought into the laboratory (May) maintain this rhythm under constant conditions throughout the sum-

mer. Outdoors, the Veliger larvae hatch about 13 days later, again within about four days at spring tide. Laboratory experiments show that the eggs can break open within a period of 10 to 24 days after the eggs have been deposited as a reaction to being flooded four times (by the tides) within 50 hours. Presumably the Veliger larvae live approximately 14 days as plankton, before settling in the habitat of their parents at the time of the spring tides [1077]. On the relationship between these rhythms and neuroendocrinological processes compare [452].

Little Sippewisset Marsh, Cape Cod, Massachusetts, USA
The adult samples (> 5mm) of these snails living in the tidal zone are subject to semi-lunar migration so that the majority of these snails at neap tide are 2m below the place where they are at the time of the spring tides [453].

Order: Anaspidea
Family: Aplysiidae

Aplysia californica Cooper, 1863
(Californian sea hare)
Pasadena, California, USA
Diurnal and semi-lunar rhythm in an isolated nerve cell of the intestinal ganglion [454].

Order: Nudibranchiata
Family: Tritoniidae

Tritonia diomedea
Olga, Orcas Island, Washington, USA
The magnetic field of the earth or other magnetic fields as strong as the former influence turning behaviour of the animal. The reaction to the magnetic fields varies with the lunar phases [70].

Class: Bivalvia, mussels
Order: Protobranchiata, Nuculids or nut shells
Family: Nuculanidae

Yoldia sapotilla Gould
Canadian Atlantic coast
Semi-lunar spawning rhythm, at maximum at Full and New Moon [455, 456].

Order: Filibranchiata
Family: Mytilidae

Mytilus edulis Linnaeus, 1758
(common mussel)
Roscoff, Brittany, France; Cape Cod, Massachusetts, USA; Canadian Atlantic coast; Santa Monica, California, USA
Tidal rhythm of water through-flow, with animals from the Atlantic also independently of the tides [457]; in follow-up studies determined as being largely controlled by exogenous rhythms (quoted according to [459 p.361]). Lunar swarming of larvae, maximum at Full Moon, minimum at New Moon [371, 455, 456, 457].

Mytilus californicus Conrad, 1837
(Californian blue mussel)
Santa Monica, California, USA
Tidal rhythm of water through-flow (in the laboratory over 4 weeks) parallel to the tides, at maximum at high water [457, 458]; in follow-up studies determined as being largely controlled by exogenous rhythms (quoted from [459 p.361]).

Family: Pectinidae, scallops

Pecten maximus Linnaeus, 1758
(giant scallop)
Port Erin, Isle of Man
Lunar spawning rhythm, at maximum at Full Moon from April to June [460].

Placopecten magellanicus Gmelin, 1791 *(giant scallop)*
Passamaquoddy Bay, New Brunswick, Canada
Semi-lunar spawning rhythm, at maximum in the days before New and Full Moon or the spring tides in the time from July to September [461].

Pecten opercularis *(small scallop)*
Plymouth, England
Lunar spawning rhythm, at maximum at Full Moon from March to June [462].

Family: Ostreidae, oysters

Crassostrea virginica Gmelin, 1791 *(American oyster)*
Long Island Sound, Connecticut, USA
Tidal rhythm of opening and closing, lunar spawning rhythm [463, 464, 465].

Ostrea edulis Linnaeus, 1758 *(common, edible, table or European oyster)*
Netherlands, Falmouth, Cornwall, England
Semi-lunar spawning rhythm, at and shortly after Full and New Moon from end June to end August [3, 464, 465, 466, 467, 468].

Order: Eulamellibranchiata
Family: Cardiidae, cockles

Cardium (= Cerastorma) edule Linnaeus, 1758 *(common cockle)*
Burry Inlet, S Wales
The additional growth of the shell occurs once daily and shows regular bundling with a period of 29 days. Spring tide maximum and growth minimum correlate [469].

Gorad y Gyt, Menai Straits, Wales
Tidal rhythm of shell growth. The (semi-)lunar tidal rhythm is also reflected in the alternation of thicker and thinner growth bands. Since animals which are artificially submerged also show signs of periodic shell formation under constant conditions, endogenous periodicity must be considered [470, 471, 472, 473, 474, 475, compare 476; for indices arguing against a circatidal rhythm compare also 477].

Southern Odde tidal flat, Sylt, Germany
Juvenile animals which drift away can be traced at maximum frequency approx. every 15 days at the spring tides and at night [478].

Family: Tridacnidae

Tridacna gigas Linnaeus, 1758
Tridacna derasa
Great Barrier Reef, Australia
Escape Reef: spawning takes place mid to late afternoon at New and Full Moon. Myrmidon Reef: late afternoon, 3–6 days before New Moon. Lizard Island: later afternoon, 6–9 days after Full Moon [479].

Family: Veneridae

Cytherea (= Callista) chione Linnaeus, 1758 *(Venus shell)*
Tidal rhythm with activity maximum at high water [480].

Venus (= Mercenaria) mercenaria Linnaeus, 1758 *(northern quahog)*
Woods Hole, Massachusetts; Chesapeake Bay, Virginia, USA
Diurnal and tidal activity rhythm [481, 482].

Family: Tellinidae, Tellins

Macoma balthica Linnaeus, 1758 *(Baltic macoma)*
Canadian Atlantic coast
Lunar rhythm of swarming from July to September, maximum at Full Moon, minimum at New Moon [455, 456].

Southern Odde tidal flat, Sylt, Germany
Juvenile animals which drift away can be traced at maximum frequency

approx. every 10 days at night. Only approximately every third maximum fell at a spring tide [478].

Family: Semelidae

Cumingia tellinoides
Woods Hole, Massachusetts, USA
Lunar spawning rhythm, maximum from full to New Moon [483].

Family: Donacidae

Donax serra *Röding (white mussel)*
Sandags River, Algoa Bay, South Africa
Semi-lunar rhythm of migration vertical to coastline; the mussel moves from the middle tidal level during the spring tides to the level of the low tide during neap tides [484, 485].

Family: Solenidae, razor shells

Ensis directus
Southern Odde tidal flat, Sylt, Germany
Juvenile animals which drift away can be traced at maximum frequency approx. every 14 days at the spring tides and at night [478].

Family: Myidae

Mya arenaria *Linnaeus, 1758*
(soft-shelled clam)
Canadian Atlantic coast
Lunar rhythm of swarming from June to September, maximum at Full Moon, minimum at New Moon [455, 456].

Family: Teredinidae

Teredo pedicellata *Quatrefages, 1849*
(Atlantic shipworm)
Mediterranean
Lunar rhythm of swarming of larvae, maximum at Full Moon from October to March [486].

Phylum: Annelida, segmented worms
Class: Polychaeta

Hediste diversicolor
Wimereux, Boulogne, France
Tidal rhythm, leaves sand substrate during high tide [407].

Order: Phyllodocida
Family: Phyllodocidae

Eulalia punctifera *Grube*
Concarneau, Brittany, France
Lunar rhythm of swarming, last quarter in July and August [487, 488].

Family: Syllidae

Eusyllis blomstrandi
Cherbourg, Normandy, France
Semi-lunar rhythm of reproduction, with an annual one superimposed, after Full and New Moon in May [489].

Odontosyllis spec.
Bahamas
Diurnal, lunar and annual rhythm of swarming, in the hours following sunset around the last quarter in October [4].

Odontosyllis enopla *Verrill, 1900*
Flatts Island, Bermuda
Lunar rhythm of swarming with an annual and diurnal one superimposed, above all on first or second day after Full Moon, shortly after sunset in July to August [490, 491, 492].

Odontosyllis gibba *Claparede*
Concarneau, Brittany, France
Lunar rhythm of swarming with an annual one superimposed, at New Moon from July to September [488].

Odontosyllis hyalina *Grube*
Jakarta, Indonesia
Lunar rhythm of swarming with a diurnal one superimposed, during the three nights following Full Moon shortly after sunset [493].

Odontosyllis phosphorea Moore
Nanaimo, Vancouver Island, BC, Canada
Semi-lunar rhythm of swarming with an annual and diurnal one superimposed, at the time of the quadratures from end July to early October at sunset [494, 495].

Mission Bay, San Diego, California, USA
Semi-lunar rhythm of spawning, for example in the first and third quarter, i.e. around the time of the neap tides, from June to October. To do so, the animals appeared on the surface in the hour after sunset and spawning is accompanied by bioluminescence [496].

Pionosyllis lamelligera
Cherbourg, Normandy, France
Semi-lunar rhythm of swarming with an annual one superimposed, in the days after Full and New Moon in May [489, 497].

Syllis amica Quatrefages
Cherbourg, Normandy, France
Lunar rhythm of swarming with an annual one superimposed, at the first quarter in July [489, 497].

Typosyllis prolifera Krohn, 1852
Porec, Croatia
Annual, circalunar and diurnal reproductive rhythm, from March to October mostly around the days of the Full Moon at sunrise. The prostomium of a male animal completes a spontaneous lunar cycle of hormonal activity. Hormone secretion always takes place about 13 days after the previous release of stolons. With a light-dark alternation of 16h:8h and 20°C in the laboratory, the animals continue their reproductive cycle in individualised form. With artificial moonlight provided for a few days every 30 days as stimulation of the natural moonlight cycle, a synchronisation of free rhythms can be produced [498, 499, 500, 501, 502, 503].

Family: Nereidae

Leptonereis glauca Claparede
Concarneau, Brittany, France
Lunar rhythm of swarming, in last quarter from May to July [488, 504, 505].

Nereis irrorata Malmgren, 1867
Concarneau, Brittany, France
Lunar rhythm of swarming with an annual one superimposed, in the last quarter and at New Moon from August to September [488].

Nereis japonica Izuka, 1908 *(Japanese palolo worm)*
Japan
Lunar rhythm of swarming with an annual and diurnal one superimposed, either at New or Full Moon in the second half of December at midnight after high tide [506].

Nereis linbata (= *succinea*) Frey & Leuckart, 1847
Caen, Normandy, France; Woods Hole, Massachusetts, USA
Lunar rhythm of swarming, in third quarter [507, 508, compare 509 for induction of swarming in the laboratory through temperature changes].

Nereis longissina Johnston, 1840
Wimereux, Boulogne, France
Lunar rhythm of swarming, in the last quarter [510].

Nereis pelagica Linnaeus, 1761
Concarneau, Brittany, France
Lunar rhythm of swarming with an annual one superimposed, in the third quarter in January and February [488, 489].

Perinereis cultrifera Grube, 1840
Cherbourg, Normandy; Concarneau, Brittany, France
Lunar rhythm of swarming, with a diurnal one superimposed: in the first quarter between 14.00 and 16.00 at high tide and at Full Moon and at night [488, 489, 497, 511, 512].

Perinereis marioni
Cherbourg, Normandy, France
Semi-lunar rhythm of swarming superimposed by an annual one, at Full Moon in June and July [489, 497].

Perinereis nuntia var. brevicirrus
Grube, 1857
Qingdao, China
Semi-lunar spawning rhythm, in the days after Full and New Moon in the months June to September [513].

Platynereis bicanaliculata Baird
Soquel Point, Monterey Bay, California, USA
Circalunar swarming rhythm, maximum at New Moon. The timer is probably the gradual waning of the moonlight from Full Moon to the third quarter [514].

Platynereis dumerilii Audouin & Milne-Edwards, 1834
Concarneau, Brittany, France; Naples, Italy; Split, Croatia; Algiers, Algeria
In French Atlantic semi-lunar swarming: maximum in first and third quarter, May to September; in Mediterranean lunar swarming: maximum 1 to 3 days after Full Moon, June to September. In the laboratory photoperiodic triggering of circalunar swarming with maximum at New Moon (Tübingen, Germany, with animals from Naples) [125, 126, 488, 504, 505, 515, 516, 517, 518, 519, 520, 521, 522 p.927ff, 523, 524, 525].

Platynereis megalops
Woods Hole, Massachusetts, USA
Lunar rhythm of swarming with an annual and diurnal one superimposed, between the last quarter and New Moon in the first 2–3 hours of night from June to August [526].

Platynereis spec.
Madras, Bay of Bengal, India
Lunar rhythm of swarming, with quarter yearly and diurnal one superimposed, at New Moon between 7.00 and 10.00 p.m. in March, June and September [527].

Order: Eunicida
Family: Eunicidae

Ceratocephale osawai (= Tylorhynchus chinesis) Izuka (Japanese palolo worm, 'Itome')
Tokyo, Japan
Semi-lunar rhythm of reproduction with an annual and diurnal one superimposed, only in October and November at Full and New Moon between 18.00 and 19.00 [528, 529, 530].

Eunice (= Leodice) fucata Ehlers (Atlantic palolo worm)
Lesser Antilles; Bermuda; Florida, USA
Lunar rhythm of swarming with an annual and diurnal one superimposed, between 27 June and 28 July in the 3 days after the last quarter before sunrise [522 p.929ff, 531, 532, 533, 534, 535, 536, 537].

Eunice harassi (brook lamprey, European palolo worm)
Concarneau, Brittany and Cherbourg, Normandy, France
Lunar rhythm, swarming in the last quarter in May, June and in the northern section also in July [488, 489, 497].

Eunice viridis Gray, 1847 (Pacific palolo worm)
Gilbert Islands (Kiribati), Samoa, Tonga and Fiji, Vanuatu, S Pacific
Diurnal, lunar and annual rhythm of swarming: in October or/and November 3 days at the time of the last quarter before sunrise [5, 6, 35, 36, 522 p.924ff, 538, 539, 540, 541, 542, 543, 544, 545, 546, 547, 548, 549, 550, 551, 552, 553, 554, 555, 556, 557, 558, 559, 560].

Lumbriconereis sphaerocephale
Vanuatu, Pacific
Lunar rhythm of swarming, 2–3 days at the last quarter [548].

Lumbriconereis spec.
Moluccas, Indonesia
Lunar rhythm of swarming, second and third night after Full Moon in March and April [561].

Lysidice fallax
Samoa, Pacific
Lunar rhythm of swarming with an annual rhythm superimposed, in the last quarter of the Moon in October, about 1–2 days before Palolo [34, 553].

Lysidice oele Horst (wawo worm)
Ambon, Moluccas, Indonesia
Diurnal, lunar and annual rhythm of swarming: on second, third and fourth day after Full Moon in February, March and April at sunset [34, 550, 561].

Nematonereis spec.
Moluccas, Indonesia
Lunar rhythm of swarming with an annual one superimposed, in the second and third night after Full Moon in March and April [561].

Staurocephalus rudolphii Delle Chiaje
Concarneau, Brittany, France
Lunar rhythm of swarming, in the last quarter [488].

Order: Spionida
Family: Spionidae

Prionospio cirrifera Wiren
Concarneau, Brittany, France
Lunar rhythm of swarming, in the last and first quarter [488].

Prionospio malmgreni Claparède
Concarneau, Brittany, France
Lunar rhythm of swarming, in the last and first quarter [488].

Order: Opheliida
Family: Opheliidae

Polyophthalmus pictus Dujardin
Concarneau, Brittany; Port Vendres, France
Lunar rhythm of swarming, minimum at Full Moon, maximum between New Moon and first quarter [488].

Order: Capitellida
Family: Arenicolidae

Arenicola cristata Stinpson
Moura coast, Mutu Bay, Aomori, Japan
Semi-lunar rhythm of swarming, during the quadratures, i.e. at the neap tides from mid-July to mid-September [562].

Arenicola marina Linnaeus, 1758 (lugworm)
Roscoff, Brittany, France
Tidal rhythm of oxygen consumption, maximum at high tide, minimum at low tide [371, 407, 441].

Order: Terebellida
Family: Terebellidae

Amphitrite ornata Verril
Woods Hole, Massachusetts, USA
Semi-lunar rhythm of egg deposit, at Full and New Moon or 2 days thereafter [563].

Order: Sabellida
Family: Sabellidae

Chone teres
Japan
Lunar swarming rhythm [compare 564].

Family: Serpulidae

Spirorbis borealis Daudin
Concarneau and Roscoff, Brittany, France
Semi-lunar rhythm of swarming of larvae with an annual one superimposed; maximum in first and third quarter from August to October [565, see also 566].

Class: Clitellata
Sub-class: Oligochaeta
Order: Opisthopora
Family: Lumbricidae

Lumbricus terrestris Linnaeus, *1758*
(earthworm)
Laboratory experiments under constant conditions, Burlington, North Carolina, USA
The time required by the animals to creep 10 cm was measured at midday (12.00) and in the evening (7.00 p.m.). The worms were always faster in the evening than at midday. The difference between both speeds varies throughout the year and was greatest in August and smallest in January. In addition, in spring (February to May) and winter (December) the difference between the midday and evening creeping speed was greater in the week before and after Full Moon than in the two weeks either side of New Moon [567].

Evanston, Illinois, USA
Diurnal, lundian (semi-)lunar rhythm of movement activity and oxygen consumption. Maximum movement activity in the evening and at night as well as at the syzygies, particularly at New Moon, minimum between 7.00 and 11.:00 a.m. in the morning as well as at the quadratures. The rhythms of oxygen consumption do not just reflect movement activity, as the curves for both are not always the same [568].

Phylum: Arthropoda, arthropods
Sub-phylum: Arachnata
Class: Xiphosura, horseshoe crab

Order: Linulida
Family: Linulidae

Linulus polyphemus Linnaeus, *1758*
(Atlantic crab)
Madison, Wisconsin, USA
Lunar rhythm of light sensitivity of the ventral eye. This is at maximum at New Moon and at minimum at Full Moon [569, on the relationship of the ventral eye to the other eyes of Limulus compare 570, 571].

Apalachee Bay, Florida, USA
Circadian, tidal and semi-lunar rhythm of egg deposit, with maximum activity during nocturnal high water, particularly at Full Moon [572].

Seahorse Key, Florida, USA
Maximum egg deposit by the female animals on the beach in spring and summer at high tide and with a largely Full Moon, but also New Moon, more frequently during the day than at night [573].

Delaware Bay, New Jersey, USA
Semi-lunar spawning rhythm [574, and information in 575].

St Joseph Bay, Florida, USA
No semi-lunar or lunar rhythm. For certain behaviours (search for food) the depth of water appears to be more important than the changing phases of the Moon. In contrast to the sections of coast above, we do not here have the mixed, largely half-day form of tide but the whole-day tide[575].

Mashes Sands Beach, Apalachee Bay, Florida, USA
The larvae creep to the surface at the time of the nocturnal spring tide at Full Moon from the nests deposited in the

sand on the beach. In the laboratory the larvae are active only at night both under natural photoperiods and in permanent darkness, reaching maximum at Full and New Moon, with positive phototaxis in all lunar phases except after New Moon [576].

St James, Franklin County, Florida, USA
Diurnal and circatidal activity rhythm. In the laboratory: diurnal activity during natural photoperiods and constant water depth [577].

Class: Arachnida, arachnids
Order: Acari, mite
Family: Bdellidae, chiggers, mites, ticks

Bdella interrupta Evans
Hut Marsh, Scolt Head, Norfolk, England
Tidal, approximately 12.5 hour rhythm of activity on the surface of the sand if the environment (salt marsh) is flooded by water from the tides. If it is not flooded during other periods, an 11.5 hour activity rhythm occurs [578].

Family: Trombidiidae

Schöngastia (= Thrombidium) vandersandei
Opposite Yule Island, Papua New Guinea
Only appears in the days after New Moon as parasitic form of larva [579, 580].

Family: Phytoseiidae

Typhlodromus pyri Schleuten
Laboratory culture
The predatory activity of these animals culminates on the seventh and twenty-fourth days of the synodic cycle and is least about the time of the Full and New Moon [581].

Sub-phylum: Mandibulata
Super-class: Diantennata
Class: Crustacea, crustaceans

Various species of planktonic freshwater crustaceans
Gorge Basin, Lake Cahora Bassa, Mozambique
Despite constant reproduction rates, the population density of zooplankton displays a fluctuation in connection with the lunar phase. Population density increases from the last quarter through New Moon and the first quarter to Full Moon and then drops abruptly to its lowest level around the time of the last quarter. This fluctuation is induced by the feeding behaviour of a fish *(Linnothrissa miodon = Tanganyikan sardine)*. This catches the zooplankton with particular efficiency when the Full Moon rises after sunset and the zooplankton came to surface in the preceding darkness. After the last quarter the population density is low, the fish looks for other sources of food until the next Full Moon [582, 583].

Zooplankton
Moreton Bay, Queensland, Australia
Species from 43 taxa rising to the surface water were captured and counted. 29 taxa comprised adult forms, including 17 groups which sink down to the ocean floor during their vertical migration. The remaining taxa comprise larval forms. The quantity of caught zooplankton varies with the season and synodic month. The summer samples were larger in taxa comprising 25 of the adult forms and 10 of the larval forms. 10 of the 17 groups sinking as far down as the ocean floor, 5 other groups of the adult forms and 5 groups of the larval forms were encountered in larger numbers either at the first quarter or third quarter or at New Moon. Significantly larger numbers at Full Moon resulted for 9 of the larval taxa,

5 of 17 of the groups sinking to the ocean floor and 5 other of the adult taxa [584]. Results differentiated by the various groups of zooplankton in respect of behaviour during different lunar phases were produced by similar studies on Heron Reef, Capricorn Group, South End of the Great Barrier Reef (compare [585]).

Litoral and limnic zones of Mahanadi River near Sambalpur, Orissa, India,
Rhythmical rising and falling of zooplankton in parallel to the lunar cycle with maximum density in surface waters at the time of the Full Moon. Associated morphological changes of the organisms by size and form [586].

Gulf of California, Mexico
Animals of the groups amphipoda and isopoda: the vertical upwards migration of the species from this group takes place to avoid moonlight as far as possible, i.e. surfacing is postponed until the Moon has set, and when the animals have risen they return to the benthos before the Moon rises [587].

Gulf of Nicoya, Costa Rica
Larval forms of crabs from various families: zoea stage I of the species from the families: Grapsidae, Xanthidae, Ocypodidae *(Uca spp.),* Porcellanidae *(Petrolisthes spp.)* and Pinnotheridae *(Pinnotheres spp.)* were significantly more frequent during the period of low tide than at high tide. Later larval stages and megalopae were more frequent during high tide. This rhythm has a day and lunar rhythm superimposed. Some stages were encountered with conspicuous frequency during the nocturnal high tide and particularly frequently during the nocturnal spring tide [588].

Sub-class: Copepoda, copepods
Order: Gymnoplea
Family: Temoridae

Eurytemora affinis Poppe

Conwy Estuary, N Wales
Circatidal rhythm of upwards swimming movement [589]. Semi-lunar rhythm of frequency of copepods along the river mouth: during the spring tide the largest part of the population is found further upriver (inland) than at neap tides [590].

Sub-class: Malacostraca
Order: Stomatopoda, mantis shrimps

A semi-lunar rhythm of egg deposit (E) and moulting (H) is shown by all the following mantis shrimps:

— H only at neap tides: ***Gonodactylus zacae*** (Gulf of California, Mexico), ***Haptosquilla glyptocercus*** (Eniwetak, Marshall Islands, Pacific), ***Gonodactylus graphurus*** (Australia)

— H only at spring tides: ***Gonodactylus falcatus*** (Hawaii, USA), ***Gonodactylus chiragra***

— E only at neap tides: ***Gonodactylus zacae*** (Gulf of California, Mexico), ***Gonodactylus falcatus*** (Hawaii, USA)

— E at rising neap tides: ***Pseudoquilla ciliata*** (Hawaii and Florida, USA)

— E at receding neap tides: ***Pseudosquilla ciliata*** (Thailand)

— E only at spring tides: ***Gonodactylus graphurus*** and further 4 spp. of Gonodactylidae (Australia).

A lunar moulting rhythm (always during the last lunar phase) is displayed by ***Pseudosquilla ciliata*** [591].

Order: Mysidacea
Family: Mysidae

Archaeomysis maculata
La Jolla, San Diego, California, USA
Tidal rhythm of swimming activity [67].

Neomysis integer Leach, 1814
Conwy Estuary, N Wales
Circatidal swimming rhythm [592].

Order: Amphipoda, amphipods
Family: Gammaridae

Gammarus chevreuxi

Gammarus zaddachi Sexton, 1912
Dourduff Estuary, Brittany, France
Circatidal and circa-semi-lunar swimming rhythm, maximum at the time of the spring tide high water [593].

Gammarus zaddachi Sexton, 1912
Conwy Estuary, N Wales
Circatidal swimming rhythm. The maximum activity in relation to expected high water varies in the course of the year and is dependent on the position in the area of the river mouth [594].

Marinogammarus marinus
Strangford Lough, Ireland
Tidal rhythm of swimming activity with a diurnal and semi-lunar rhythm superimposed, maximum activity after nocturnal neap tide high water [595].

Family: Talitridae, Sandhoppers

Orchestia cavinana Heller, 1865 (shorehopper)
Medway Estuary, Chatham, Kent, England
Circatidal rhythm of movement activity with a diurnal one superimposed, maximum after each nocturnal high water [596].

Orchestia mediterranea (shorehopper)
Sea and brackish water in the Medway Estuary, Chatham, Kent, England
Circatidal rhythm of movement activity with a diurnal one superimposed, maximum after each nocturnal high water [596].

Orchestoidea corniculata Stout
Woods Hole, Massachusetts, USA
Orients itself at night with the assistance of moonlight, no endogenous rhythm [597].

Talitrus saltator Montagu, 1808 (sandhopper)
Tyrrhenian coast near Castiglione della Pescaia, Grosseto, Italy
Diurnal rhythm of orientation behaviour, in the day by the sun and at night by the Moon, also functions after more than 10 hours of artifical darkness [598, 599, 600, 601, compare 602].

Skate Bay, Great Cumbrae, Firth of Clyde, Scotland
Circa-semi-lunar rhythm of moulting, always during the 5–7 days before Full or New Moon [603].

Derbyhaven Beach, Isle of Man
Circa-semi-lunar activity rhythm, at maximum 5–7 days after Full and New Moon. In parallel: rhythm for moulting of the same length of period, causes minimum activity 5–7 days before Full and New Moon [154].

Talorchestia quoyana
Takapuna Beach, Auckland, New Zealand
Circadian, Circatidal activity rhythm [604].

Family: Oedicerotidae

Synchelidium spec.
La Jolla, San Diego, California, USA
Circatidal rhythm of swimming activity, maximum at high water [67, 515]. Circatidal phototactic rhythm, more strongly positive at low tide [605].

Family: Pontoporeiidae

Bathyporeia pelagica Bate
Sand coast near Port Erin, Isle of Man
Circatidal rhythm of swimming activity with a diurnal one superimposed, maximum at early low tide – at night [606].

Bathyporeia pilosa Lindström, 1855
Ynyslas beach, Dyfi Estuary, Dyfed, Wales
Semi-lunar rhythm of swimming activity with a tidal and diurnal one superimposed, maximum at early low tide, specially at night and at spring tides [607].

Bathyporeia pelagica Bate
Dyfi Estuary, Dyfed, Wales

Bathyporeia pilosa Lindström, 1855
Ynyslas Beach, Dyfi Estuary, Dyfed, Wales
Diurnal, tidal, semi-lunar swimming rhythm, maximum at nocturnal low tide at the time of the receding spring tides. Copulation also takes place at this time so that the whole cycle from development, hatching and incubation takes place in correlation with the phases [608].

Family: Corophiidae

Corophium arenarium Crawford
Dyfi Estuary, Dyfed, Wales
Semi-lunar rhythm of incubation activity in female animals [609].

Corophium volutator Pallas, 1766
(mud shrimp), (Plate 8)
Dyfi Estuary, Dyfed, Wales
Semi-lunar rhythm of incubation activity in female animals [609].

Burry Inlet, S Wales
Circatidal activity rhythm, with greatest activity at time of low tide [610, compare 611, 612, 613].

S Japan
This crab displays a semi-lunar reproductive cycle in the summer which – as laboratory tests show – is transformed into a lunar cycle as is found in winter below 17.5° C [614].

Family: Hyperiidae

Parathemisto gaudichaudii Guérin-Ménéville, 1828
South Georgia (S Atlantic)
Lunar rhythm of reproduction, maximum in first quarter, slight subsidiary maximum in last quarter [615].

Order: Isopoda, isopods
Family: Cirolanidae

Eurydice longicornis Stüder
Excirolana natalensis Vanhoffen
Pontogeloides latipes Barnard
Sand coasts of South Africa
Circatidal rhythm of swimming activity varied through a circadian rhythm. The greatest swimming activity can be observed in immature animals by high water during the day, with adult animals by high water at night. Adult and immature animals of Eurydice natalensis: semi-lunar rhythm of swimming activity, increased activity always about the time of the spring tides [616].

Eurydice pulchra Leach, 1815
Isle of Man; Dyfi Estuary, Dyfed; Newborough, Anglesey; Oxwich Bay, Swansea, Wales
Adult animals possess a circatidal swimming rhythm: active in biotope at high water, digging in before the full ebbing flow begins. The circatidal rhythm is varied by a circa-semi-lunar one; during the spring tides (a few days after Full and New Moon) activity is at its highest.

O_2 intake also fluctuates circatidally, where the highest values are reached at high water. However, this rhythm is only manifest at spring and not neap tides [123, 617, 618, 619, 620, 621, 622, 623, on circatial variability of phototactic behaviour, compare 624]. Crabs caught at the coast of Newborough and Rhosneigr, Anglesey, maintain their semi-lunar rhythm of swimming activity also under artificial conditions (constant darkness, unchanging water temperature and turbidity) for a period of 50-60 days [153].

Excirolana chiltoni Richardson, 1905
La Jolla, San Diego, California, USA
Induction of circatidal activity rhythm through simulated wave movement [625].

During a 2 month study the animals, kept under constant conditions, maintained their circatidal rhythm of swimming activity. The peaks of swimming activity were initially synchronised with maximum high water. Since the average period of the rhythm in the wild at 24 hours 55 minutes was approx. 5 minutes longer than the average tidal period, the cycles gradually drifted apart. The circatidal rhythm of swimming activity has a circalunar rhythm imposed on it with a period duration in the wild between about 26 and 33 days. Both rhythms enable the crabs to copy the mixed, largely halfday tides of the Californian coast in detail [87, compare also 130, 626, 627].

Semi-lunar rhythm for moulting, maximum in the week before New and Full Moon, semi-lunar rhythm of occupation with parasites, semi-lunar rhythm of release of young forms [628, 629].

Family: Sphaeromatidae

Naesa bidentata Adams
Plymouth, England
Tidal rhythm of oxygen consumption, with maximum at high water, mininum at low water, particularly in the females [630].

Family: Tylidae

Tylos granulatus Krauss, 1843
South African coast
The respiration rate (O_2 consumption) shows a 24.8 hour rhythm (lundian), and rises dramatically during the nocturnal low tide starting from a low base value. If the nocturnal low tide occurs in the early hours of the morning, the rhythm changes and the highest respiratory values occur in the following afternoon low tide. This phenomenon escalates during spring and neap tides so that the lundian rhythm has a semi-lunar one superimposed. The activity rhythm accompanies the respiratory rhythm [631].

Order: Euphausiacea, krill
Family: Thysanopodidae

Euphausia frigida
South Georgia (S Atlantic)
Lunar rhythm of reproduction, maximum at first quarter, subsidiary maximum at last quarter [615].

Euphausia superba Dana, 1852
South Georgia (S Atlantic)
Lunar reproductive rhythm, maximum at Full Moon [615].

Thysanoessa spec. Brandt, 1851
South Georgia (S Atlantic)
Lunar reproductive rhythm, maximum at Full Moon [615].

Order: Decapoda, decapods

Decapoda
Mira Estuary, SW Portugal
Most of these species have a semi-lunar rhythm of larva release, centred around the first and third quarter [632].

Decapoda
Lunar and semi-lunar rhythm of larvae release in various decapod crab species ranging to the sublitoral area (overview with tables [compare 633]).

Anchistoides antiguensis Schmitt
Bermuda
Lunar rhythm of swarming, 2 maximums: in the first night after New Moon and between the last quarter and New Moon [641, 642].

Family: Penaeidae

Metapenaeus dobsoni
Penaeus merguiensis
Kali Estuary, Karwar, Karnataka, India
The frequency of post-larval and juvenile stadia varies with the phase of the Moon. Samples taken at New Moon (A), Full Moon (B) and Half Moon (C) produced particularly high values for A (67% of all catches in total), follwed by B (30%) and C (4%) [638].

Metapenaeus endeavouri
Penaeus merguiensis
Embley River, N Queensland, Australia

Panaeus esculentus
Moreton Bay, S Queensland, Australia
If juvenile forms of these three species are subject either to the artificial alternation of light and dark or to an aratificial change of tides, all species were more active at night or in the first hours of high water. If the animals were subject to both specified rhythms, the behaviour of the three species varied: *M. endeavouri* was no longer active at all during the day but only at high water; *P. merguiensis* and *P. esculentus* kept their daytime activity but the maximums fell in the period of the nocturnal high tide [639, compare 640].

Penaeopsis goodei
Bermuda
Lunar rhythm of swarming, maximum between the last quarter and New Moon and subsidiary maximum in the first night after New Moon [641, 642].

Penaeopsis smithi
Bermuda
Lunar rhythm of swarming, 2 maximums: in the first night after New Moon and between the last quarter and New Moon [641, 642].

Penaeus duorarum Burkenroad
Buttonwood Canal (near Flamingo), Everglades, Florida, USA
Juvenile animals collected at the time of the Full or New Moon show a daytime rhythm in the laboratory under constant conditions (faint light, water flow, water level): buried in the sand during the day, they come out at night and are active until midnight. A second activity phase occurs in the hours before dawn. The delayed nightly activity phase in accordance with the tides indicates an additional circatidal component [643, 644, 645, compare 646].

Buttonwood Canal, Everglades, Florida, USA
Circatidal swimming rhythm in bei post-larval animals. Depending on whether they were collected at Full or New Moon or in the first or third quarter, juvenile animals showed different endogenous movement patterns with tidal and circadian elements [647].

Lunar activity rhythm: at New Moon and in the course of the first quarter the crabs showed increased activity under UV light in relation to the time at Full Moon and in the last quarter. An endogenous rhythm for moulting is semi-lunar [648].

Penaeus indicus Milne-Edwards
(prawns)

Penaeus monodon Fabricius, 1798
Vellar estuary, Tamil Nadu, India
The movement activity of the animals follows a tidal and daily rhythm. The maximum activity of the tidal rhythm is synchronised with the times of high tide and correspondingly the greatest activity of the daily rhythm with the

night. Both rhythms persist as circatidal and circadian rhythms respectively for some days under constant conditions when the animals are brought into the laboratory from their natural environment. In addition, the movement activity also varies in permanent darkness with the lunar phases, with increased activity at Full and New Moon and reduced activity in the first and third quarter [649, 650].

Penaeus merguiensis
Norman River, Gulf of Carpentaria, Queensland, Australia
Tidal rhythm of catch rate, maximum at low tide [651].

Penaeus schmitti Burkenroad
Salvador, Bahia, Brazil
Mating between the animals kept in tanks takes place above all during Full and New Moon, moulting generally occurs between these periods [652].

Penaeus semisulcatus (tiger prawns)
Embley River, north-eastern Gulf of Carpentaria, Queensland, Australia
Higher catches of juvenile prawns are influenced by the day-night cycle and the tidal cycle: higher catches occur during the night and at low tide [653].

Penaeus spec. Fabricius, 1798
(penaeid shrimps)
Ras At Tannura, the Gulf, Saudi Arabia
The quantity of prawn eggs shows – like plankton – a lunar cycle: they accumulate in the water in the first quarter [654].

Penaeus stylirostris Stinpson
Guaymas, Sonora, Mexico
The reproductive activity (mating and spawning) of these animals, kept in hypersaline water, is greatest 3 to 4 days after New Moon [655].

Penaeus vannamei Boone
Commercial breeding stations in Colombia and near Guayaquil, Ecuador
Semi-lunar rhythm of weight increase, maximum at the time of the syzygies [656].

Larval forms of 8 species of the classes Metapenaeus, Parapeneopsis, Panaeus
Mandovi Estuary, Goa, India
The total number of larval forms carried into the estuary zone was greater at full than at New Moon [657].

Family: Palaemonidae

Palaemon elegans Rathke
(common prawn)
Tidal rock pools, Oxwich, West Glam., Wales
Only at spring tides: circatidal rhythm of migrating activity, maximum at low tide; at neap tides: change to a semi-diurnal rhythm of migrating activity [658].

Palaemon serratus Pennant, 1777
(common prawn)
Tidal rock pools, Oxwich, West Glam., Wales
Tidal rhythm of migrating activity, maximum at low tide [658].

Palaemonetes pugio Holthuis
Salt marshes along Duplin River, west of Sapelo Island, Georgia, USA
The number of animals in the post-larval stage was particularly large every 2 weeks within a period of 3 days befor to 3 days after Full or New Moon, in parallel to the monthly spring tides [659].

Family: Crangonidae

Crangon crangon Linnaeus, 1758
(brown shrimp)
Carrick Bay, Isle of Man
Circatidal rhythm of surfacing, maximum at high water, and in parallel of swimming activity [660].

SPECIES CATALOGUE

Family: Palinuridae, spiny lobster

Panulirus ornatus Fabricius
Area between Warrior Reefs and Mabuiag Islands, Torres Strait, N Australia
After moulting when reaching maturity, a section of the two-year-old animals (above all the females) migrates towards the east in the spring of each year. The preceding moulting of these animals always takes place in the third quarter and coincides with the general lunar rhythm for moulting of the whole population [661].

Family: Paguridae

Clibanarius misanthropus Risso
Arcachon, Biscay, France
Tidal rhythm in reaction to the light also under constant laboratory conditions if originating from coasts with clear tides [662].

Family: Coenobitidae

Coenobita clypeatus Fabricius, 1787
Curaçao, Caribbean
Female animals with mature eggs spawn near water line at low tide. Fresh eggs can always be found there at the time of the Full Moon [663].

Family: Hippidae

Emerita analoga Stinpson, 1857
La Jolla, San Diego, California, USA
Tidal rhythm of swimming activity [67].

Emerita asiatica
Circatidal and circadiurnal activity rhythm and rhythm of oxygen consumption [664].

Family: Cancridae, common crabs

Cancer pagurus Linnaeus, 1758 *(rock-dwelling crab)*
Dublin, Ireland
Semi-lunar, circatidal and circadian rhythm of O_2 absorption rate both in normal crabs and those made artificially immobile [665].

Cancer novaezelandiae Jacquinot
Otago Harbour, Dunedin, New Zealand
The search for food *in situ* is influenced by variables of daytime and tides. Under constant laboratory conditions only a circadian rhythm was to be observed, where the maximum movement occurs in the night. There is little indication of an endogenous tidal component [666].

Family: Portunidae, swimming crabs

Callinectes sapidus Rathbun, 1896
(blue crab)
Lake Pontohartrain at New Orleans, Louisiana, USA
Diurnal rhythm for change of body colour (dark during the day, light at night) with a tidal rhythm superimposed [667].

Chesapeake Bay, Maryland, USA
Lunar rhythm of moulting activity. This is greatest at Full Moon, intermediate at Half Moon and least at New Moon [668].

Carcinus maenas Linnaeus, 1758
(shore crab)
Dublin, Ireland
Semi-lunar, circatidal and circadian rhythm of O_2 absorption rate both in normal crabs and those made artificially immobile [665, 669, 670].

Circadiurnal and circatidal rhythm of movement activity, the diurnal rhythm can be transformed into a tidal one through cold shock [123, 142, 671, 672, 673, 674, 675, 676, 677, compare 678, 679, 680, 681].

Keppel Pier, Millport, Great Cumbrae, Firth of Clyde, Scotland
The endogenous rhythm of the heartbeat rate shows a seasonal variation in *in situ* experiments. It is only expressed

weakly in winter – if at all – and is more clearly evident in late spring and summer. The position of the studied animals on the coast is also of importance in this respect. In the largely younger and smaller animals of the eulitoral the rhythm was more pronounced than in the older and larger animals of the sublitoral [682].

Canal de Mira, Ria de Aveiro, Portugal
Tidal, diurnal and semi-lunar rhythm of appearance of the first larval stages, maximum at nocturnal low tide in the first and third quarter [683].

Charybdis cruciata
Portunus pelagicus
Portunus sanguinolentus
Coast of Goa, India
Particularly large trawl net catches at Full and New Moon [684].

Family: Xanthidae

Cataleptodius taboganus
Xanthodius sternberghii
Naos Island, Panama, (Pacific side)
Semi-lunar rhythm of larval release, maximum 1 to 4 days before the first and third lunar quarter [685].

Rhithropanopeus harrisii Gould, 1841
Neuse River and Newport River, North Carolina, USA
Circatidal or circadian rhythm of larval release, the former can be induced through cyclical change of alt concentration [634, 635]. Circatidal or circadian vertical migration rhythm of the zoea larvae [636, 637]. Circatidal rhythm of phototaxis [635].

Xantho floridus
Roscoff, Brittany, France
Tidal rhythm of oxygen consumption, maximum at high tide, mininum at low tide [371].

Family: Ocypodidae

Ilyoplax gangetica Kemp, 1919
Phuket Island, S Thailand
Semi-lunar rhythm of male courting and female incubation behaviour [686].

Ilyoplax delsmani
Ilyoplax gangetica Kemp, 1919
Ilyoplax orientalis
Malaysia and Thailand
Semi-lunar waving rhythms in the male animals [687].

Megalopae of Uca pugilator, Uca minax, Uca pugnax
Newport Estuary, Beaufort, North Carolina, USA
Circatidal swimming rhythm, with maximum activity at the time when high water is to be expected in the field [688, 689]. These results do not accord with field studies which produced particularly large catches for the megalopae during the nightly high tide before the high water period [compare 690].

Macrophthalmus boteltobagoe
Sakai, 1939
Horikawa, Okinawa, Japan
Most of the femal animals stick the eggs to the rear extremities of the body at the time of the New Moon; the larvae are then released about 14 days later at the time of Full Moon. Then moulting takes place: in the female animals there was only one moulting maximum per month from May to September always in the time between Full Moon and the last quarter.

In male animals: in May, the moulted carapaces were frequently found in the first quarter, with a further maximum at Full Moon, in July and August always at the first quarter and between Full Moon and the last quarter, in September in the first quarter. In the laboratory, the female animals occasionally took on the next brood immediately after larvae release without a moulting period in between [691].

Macrophthalmus hirtipes
Circatidal activity rhythm [88].

Macrophthalmus japonicus De Haan
Tatara-Umi River, east of Fukuoka, Japan
The animals live in caves which they leave during the day at low tide to search for food near the cave. But during the summer proper migrations occur in some places. The animals leave the cave area and migrate at low tide to the water's edge to take in food there. The number of these migrating animals follows a semi-lunar periodicity with maximum values 2 or 3 days after spring tides [692].

Ocypode ceratophthalmus Pallas, 1772
Cousin Island, Seychelles
The number of copulation caves created by the male animals reaches its maximum at or shortly before New Moon [693].

Uca beebei Crane, 1941
Uca latinanus
Uca musica terpsichores Crane, 1941
Canal Zone, Panama
Lunar rhythm of courting behaviour of the male animal, above all in the week of Full Moon. The construction of a clay hood above the entrance to their caves only occurred in the courting male animals and climaxed at the time of Full Moon. The whole reproductive behaviour of male and female animals appears to be co-ordinated in relation to this period [696, compare 697].

Uca crenulata
Aqua Hedionda Lagoon, Carlsbad, California, USA
Circatidal activity rhythm [698].

Uca lactea lactea
Taiwan
Semi-lunar rhythm of pairing behaviour, maximum 1 to 3 days after New and Full Moon [699].

Uca latinanus
La Boca, Panama
Semi-lunar activity rhythm, maximum during the spring tides [694, 700].

Uca maracoani
Trinidad
Diurnal and tidal activity rhythm, maximum at low water between 8.00 and 10.00 a.m. [694, 695].

Belém and Salinópolis, Paráa State, Brazil
Circatidal activity rhythm, maximum at low water [701].

Uca minax
Biloxi Bay, Ocean Springs, Mississippi, USA
Diurnal, circatidal and semi-lunar rhythm of the activity phases [91, 702, 703].

Uca minax
Uca pugilator Bosc, 1802
Uca pugnax
North Inlet estuary, Georgetown, South Carolina, USA
Semi-lunar rhythm of larval release, maximum at Full and New Moon ±1.5 days [704, compare also 705] at nocturnal high water [706].

Uca mordax
Caribbean coast of Costa Rica
Eleven male animals taken from non-tidal river; taken 110 km to the Pacific coast; there subjected to the tides in a cage for 5 days; in the laboratory 2 animals in permanent darkness displayed semi-diurnale and tidal activity patterns [694, compare 701].

Uca princeps
Uca stylifera
Pacific coast of Costa Rica
Diurnal and tidal activity rhythm, maximum at high water between 07.00 and 12.00 [694].

Uca pugilator Bosc, 1802
Tidal rhythm of the activity phase tied to intact eyes in contrast to the diurnal

rhythm; a semi-lunar rhythm superimposed [703, 707, 708].

Uca pugilator Bosc, 1802
Cayo Pelau, Charlotte Harbor, Florida (west coast), USA
Semi-lunar reproductive rhythm (courting behaviour of the male animals, number of female animals ready for fertilisation). The female animals mate once every month 4–5 days before one of the half-monthly syzygies. Larval release takes place about 13 days later, i.e. approx. one week before the spring tides [709].

Uca pugnax
Cape Cod, Massachusetts, USA
Diurnal and circatidal rhythm of migrating activity [109, 707, 710, 711, compare 712, 713, 714, 715].

Georgetown, South Carolina, USA
Semi-lunar reproductive rhythm (number of courting male animals, number of caves with domes, number of female animals with decalcified sexual openings, number of female animals carrying incubated eggs), maximum at the spring tides [716].

Woods Hole, Massachusetts, USA
Animals kept under constant environmental conditions (permanent darkness or low light) show ultradian rhythms with periods ranging from 2.5 to 8.0 hours [717].

Uca speciosa
Tidal rhythm of colour change [708].

Uca tangeri Eydoux, 1835
(fiddler crab)
Guadalquivir Estuary, S Spain
Semi-lunar rhythm of reproduction, eggs deposited at New Moon. At non-tidal pools without lunar rhythm [694, 718, 719]. Circatidal activity rhythm [720].

Ramalhete Station, Ria Formosa, near Faro, Portugal
Tidal activity rhythm (food intake, digging of caves, courting or migrating to other tidal flat), at low tide irrespective of day or night [721].

Family: Grapsidae

Cyclograpsus lavauxi Milne-Edwards, 1853
Kaikoura Peninsula and Governors Bay, Christchurch, New Zealand; Laboratory study
Circadian-circatidal activity rhythm, maximum at night during high water [722].

Helice crassa Dana, 1851
Circatidal activity rhythm [88].

Hemigrapsus edwardsi Hilgendorf
Dunedin, New Zealand
Circatidal activity rhythm with weak circadian component [723].

Sesarma haematocheir
Sesarma intermedium
Japan
These terrestrially living crabs release their larvae, sensitive to fresh water, in the lower reaches of the river close to the sea only at New and Full Moon (semi-lunar rhythm), i.e. during the spring tides. Crabs from larvae which are tolerant of fresh water, e.g. S. dehaani, do not display this rhythm [724].

Ogamo River, Izu Peninsula, Japan
Sesarma haematocheir and Sesarma intermedium live terrestrially. But their larvae develop in the sea and therefore the adult animals migrate to rivers or waters close to and flowing into the sea for spawning. The number of crabs which release zoeae larvae reaches its maximum at the time of each syzygie and its mininum in the first and third quarter (semi-lunar rhythm) [725]. Induction with aritificial moonlight as

time-giver is possible in animals in the laboratory [127, 726, 727], compare also [728].

Sesarma haematocheir
Kasaoka, Okayama, & Gokasho, Mie, Japan
Circatidal and semi-lunar rhythm of larval release. Induction with aritificial moonlight as time-giver is possible in animals in the laboratory [729, 730, 731].

Sesarma reticulatum Say, 1817
Woods Hole, Massachusetts, USA
Circatidal rhythm of migrating activity with a circadian one superimposed [732].

Sesarma spec. (nr. reticulatum)
Coast of Louisiana, USA
Semi-lunar rhythm of egg deposit and larval release, for the whole season (approx. April to October) [733, 734].

Sesarma rotandata Hess.
Samoa (Pacific)
Lunar migrating activity: 10 days before the appearance of the Palolo just before Full Moon, it climbs from the Inocarpus trees, migrates to the beach to lay its eggs in the sea water [553, 735].

Super-class: Antennata
Class: Diplopoda, diplopods
Family: Paradoxosomatidae

Orthomorpha coarctata *(grassland millipede)*
India
The mating rate is at maximum respectively in the tenth to twelfth night after Full Moon or New Moon. It then rapidly reduces to a minimum at Full and New Moon and remains at this level for the next 8 or 9 nights [736].

Class: Insecta, insects
Insects
Kawanda, Uganda; Kwadaso, Ghana
Light trap study: the nightly catch rates of isoptera and Bostrychidae increase with moonlight, those of Marasmia trapezalis *(Pyralidae), Lampyridae* and *Dorylus spp. (Formicidae)* decrease [737].

Muguga, Kenya
Two-year study in the wild using suction traps at heights of 1.5 m and 9 to 15 m: the vertical distribution of all insects follows a lunar cycle: the insects fly higher in the first quarter than in the third quarter. The total number of moths as well as their numbers at both flight levels followed a lunar cycle in one of the two years of the study, with maximums at New Moon and Full Moon and minimums in the first and third quarter [738].

Great Scott Cave, Washington, Missouri, USA
Bats returning to the cave from feeding were studied with regard to their food (insects of various orders). Light trap catches were carried out in parallel. The catches of Lepidoptera by means of light traps went down with increasing illumination from the Moon, and the reverse applied to catches from the order of Homoptera. Most of the orders of insects determined via food displayed a correlation with the lunar cycle even though the percentage of the food which can be explained in this way was relataively small. In the determination via food, representatives of the following orders revealed themselves as photopositive: Diptera, Plecoptera, Homoptera, and seen as a whole, the orders related to water. The representatives of the orders Lepidoptera, Coleoptera and, and seen as a whole, the terrestrial insect groups revealed themselves as photonegative [739].

Kenya
Light trap study: in the period before Full Moon, when the Moon rises early and sets well before dawn, more insects are caught in the late phases of the night than in the period after Full Moon when the Moon rises late and is still visible in the morning [740].

Night insects
Haut-Richelieu Valley, Quebec, Canada
Light trap catches in an oak clearing (*Quercus bicolor*). The activity of the insect community was at maximum between Full Moon and the last quarter. 60% of all catches fell in this period. The principle limiting factor was temperature, moonlight only took second place [741].

Order: Ephemenoptera, mayflies

Mayflies
Kangra Valley, Punjab; Nerbadda River, Central India
Swarming at Full Moon [742].

Hexagenia bilineata
India
Swarms between first quarter and Full Moon and between last quarter and New Moon [742].

Oligoneuria rhenana Inhoff
Basel, Switzerland
Swarms at Full Moon, July to September [742, 743].

Palingenia robusta Eaton
India
Swarms between last quarter and New Moon [742].

Polymitarcis (= Epheron) virgo Olivier
Swarms at Full Moon and the last quarter [742].

Povilla adusta Navas (tropical mayfly)
N Lake Victoria, Kaazi and Jinja, Uganda
Circalunar reproductive rhythm: hatching and swarming takes place from approx. 7.30–9.00 p.m. in the 5–6 days before and after Full Moon, particularly on the second day after Full Moon. The individual animals have a lifespan of only about 1.5 hours [744, 745, 746, 747].

Barombi Mbo Crater Lake, Cameroon
Lunar reproductive rhythm, most adult animals surface on the fourth day after Full Moon [748].

Order: Blattodea, cockroaches

Periplaneta americana Linnaeus
(American cockroach)
Wichita, Kansas, USA
Ethanol extracts with cordial activity (increased heart rate in cockroaches) can be found in the blood of stressed and unstressed cockroaches, laboratory mice and human beings. The stimulating effect of blood extracts of stressed animals reached very high values on the second day after Full and New Moon. On these days the extracts of unstressed animals remained without effect [333, compare also 332, 335, 336, 337, 338 and under Coleus for angiosperms].

Extracts with cordial activity (increased heart rate in cockroaches, comparable to the effect of neurotransmitters) can be won from various nervous tissues of cockroaches. The effect of the extracts varies from day to day and was greatest when the extract was won after Full or New Moon [335, compare also 332, 333, 336, 337, 338].

The effect of acetylcholine and noradrenalin on the heart of the cockroach was studied daily over a longer period. Both substances always produced an increase in the heart rate. Here the reaction of the heart to acetylcholine varied in a semi-lunar rhythm, as opposed to noradrenalin which varied in a lunar rhythm. The reaction to both substances changed at the spring and autumn equinox and at the summer solstice [336, also compare 332, 333, 335, and for antagonists 337].

SPECIES CATALOGUE   189

Order: Hemiptera, hemiptera
**Family: Belostomatidae
(giant water bugs, fish killers)**

Diplonychus nepoides Fabricius
Ivory Coast
The number of animals caught using light traps reaches its maximum (approx. 80%) 1 week before and after Full Moon, and its minimum (approx. 20%) 1 week before and after New Moon [749].

Family: Cicadidae

Nephotettix spec. (rice green leafhoppers)
W Bengal, India
Light trap catches at seven locations: the catch rate increases towards New Moon and decreases towards Full Moon [750].

Nephotettix nigropictus Stal
Nephotettix virescens Distant
Kalyani, W Bengal, India
The maximum catch rates fell in the time between the first quarter and Full Moon [751].

Order: Planipennia
Family: Myrmeleontidae, ant lion

Myrmeleon obscurus
Grahamstown, South Africa
Circalundian, circalunar activity rhythm of the larvae, determined by the volume of the conical pit traps built by them, maximum at Full Moon with a smaller maximum at New Moon always 4 hours after Moon rise [752].

Order: Hymenoptera,
    hymenopterous insects

Armigeres subalbatus
Madurai, India
The population density of female, host-seeking mosquitos increases and decreases in a lunar rhythm, being greater at Full Moon and lesser at New Moon [753].

Family: Sphecidae

Sphecodogastra texana Cresson
(North American bee)
Kingman County State Park, Kansas, USA
This solitary-living bee builds underground nests in the summer to bring up its young where it stays during the day. Only with the onset of dusk do the females leave their nests to search for food – exclusively slender evening primrose *(Oenothera rhombipetala)* – for their young. The bees only extend their flights beyond dusk into the night if moonlight follows the dusk phase (approximately from first night after New Moon to approximately the third night after Full Moon). As the Moon sets, they end their search for food. But if the Moon only sets when dusk has passed, the collection acitivity is restricted to dusk. This results is a rhythmical alternation of pronounced night activity and phases in which the bee flies only during the period of dusk [754, 755].

Family: Apidae, true bees

Apis mellifera carnica Linnaeus, 1758
(Carnica honey bee)
Lunar rhythm of flying activity, dependent on the direction of the honeycombs and the flight hole. Honeycomb in east-west direction and flight hole towards east: maximum at New Moon. Honeycomb in north-south direction and flight hole towards north: maximum at Full Moon. At the start of the winter rest period the lunar rhythm changes into a semi-lunar one, maximum at Full and New Moon [756].

Apis mellifera intermissa
(Moroccan honey bee)
Morocco
Semi-lunar rhythm of flight activity between 25 June and 8 August 1968:

maximum between New Moon and first quarter and between Full Moon and last quarter; no rhythm from 23 August to 21 September 1968. Semi-lunar rhythm between 21 September and 10 October 1968. Maximum reversed in relation to July. Change of location from Morocco to Frankfurt: change of semi-lunar rhythm into a lunar one [756].

Temera, Morocco
Semi-lunar rhythm of flying activity, maximum at time of syzygies [757].

Apis mellifera mellifera Linnaeus
(worker honey bee)
Cyclical fluctuation of sugar content in blood of hatching worker honey bees. Maximum values both for trehalose and glucose at Full and New Moon auf. A lower maximum value occurs in the period of the third quarter. For the trehalose sugar the values allow the assumption of a circa-seven-day rhythm [758].

Order: Trichoptera, caddis fly

Athripsodes stigma Kinmins
Lake Victoria at Jinja, Uganda
Lunar rhythm of hatching and swarming, maximum shortly before New Moon [744].

Athripsodes ugandanus Kinmins
Lake Victoria at Jinja, Uganda
Lunar rhythm of hatching and swarming, maximum shortly after first quarter [744].

Order: Lepidoptera, butterflies

Lepidoptera
Mount Glorious State Forest, Queensland, Australia
The lunar cycle has a weak influence on the activity of arboreal imagines [759].

Lepidoptera, *9 species*
Kenya, Uganda, Tanzania
Light trap catches of all 9 species were reduced in moonlight, albeit to different degrees [760].

Lepidoptera, *various types of cotton pests*
Cairo, Egypt
Light trap catches north of Cairo: the average catches in the week before and during New Moon were more than twice as large as at Full Moon [1064].

Heliothis zea Boddie *(bollworm moths)*
Texas, USA
The number of deposited eggs fluctuates in connection with the lunar cycle. Large egg deposits take place immediately after Full Moon, minimal ones at times of New Moon. The number of animals caught in light traps varies in similar form [1065, compare 1066].

Scirpophaga nirella
Paddy fields, W Bengal, India
The catch rate for immature and mature females respectively were always highest at Full Moon and New Moon respectively [1067].

Spodoptera litura Fabricius *(tobacco caterpillar)*
Raipur, Madhya Pradesh, India
The catch rate was higher at New Moon ± 3 days than at Full Moon ± 3 days [1068].

Prays citri
Bet Oagan, Israel
Significantly more male animals are caught at Full Moon than in other phases [1069].

Family: Tortricidae, bell moth

Epiphyas postvittana
Australia
These moths display a semi-lunar periodicity in their flying activity: shortly before and shortly after New Moon and

at Full Moon it is at its most pronounced [1070].

Family: Plutellidae

Plutella xylostella
Victoria, Australia
Lunar rhythm of flying activity, maximum at Full Moon, 2 further small maximums a few days before and after Full Moon [1071].

Family: Pyralidae, pyralid

Galleria mellonella Linnaeus
(wax moth)
Fort Collins, Colorado, USA
These darkness-adapted larvae are negative phototactic. The reaction is dependent on wave length and temperature. The larvae react with particulary sensitivity at about 300, 500 and 600 nm. The energy required for a specific threshold value (95 of 100 larvae react) at a given wavelength falls with temperature (21° C > 24° C > 27° C > 32° C). The threshold energy also varies with the lunar phases. At 393 nm the threshold energy at the time of the Full Moon is higher than at New Moon. The reverse applies at 503 nm and 653 nm. The temperature dependence of threshold energy does not vary with the phases [1072].

Family: Sphingidae

Seventy-five species
Trinidad
Catch rates vary with the linar light change of the Moon: they are significantly reduced as the light increases [761].

Family: Noctuidae, owlet moth

Noctuidae
S Queensland, Australia
Lunar periodicity of insect catches by means of light traps. Most of the species studied could be caught at maximum rates at the time of New Moon [762].

Agrostis ipsilon Hufnagel (cutworm)
Varanasi, India
Studies in the wild using light traps: catches were lowest in light nights around Full Moon [763].

Order: Diptera, dipterous insects

Family: Culicidae, mosquitos

Anopheles annulipes Walker
Griffith, NSW, Australia
Study with traps which preferentially bait female animals looking for a host: lunar rhythm of catch rate, maximum at Full Moon [764].

Anopheles bellator Dyar & Knab
Platanal, Cumaca, Trinidad
Of a total of 2314 female animals, 914 are caught at Full Moon, 852 at the last quarter, 409 at the first quarter and 364 at New Moon. The maximum activity occurred at Full Moon in the period from 6.00 to 8.00 p.m., at the other three times from 4.00 to 6.00 p.m. [765].

Anopheles culicifacies
Kheda District, Gujarat, India
The activity was greater in hours without Moon than with Moon [766].

Anopheles farauti Laveran
Coastal region of Papua New Guinea
The number of biting mosquitos was less during moonless nights than in nights with waxing Moon or Full Moon [767].

Culex (= Melanoconion) candelli
Dyar & Knab
Trinidad
The mouse bait trap was approached by the female mosquitos all night. But the peak values occur at different times during the night depending on phase: at Full Moon in the period from 22.00 to

24.00, in the first quarter from 22.00 to 04.00, at New Moon and in the last quarter from 24.00 to 02.00. Here the number of mosquitos also varies with the phase. Most of the mosquitos are collected in the first quarter, the least number in the last quarter [768, compare also 769].

Culex (= Melanoconion) portesi
Senevet & Abonnenc

Culex (= Melanoconion) taeniopus
Dyar & Knab
At sea east of Trinidad
Study using baited and light traps: maximum catches at dawn and dusk at New Moon. At Full Moon biting activity moves to the time of Moon rise and into the middle of the night [770].

Culicoides, 4 different species
Yankeetown, Levy Countee, Florida, USA
The greates flight activity occurred in the periods of dawn and dusk, and here there were clearly more individuals hunting for food during Full Moon than at any other phase [771]. The flight activity of male and female animals fell after sunset, but the female animals nevertheless remained flying relatively actively in the nights with Full Moon [772].

Culicoides peliliouensis
Palau Islands, Caroline Islands, Pacific
Semi-lunar rhythm of swarming, some days at New and Full Moon from midday to midnight [773].

Culicoides spec.
Brazil
Semi-lunar rhythm of swarming, at New and Full Moon [774].

Psorophora confinnis
St Lucie County, Florida, USA
Prolonged evening swarming of males by moonlight shortly before Full Moon [775].

Family: Chaoboridae

Chaoborus anomalus Lichtenstein (= *Corethra* Metg.)
Ekunu Bay, Lake Victoria at Jinja, Uganda
Lunar hatching and swarming rhythm, starts with waning Moon, maximum 2 to 3 days after New Moon [776].

Chaoborus edulis Edwards (= *Corethra* Metg.)
Lake Victoria at Jinja, Uganda
Lunar rhythm of hatching and swarming, maximum 5 to 2 days before New Moon [744].

Chaoborus spec. Lichtenstein (= *Corethra* Metg.)
Lake Kariba & Lake McIlwaine, Zimbabwe
The appearance of the adult forms occurs synchronously with the lunar cycle. If there is great population density at certain times, hardly any animals can be seen at other times. The lunar phase associated with maximum occurrence is variable. In the studied populations this initially ran in parallel with the period of the New Moon. This changed in the course of the study and the Full Moon accompanied the periods of greatest frequencyof the adult forms. Artificially kept animals do not display any occurrence synchronised with the lunar cycle [777].

Family: Chironomidae, chironomids

Chironomus (= Nilodorum) brevibucca Kieffer
Lake Bangweulu, Zambia
From June or July to October or November: mass occurrence of imagines. First appearance shortly after Full Moon, maximum occurrence 3 to 5 days after Full Moon [778].

Clinotanypus claripennis Kieffer
Lake Victoria, Jinja, Uganda
Lunar rhythm of hatching and swarming, maximum at first quarter [744].

Clunio marinus Haliday *(marine midge)*
Heligoland, Germany; Normandy, France; Santander, Spain
Circa-semi-lunar rhythm of pupation and subsequent hatching and swarming, with a circadian and annual rhythm superimposed: swarming only takes place shortly after the syzygies at deep low water in the evening from end of April to start of October only. The semi-lunar rhythm of pupation and hatching of the chironomid population of Heligoland and other northern regions (England, Ireland) can be induced through the interaction of artificial ligh-dark alternation in a 24-hour rhythm (LD 12:12) and a 12.4-hour cycle of simulated tides. Turbulence, underwater noise and floor vibrations as components of mechnical water movement were shown to be effective factors in artificial tides. In other marine midge populations, e.g. at Port-en-Bessin, Normandy, a semi-lunar rhythm can be induced in the laboratory through LD 12:12 and artificial moonlight (0.4 lux during 4 subsequent nights per month) [26, 27, 29, 30, 31, 32, 33, 128, 148, 149, 150, 151, 152, 522 pp.922ff, 779, 780, 781, 782, 783, 784, 785, 786, 787, 788, 789, 790, 791, 792, 793, 794, 795, 796].

Clunio pacificus Edwards
Japan
Lunar rhythm of swarming with an annual one superimposed, at New Moon in March and in the first half of April [797, 798].

Tanytarsus balteatus Freeman
Lake Victoria, Jinja, Uganda
Lunar rhythm of hatching and swarming, maximum 2–5 days after New Moon [744].

Family: Muscidae, true flies

Glossina pallidipes *(tsetse fly)*
Lake Tanganyika, Tanzania
Lunar rhythm of daytime flying frequency, maximum at New Moon, mininum at Full Moon, presumably reversed activity at night [799].

Family: Tachinidae, tachina fly

Euphasiopteryx ochracea
N Florida, USA
The number of insects caught correlated with the phases with greates catch rates a few days after New Moon [800].

Phylum: Echinormata, echinoderm

Class: Crinoidea, sea lilies and feather stars

Order: Comatulida, feather stars

Family: Comasteridae

Comanthus japonicus Müller *(Japanese feather star)*
Spawning at first and last quarter of the Moon in the first half of October within two hours (also single severed arms) [1073].

Class: Echinoidea, sea urchin

Hemicentrotus pulcherrinus Agassiz
Seto, Japan
Semi-lunar spawning rhythm on a few days at Full and New Moon [801].

Sub-class: Cidaroida, pencil urchins

Family: Cidaridae

Eucidaris tribuloides Lamarck
San Blas Archipelago, Panama (Caribbean coast)
Lunar rhythm of spawning, at around Full Moon [802].

Sub-class: Euechinoidea
Family: Diadematidae, Diadem sea urchin

Centrostephanus coronatus Verrill
Southern Californian coast, USA
Circalunar rhythm of gamatogenesis and spawning. The rhythm coincides with the lunar phases (maximum spawning in the days of the third quarters until close to New Moon [803, 804], but not with the rhythm of occurrence of stronger spring tides (first observation year at Full Moon, second observation year at New Moon) [805].

Diadema antillarum Philippi, 1845
(long-spined sea urchin)
Florida Keys, USA
Lunar rhythm of spawning with a diurnal one superimposed, nocturnally at Full Moon in summer [806, compare 807, 808].

Tropical Atlantic, Bermuda
Lunar rhythm of gamatogenesis. Oocytes grow between first and third quarter, spawning takes place at time of New Moon [809].

Diadema (= Centrechinus) setosum
Leske, 1778 (Indo-Pacific diadem sea urchin)
Suez, Red Sea, Egypt
Lunar rhythm of swarming in summer, maximum at or shortly before Full Moon [1, 810, 811, 812, 813, 814].

Lunar reproductive rhythm. Gonade maturity of various globally distributed populations falls after various phases of the lunar cycle: Island: first quarter; NE Borneo: New Moon; Vanuatu: third quarter [804].

Family: Toxopneustidae

Lytechinus variegatus Lamarck, 1816
(West Indian white sea urchin)
San Blas Islands, Panama (Caribbean coast)
Semi-lunar spawning rhythm, at Full and New Moon [802].

Bermuda; Miami, Florida, USA
Semi-lunar spawning rhythm, maximum at syzygies [815].

Toxopneusthes (= Lytechinus) variegatus Lamarck, 1816 (West Indian white sea urchin)
Coast of North Carolina, USA
Lunar rhythm of swarming, maximum at Full Moon [816].

Family: Echinidae

Paracentrotus lividus Lamarck, 1816
Roscoff, Brittany, France
Tidal rhythm of oxygen consumption, maximum at high tide [371].

Family: Echinometridae

Heterocentrotus mammillatus
Linnaeus
Northern Red Sea
Lunar or semi-lunar spawning rhythm in June and July [817].

Class: Holothuroidea, sea slugs, sea cucumbers

Bohadschia argus
Davies Reef, Great Barrier Reef, Australia
Lunar rhythm of swarming: in the first 3 nights after Full Moon in the period from 19.30 to 22.00 [818].

Euapta godeffroyi
Davies Reef, Great Barrier Reef, Australia
Lunar rhythm of swarming: at Full Moon and in the first night thereafter, from October to December towards sunset [818].

Stichopus chloronotus
Davies Reef, Great Barrier Reef, Australia
Lunar rhythm of swarming: in the second and third night after Full Moon between 19.30 and 23.00 [818].

Stichopus variegatus
Davies Reef, Great Barrier Reef, Australia
Lunar or semi-lunar rhythm of swarming: 5 of 6 spawning events were

observed in the 3 days at New Moon, the sixth spawning event 3 days after Full Moon, in the period from 16.45 to 21.45 [818].

Order: Apoda
Family: Chiridotidae

Polycheira rufescens
Kagoshina, Japan
Semi-lunar spawning rhythm. The release of gametes takes place within 1–2 days and always starts 1–2 days before each Full and New Moon [819].

Phylum: Chordata, chordates
Sub-phylum: Tunicata, tunicates
Class: Thaliacea, salpa, pyrosomes
Order: Desmomyaria, salpa
Family: Salpidae

Salpa fusiformis
South Georgia, S Atlantic
Lunar reproductive rhythm, strongest reproduction at Full Moon [615].

Sub-phylum: Vertebrata, vertebrates
Super-class: Gnathostomata, jawed vertebrates
Class: Chondrichthyes, cartilaginous fish
Sub-class: Elasmobranchii, sharks and skate
Order: Carcharhiniformes
Family: Carcharhinidae

Negaprion brevirostris (lemon shark)
The growth of vertebral body centres takes place in a lunar rhythm [820].

Class: Osteichthyes, bone fish
Sub-class: Actinopterygii, ray-finned fish
Teleostei, true bone fish

Reef-inhabiting fish
Tresher provided the following information in 1984 [821] on the occurrence of lunar-periodic spawning rhythms in reef fish (? = only indirect or unconfirmed references; * = based on observation in aquarium). Semi-lunar spawning rhythm in at least one or several species from the families: Apogonidae, Balistidae, Blenniidae (*), Pomacentridae, Pseudochromoids (*), Pteroidae, Ophistognathidae (*). Lunar spawning rhythm in at least one or several species from the families: Acanthuridae, Apogonidae, Balistidae, Carangidae (?), Chaetodontidae (?), Epinephelinae, Labridae (?), Lutjanidae, Mugilidae (?), Mullidae (?), Pomacentridae, Scaridae (?), Siganidae (?), Sphyraenidae (?), Sparidae (?). Several species from these families are named in the catalogue of species in the corresponding place.

Fish
Lake Kainji, Nigeria
Two to four days before and at Full Moon larger fish are caught in the open surface waters, 2–4 days after Full Moon in waters close to the coast [961].

Order: Anguilliformes, eels
Family: Anguillidae, eels

Anguilla anguilla (= vulgaris)
Linnaeus, 1758 (fresh-water eel)
Upper Rhine between Maxau/Karlsruhe and Speyer, Germany; east coast of Rügen, Baltic Sea, Germany
Maximum catadromous (towards the sea) migration intensity at waning Half Moon and minimum at waxing Half Moon. Approximately same intensity

at New and Full Moon [822, see also 823, 824].

Six places in tidal zone, Cumbria, England
Most eels migrate when high tide coincides with moonless nights. Only very few eels migrate at Full Moon [825].

Norway
Catadromous migration: most eels in the first quarter of the Moon, only a few at Full Moon [826].

Coast of Netherlands
The elvers entering the rivers from the sea enter the river estuaries with the high tide [827, see also 828].

Anguilla australis (shortfin)
Anguilla dieffenbachii
A: Lake Ellesmere (Canterbury), B: Lake Onoke (Wellington), C: Makara Stream (Wellington), New Zealand
A: Commercial catches of eels gathering for the migration to the sea display lunar periodicity: largest catches shortly after the third quarter and smallest catches during Full Moon.
B: Similar to A.
C: Greatest movement of eels during last half of lunar cycle and maximum activity shortly before third quarter [829].

Anguilla rostrata LeSueur, 1821
(American eel)
Rhode Island, USA
The migration of eels to the sea takes place mainly during the months of September to November and increases when there has been significant rain as well as during the third quarter and New Moon. Most animals migrate in the third quarters in the approximate 2 hours of darkness between sunset and Moon rise [830, 831].

Penobscot estuary, Maine, USA
The still transparent elvers migrate with the high tide into the river estuaries. Caught animals maintain this rhythm in the laboratory if there is a water current and otherwise constant conditions. Animals which have completed their migration through the tidal zone no longer display this circatidal activity rhythm [828].

Order: Clupeiformes, herring fish
Family: Clupeidae, herring

Alosa aestivalis (blueback herring)
Connnecticut River, Massachusetts, USA
Juvenile fish migrate from September to early November into downriver zones of the river. With regard to maximum movement compare Alosa sapidissina [832].

Alosa sapidissina Wilson, 1811
(American shad)
Connecticut River, Massachusetts, USA
Juvenile fish migrate end of September/early October into downriver zones of the river. Maximum movement is mostly focused in the first and third quarter. Some movement at New Moon, none at Full Moon [832].

Clupea pallasii Valenciennes, 1847
(Pacific herring)
Coast of British Columbia, Canada
Most spawning events occur in the period of the neap tides, particularly the ones in the third quarter [833].

Clupea harengus Linnaeus, 1758
(Atlantic herring)
East coast of England
Lunar rhythm (1921–32) of herring catches, greatest in September to December at Full Moon [834, 835, 836, 837, see also 823].

Herklotsichthys castelnaui
Townsville, Queensland, Australia
Semi-lunar spawning rhythm (associated with first and third quarter) probable [838].

Linnothrissa miodon *(Tanganyikan sardine)*
Gorge Basin, Lake Cahora Bassa, Mozambique
Despite constant reproductive rates, the population density of various species of planktonic fresh water crustaceans displays fluctuation connected with the phases. Population density increases from the last quarter through New Moon and the first quarter to Full Moon and then abruptly falls to its lowest level in the time of the last quarter. This fluctuation is induced by the feeding behaviour of the fish. This catches the zooplankton particularly efficiently when the Full Moon rises after sunset and the zooplankton rises to the surface in the preceding darkness. After the last quarter the population density is low and the fish seeks other sources of food until the next Full Moon [582, 583].

Spratelloides delicatulus
Spratelloides gracilis
Solomon Islands (Melanesia)
Lunar spawning rhythm, at Full Moon [962].

Spratelloides lewisi
Solomon Islands (Melanesia)
Lunar spawning rhythm, at New Moon [962].

Family: Engraulidae

Encrasicholina devisi
Encrasicholina heterolobus
Solomon Islands (Melanesia)
Lunar spawning rhythm, at Full Moon [962].

Order: Cypriniformes, cyprinids
Family: Catostomidae, sucker

Catostomus commersonii Lacépède, *1803 (white sucker)*
Canada
Lunar rhythm of thermoregulation: at New Moon, the fish heads for water with higher termperature than at Full Moon [839].

Order: Salmoniformes, salmonids
Family: Salmonidae, salmon

Salmo salar Linnaeus, *1758 (Atlantic salmon)*
Pitlochry, Perthshire, Scotland
The advancement rate is slower overall at Full Moon than in other phases [840].

Le Conquet, Brest, France
The blood plasma concentration of thyroxin was low from January to March. In April this value rose abruptly and reached its maximum in mid- April, subsequently falling again in the following 2 weeks back to the starting value. The maximum coincided with New Moon [841]. (With regard to the chemical and artificial stimulation of the thyroxin concentrations compare [842].)

Girnock Burn, near Ballater, Aberdeenshire, Scotland
The emigration of juvenile salmos in autumn and early winter and in spring from fresh water into sea water is less at the time of the Full Moon [843].

Salmo gairdneri Richardson, *1836 (rainbow trout)*
Guelph, Ontario, Canada
Semi-lunar rhythm of growth rate under laboratory conditions, in the case of longitudinal growth maximum at New and Full Moon [844].

Salmo trutta lacustris Linnaeus, *1758 (lake trout)*
Lac de Neuchâtel, Switzerland
Lunar rhythm of catch rate, maximum at New Moon, minimum at Full Moon [845].

Oncorhynchus kisutch Walbaum,
1792 (silver salmon)
California, Washington, Oregon, USA
Together with the catadromous emigration of the juvenile fish, other physiological changes take place (on morphological changes compare [846]). A particularly high value of the blood plasma concentration of thyroxin occurs on the 5 days of the New Moon closest to the spring equinox (California) or the following one (Washington and Oregon) [847, 848, compare 849].

Fish farm, Ontario, Canada
Artificial light L:D 12h:12h: semilunar rhythm the growth rate of juvenile fish, minimum at the time of the New and Full Moon [850, compare also 851].

Fish farm, Ontario, Canada
Artificial light L:D 12h:12h: semilunar rhythm in juvenile fish of the
— liver RNA:DNA ratio (minimum 1 day before New and Full Moon),
— muscle RNA:DNA ratio (maximum 1 day after New and Full Moon),
— animal carcass water content (minimum about 4 days before New and Full Moon),
— haematocrit concentrations (maximum 1 day before New and Full Moon)
— plasma triglyceride quantity (significantly lower before New and Full Moon than after)
— glucose content (significantly lower 7 days after New and Full Moon)
— plasma cholesterol content (significantly higher about 4 days before New and Full Moon)
— plasma-L thyroxin content in months with generally relatively low content (March), with significant minimums (4 days) and maximums (3 days) before New and Full Moon [851].

Capilano Hatchery, North Vancouver, Canada
The concentration of glucose and amino acid nitrogen varied in juvenile fish in a lunar cycle with the maximum at or near Full Moon [852].

Oncorhynchus masou Brevoort, *1856 (Masu salmon)*
Oippe River, Aomori Prefecture, Japan
The start of migration down river occurred (March 1982 and 1983) in the period of the New Moon. Most animals migrated directly after rain which followed the New Moon. The serum concentration of thyroxin in free-ranging fish was particularly high at New Moon (end April) [853, compare 849].

Oncorhynchus mykiss Walbaum, *1792*
Guelph, Ontario, Canada
Food consumption, plasma thyroid and growth hormone concentration vary in 14 day periods [854].

Oncorhynchus tschawytscha
Walbaum, *1792 (Chinook salmon)*
New Zealand
The plasma concentration of thyroxin (T4) in juvenile fish of a salmon research institute fluctuates in cyclical form, with maxiumu values around the time of each New Moon [855]. Marked juvenile animals released into a river (Rakaia River) at different stages of the lunar cycle were later captured again. Those adult animals were captured at the greatest rate which had been released as juveniles 3–6 days before New Moon [856].

Family: Thymallidae, grayling

Thymallus thymallus Linnaeus, *1758 (European grayling)*
Dunajec, Lesnica, Niedziczanka, Poland
The adult animals move upstream for spawning. The majority of fish move to the spawning grounds at the time of the New Moon [857].

Family: Galaxiidae, whitebait

Galaxias attenuatus (= maculatus)
Jenyns, 1842 (New Zealand whitebait)
Australia and New Zealand
Semi-lunar spawning rhythm, spawning at the time of the spring tides, hatching during the following spring tide [858, 859].

Order: Gadiformes, codfish
Family: Gadidae

Enchelyopus cinbrius *Bloch & Schneider, 1801 (four-bearded rockling)*
Canadian Atlantic coast
Semi-lunar spawning rhythm with a diurnal one superimposed, maximum at Full and New Moon early in the morning [860].

Lota lota *Linnaeus, 1758 (eel pout, burbot)*
Ängeran River, Sweden; Gulf of Bothnia
Lunar periodicity of the catadromous migration of juvenile fish, maximum at New Moon even when the river is covered with 40 to 50 cm ice and 30 cm snow [861].

Family: Merlucciidae, hakes

Macruronus novaezelandiae *(blue grenadier)*
South Australian waters
Lunar rhythm of spawning activity. This is greatest 7–10 days after Full Moon [862].

Order: Cyprinodontiformes, cyprinodonts
Family: Cyprinodontidae, cyprinodont

Adinia xenica
Fandulus pulvereus
Coast of Alabama, USA
Building on other findings, the results indicate a semi-lunar rhythm of spawning activity [863].

Fandulus grandis *Baird & Girard (gulf killifish)*
Louisiana, USA
Fandulus grandis displays a semi-lunar rhythm of spawning activity. This rhythm is maintained also in the laboratory under constant conditions: for 4 months [68, 69]. Studies carried out on fish kept in aquaria makes it seem likely that it is the maturing of the oocytes and ovulation which are the deciding factors leading to increased spawning activity during the circa-semi-lunar cycle. In the laboratory, the circa-semi-lunar rhythm of reproductive readiness could be induced through simulation of tidal changes of water temperature [864, 865, 866, compare 867, 868].

Gulf of Mexico
The semi-lunar spawning rhythm is accompanied by a change in the concentration of various hormones (E2, progesterone, testosterone, thyroxin, triidothyronin) in female and male animals [869].

Coast of Alabama, USA
Cyclical spawning activity, maximum in the days of the spring spring tides. The periodicity of the latter is 13.6 days in the studied section of coast (lunar declination of greater significance than the phases) [870].

Coast of Alabama, USA
Semi-lunar rhythm of spawning and gamete development, most pronounced from mid-April to late June, only during rising tides or spring tides [871].

The plasma concentrations of oestradiol-17β and testosterones in female animals vary in a semi-lunar rhythm. They are greatest shortly before and during spawning at each spring tide [872].

Fandulus heteroclitus Linnaeus, 1766
(common mummichog)
Maryland, USA
Free-living populations studied during the first 45 days of the brooding season display a semi-lunar cycle of sperm index (maturing). 6 days before the peak value of the sperm index, the serum testosterone concentration reaches its highest value [873].

Delaware Bay, Atlantic Ocean, USA
The semi-lunar rhythm of spawning is accompanied by fluctuations in the concentration of various substances in the serum and in the ovarial liquid (oestradiol-17β, cortisol) [874, 875, compare also 876].

Sapelo Island, Georgia, USA
Semi-lunar and lunar spawning activity depending on the size of the fish. Large fish (⩾ 56 mm SL = Standard Length) display a pronounced semi-lunar periodicity. They spawn at every spring tide during the two reproduction cycles of the year (first end of February to mid-May; second early July to mid-September). Medium sized fish (46–55 mm SL) spawn in a semi-lunar rhythm; but occasionally a larger group of these fish jumps a spawning event, thus displaying a lunar periodicity. Small fish (36–45 mm SL) overhwlemingly spawn in a lunar rhythm, above all during the first reproduction cycle. Spawning during the two reproduction cycles starts in the small fish later and ends earlier than in the larger fish. In fish of all sizes the maximum spawning activity during the first reproductive phase lies at or shortly before the spring tides which belong to the Full Moon. During the second phase it coincides with the spring tides of the New Moon, or follows shortly thereafter [877].

Sapelo Island, Georgia, USA
Circa-semi-lunar spawning rhythm, in comparison to the fish *in situ* with delayed phases but with the same transparency [878].

Delaware Bay, USA
Circa-semi-lunar spawning rhythm [865].

Canary Creek Marsh, Delaware Bay, USA
Circa-semi-lunar spawning rhythm, gonad maturity (GSI) and spawning readiness is greatest for some days around the syzygies. The spawning cycle is accompanied by cyclical changes of the water content of the eggs, egg size and the developmental stage of the follicles. The maximums of these parameters are achieved synchronously with the maximums of the spaewning activity at the spring tides [879, 880, 881, 882].

Muddy Creek, Rhode River, Chesapeake Bay, S of Annapolis, Maryland, USA
Semi-lunar spawning rhythm, at Full and New Moon [883].

Fandulus sinilis (longnose killifish)
Coast of Alabama, USA
Semi-lunar spawning rhythm; spawning takes place at the high tide in the 3–6 days before and including the sping tide [884, 885].

Family: Poeciliidae, live-bearers

Lebistes (= Poecilia) reticulatus
Peters, 1859 *(guppy)*
Venezuela and Caribbean islands
Lunar rhythm of colour sensitivity of vision (light reflex): yellow sensitivity is greatest shortly before or at Full Moon and red and purple sensitivity is at its lowest; at New Moon the opposite applies. Measurements in yellow light showed that the position of the extreme values and the length of period does not change in animals either which have been born and raised in permanent light [199, 200, 201, 202, 203, 204, compare also 206].

SPECIES CATALOGUE

Order: Atheriniformes, atherines
Family: Atherinidae, silversides

Hubbsiella sardina Jenkins & Evermann, 1888 (sardine silverside)
Northern Gulf of California and west coast of Mexico
Semi-lunar spawning rhythm with a tidal one superimposed, 3–4 days after Full and New Moon, 1–2 hours after high water [886].

Leuresthes sardina Jenkins & Evermann (California grunion)
Northern Gulf of California, Mexico
Spawning takes place 3.5 days after Full Moon and 4 days after New Moon [887].

Leuresthes tenuis Ayres, 1860 (grunion)
Sea coast from southern California to Oregon, USA
Semi-lunar spawning rhythm, over 3–4 days after New and Full Moon from late February to early September, at night 1–3 hours after high water [24 pp.100ff., 522 pp.919ff., 886, 888, 889, 890].

Menidia beryllina Cope, 1866 (inland silverside)
Robinson Point, Blackwater Bay, Santa Rosa County, Florida
The spawning rhythm, also continued under constant conditions, follows the synodic and not the tropic month, despite the unimportant role which the former has locally in relation to the tides where the fish were caught. Spawning both in the natural environment and in the laboratory takes place at Full and New Moon at the time of high water during the day [891].

Menidia menidia (Atlantic silverside)
Edisto Estuary, South Carolina, USA
The fish spawns in a semi-lunar rhythm, at New and Full Moon [892].

Salem Harbor, Massachusetts, USA
Spawning in the wild coincides with the New Moon and Full Moon (= semi-lunar). Fish kept artificially in outside tanks under daylight spawn every 1–3 days in contrast [893, 894].

Menidia peninsulae Goode & Bean (tidewater silverside)
Santa Rosa Island, Florida, USA
No connection was shown between the reproductive cycle and the cycle of the spring and neap tides. Nevertheless the four maximum values of the eggs, classified as mature, occurred at the time of the equatorial transit of the Moon in the period from March to July [895]. In the laboratory, spawning activity can be synchronised through low intensity artificial tidal currents and light dark alternation [896].

Order: Scorpaeniformes, scorpaenoids
Family: Scorpaenidae, sting fish

Sebastodes (= *Sebastes*) ***taczanowskii*** Steind.
Zaliv Petra Velikogo (Peter the Great Bay) Sea of Japan, Russia
The fish swims actively both at dawn and at dusk. Only the evening activity follows a lunar rhythm: it increases from new to Full Moon [897].

Order: Perciformes, perchifish
Family: Centropomidae, giant perches

Lates calcarifer Bloch, 1790 (sea bass)
Tigbauan, Iloilo, Philippines
Studies in net cages: spontaneous spawning, apparently in (semi-)lunar rhythms: of 26 observed egg deposits 17 took place within 4 days before or after the first quarter, 9 occurred within 5 days before or 3 days after the third quarter [898, compare 899].

Family: Serranidae, sea bass

Cephalopholis boenack
Horseshoe Reef, Papuan Barrier Reef, Papua New Guinea
Lunar or semi-lunar reproductive rhythm, at Full and New Moon, shortly before sunset [900].

Epinephelus guttatus
Epinephelus merra Bloch, *1793*
Epinephelus striatus
Society Islands (Pacific); shelf edge southwest of Puerto Rico and eastern end of the Cayman Islands (Caribbean)
Lunar spawning rhythm, always at the time of the Full Moon in January and February [901, 902, 903, 904].

Mycteroperca microlepis
Coast of South Carolina, USA
Semi-lunar rhythm of spawning activity, highest spawning rate in the period from new to Full Moon [905].

Seven species of the class **Plectropomus**
Arlington Reef and Green Island, Cairns Section, Great Barrier Reef, Australia
In studies on the return of pelagic youthful forms to the reef during the 1990/91 brooding season, undertaken by means of light traps, the majority were caught in the 17 days around the November New Moon 1990. Since the larval stage lasts about 25 days, it may be assumed that the new settlement of young forms in the reefs in this season can be traced back to an approximately two-week spawning period around the time of the preceding New Moon [906].

Family: Lutjanidae, snapper

Lutjanus fulvus (= variegiensis)
Schneider, *1801*
Society Islands (Pacific); Puerto Rico (Caribbean)
Lunar spawning rhythm [903].

Family: Pomadasyidae, grunt

Haemulon flavolineatum
(French grunts)
Tague Bay, Saint Croix US Virgin Islands (Caribbean)
The post-larval stages return to the coastal areas after having lived through a plankton phase. The return takes place in a semi-lunar rhythm with additional higher return rates, although not as high as the first one, 1 week before and after. An estimate of the time of fertilisation, studied in the post-larval stages through determination of the age of the otoliths, also indicated a 15-day periodicity with subsidiary peaks 1 week before and after. The maximum rates for the return to the coast as well as for fertilisation were associated with the first and third quarter, the weaker maximums with New and Full Moon [907].

Family: Sciaenidae, barfish

Plagioscion monti
Plagioscion squamosissimus Heckel, *1840 (pescada)*
Lago do Janauacá, right side of the Rio Solim es, near Manaus, Amazonas, Brazil
During the juvenile phase growth in weight occurs exponentially, longitudinal growth in a straight line (in the range 0.4–15 cm). Growth is not constant, however, but interruptions in growth occur dependent on the lunar phases. With *P. monti* growth stagnates at New and Full Moon (semi-lunar rhythm), with *P. squamosissimus* at New Moon (lunar rhythm). Age is determined by day rings on the otoliths [908].

Family: Pomacanthidae

Centropyge potteri
Coral reefs, Hawaii, USA; Enewetak, Marshall Islands (Pacific)
The fish spawns between December and May each evening during the week

before Full Moon (lunar periodicity) [909]. Other species *(Pomacanthus, Centropyge spp.)* from this family, observed at the Enewetak Atoll, Marshall Islands, spawn for the whole of the synodic month always at sunset with the exception of the half-monthly occurrence of the phase of evening slack water [910].

Holacanthus ciliaris
Punta de San Blas, Panama (Caribbean coast)
Lunar rhythm of recolonisation of the reef through the previously pelagic larval forms, maximum at first quarter [911].

Family: Chaetodontidae

Chaetodon capistratus Linnaeus, 1758
(butterflyfish)
Korbiski Reef, San Blas Archipelago, Panama (Caribbean coast)
Chaetodon capistratus prefers to graze on reef forming polyps of the Gorgonaria group. It removes individual polyps. The grazing of the polyps of a Plexaura species occurs in a lunar cycle; the colonies are grazed especially during or shortly after Full Moon. The feeding behaviour of C. capistratus appears to correlate with the spawning behaviour of the Gorgonaria [912].

Punta de San Blas, Panama (Caribbean coast)
Lunar rhythm for recolonisation of the reef through the previously pelagic larval forms, maximum 3 days at New Moon [911].

Family: Cichlidae, cichlids

Neolamprologus moorii Boulenger
Lake Tanganyika, Tanzania
Lunar rhythm of spawning, at the time of the first quarter [913].

Tilapia mariae
Ethiop River, Nigeria
Lunar rhythm of spawning; most eggs are laid during the last quarter. The river is not subject to the flow of the tides at the observation point [914].

Eight different species of
Tribus lamprologini
Lake Tanganyika, Tanzania
Lunar rhythm of spawning, all species spawn at maximum between first quarter and Full Moon. [915].

Family: Pomacentridae

Abudefduf saxatilis Linnaeus, 1758
(sergeant major)
Abudefduf troschelii *(Pacific sergeant fish)*
San Blas and Northern Perlas Island, Panama
The two 'sibling species' have developed apart since the isthmus of Panama arose. They are also different in their spawning behaviour, among other things. A. saxatilis, the Caribbean species, spawns throughout the month, the Pacific species A. troschelii in contrast spawn within a period of 9 days before and after New Moon [916].

Amphiprion bicinctus Rüppell, 1828
Coral reefs, Gulf of Aqaba, Red Sea
Spawning takes place in accumulated form in the week before and after Full Moon [917].

Amphiprion clarkii Bennett, 1830
(tropical anemone fish)
Murote Beach, Uchiumi Bay, west of Shikoku Island, Japan
1983: Continuous spawning with maximum at first and third quarter.
1984: Continuous spawning without trends.

The pelagic larval forms (called that from the time of hatching to the arrival at the reef) join the population colonizing the reef in particularly large numbers; 1983: at the time of the Half Moon, 1984: in the time from the third via the fourth to the first quarter. In both years the number of settling larval forms was larger in the period from Full Moon to New Moon than from New Moon to Full Moon. [918].

Amphiprion melanopus Bleeker, 1852 *(anemonefish)*
Guam (Micronesia, Pacific)
Maximum spawning takes place on first and third quarter of lunar cycle. Hatching reaches its peak about 1 week later close to New or Full Moon at the spring tides [919].

Chromis cyanea
Chromis multilineata
Flat Cay, St Thomas, US Virgin Islands (Caribbean)
Lunar rhythm of colonisation of reefs by larval forms, at maximum at time of Full Moon [920].

Eupomacentrus planifrons *(threespot damselfish)*
Discovery Bay, Jamaica
Lunar rhythm of spawning: new egg laying above all 4–7 days after Full Moon. The larva appear 5–8 days later accompanied by the New Moon [921].

Pomacentrus flavicauda Whitley *(tropical damselfish)*
Great Barrier Reef, Australia
Semi-lunar spawning rhythm from October to March, at maximum 6–8 days before Full and New Moon [922].

Pomacentrus nagasakiensis Tanaka
Miyake-Jina, Japan
The spawning activity of the fish increases with the Moon at maximum 2–3 days before Full Moon [923].

Pomacentrus wardi *(tropical damselfish)*
Great Barrier Reef, Australia
Semi-lunar spawning rhythm from October to March, at maximum 11–13 days before Full and New Moon [922].

Stegastes dorsopunicans Poey
Microspathodon chrysurus Cuvier & Valenciennes, 1830 *(Caribbean damselfish)*
San Blas Islands, Panama
Lunar rhythm of spawning, at maximum in the period from Full Moon to New Moon [924, 925].

Stegastes partitus
Caribbean
The generation of planktonic young forms (at maximum 6 nights after Full Moon) and their return to the reef (reaching its maximum 3 nights before New Moon) takes place in a lunar rhythm [926, 927].

Panama (Caribbean coast)
Lunar spawning rhythm, at maximum in the period between Full and New Moon [928].

Stegastes dorsopunicans C, L, N
Stegastes partitus C, L, N
Stegastes diencaeus C, L, N
Stegastes planifrons C, L, N
Stegastes variabilis C, L, N
Chromis multilineata C, L
Microspathodon chrysurus C, L
Microspathodon dorsalis P, L
Microspathodon bairdi P, L
Stegastes leucostictus C, SL, N
Abudefduf saxatilis C, SL, V
Abudefduf troschelii P, SL
Stegastes acapulcoensis P, SL
C: *Panama (Caribbean coast);*
P *Panama (Pacific coast)*
Lunar (L) and semi-lunar (SL) spawning rhythm with various maximum values in various species extending over almost the whole of the synodic cycle. In most species the cycle of hatching from the egg probably takes place in the same period, delayed by the incubation phase (3–5 days). Most species on the Caribbean side of Panama spawn at dawn [928].

Punta de San Blas, Panama (Caribbean coast)
N = lunar rhythm of re-colonisation of the reef by the previously pelagic larval forms, at maximum 3 days around New Moon.
V = semi-lunar rhythm of re-colonisation, at maximum in the first and third quarter [911].

Family: Mugilidae, marine alga

Mugilidae
Mangalore estuary, Karnataka, India
Larger catches are made in darker nights at high water and at New Moon [929].

Family: Labridae, labroids

Thalassoma bisfasciatum Bloch, 1791 (bluehead wrasse)
West coast of Barbados
Semi-lunar rhythm of spawning, more frequently at New and Full Moon (spring tides) than at neap tides [930].

San Blas Islands, Panama (Caribbean coast)
Lunar rhythm of re-colonisation of the reef by the previously pelagic larval forms, at maximum at New Moon [931]. In some observation phases also semi-lunar rhythm in first and third quarter [911].

Thalassoma duperrey
Kaneohe Bay, Hawaii
Semi-lunar rhythm of spawning, more frequently at New and Full Moon than in the first and third lunar quarter [932].
   The development of the oocytes during the autumn months is organised such that the female animals are ready for spawning – from a physiological point of view – precisely at the time of high water [933].

Family: Blenniidae, blenny

Blennius (= Lipophrys) pholis
Linnaeus, 1758 (shanny)
Rock pools on the west coast of Anglesey, Wales; Seil Island, Argyll, Scotland
Circatidal rhythm of swimming activity, at maximum during high water. Neither the semi-lunar variations nor the weak daily inequality of the local tides influence the periodicity of the activity rhythm [934, 935, 936, 937, 938, 939].

Coryphoblennius galerita Linnaeus, 1758
Plymouth, England; Roscoff, France
Circatidal activity rhythm, at maximum during high water in the natural habitat. The first maximum of each day was greater than the second, indicating a circadiane component [940].

Ophioblennius atlanticus Cuvier & Valenciennes, 1836
Barbados
Spawning activity is concentrated in the 7 days around Full Moon. Spawning takes place within the first 3 hours of daylight [941].

Panama (Caribbean coast)
Lunar rhythm of spawning, at maximum at Full Moon in the dawn period [928].

Family Gobiidae
Family Scaridae
Family Labridae

Moorea, Society Islands, Pacific
The return of the fish larvae to re-colonise the reef takes place above all at dusk and at night and is four times greater during New Moon than during Full Moon. The percentage of larva in order of the above named families was: 60.5%, 10.3% and 6.2% [942].

Family: Gobiidae, sea gudgeon

Chaenogobius castaneus
(biringo-goby)
Kanita River and Nagashita River, Mutsu Bay, N Honshu, Japan
Sea gudgeon living in the brackish water areas of river estuaries display a tidal activity rhythm both in the wild and in the laboratory under constant conditions. The fish search for benthic worms (annelids) on the river bed at falling tide and low water and migrate to brackish water ponds or stay in very shallow and hidden places along the banks when the water rises and at high tide. Like the fish, the annelids too feed (on detritus) mainly when the water is falling when parts of the work stick our of their tube. Compare also Tridentiger brevispinis [943].

Chasmichthys gulosus
Rocky coast of Asamushi, Mutsu Bay, N Honshu, Japan
A: facing the Pacific with relatively regular semi-diurnal tidal cycles
B: rocky coast of Fukaura, facing the Sea of Japan with a mixed tidal cycle
Juvenile animals from A and B were caught and kept in the laboratory under constant conditions. Only the animals from A displayed a clear approx. twelve-hour activity rhythm, the maximum of which fell into the time shortly after the predicted high water for A [944].

Coryphopterus glaucofraenum Gill
Gnatholepis thompsoni Jordan
Barbados
The colonisation of the reef by the larvae takes place in both species mainly in the third quarter. Only with G. thompsoni does spawning take place more frequently during the third quarter [945, for Lee Stocking Island, Great, Bahamas compare 946].

Tridentiger brevispinis Katsuyama, Arai & Nakamura
(numachichibu-goby)
Kanita River and Nagashita River, Mutsu Bay, N Honshu, Japan
Tidal activity rhythm for feeding; smaller fish, including Chaenogobius castaneus (compare above), are hunted with special intensity with rising water and at high tide. Bentich animals and algae also serve as food [943].

Family: Scombridae, mackerel

Scomber colias *(tuna-like mackerel)*
Fishing off NW Africa
Behavioural changes with lunar amplitude of oscillation: a more diffuse distribution of the tunalike mackerel in the dark nights of the New Moon phase, particularly in the months July to September. Increased readiness to flee of fish stocks in the Full Moon phase [947].

Family: Acanthuridae, surgeon fish

Various species from the genus
Acanthurus, Ctenochaetus, Zebrasoma and ***Paracanthurus***
Palau Islands, Caroline Islands, Pacific
Semi-lunar spawning rhythm [948].

Acanthurus bahianus
Acanthurus chirurgus
Acanthurus coeruleus
Punta de San Blas, Panama (Caribbean coast)
Lunar rhythm of re-colonisation of the reef through the previously pelagic larval forms, at maximum 3 days at New Moon. But also semi-lunar rhythm in A. bahianus and A. chirurgus during some observation phases, in first and third quarter [911].

Acanthurus triostegus sandvicensis
Streets, 1877
Hawaii, USA
Spawns only at Full Moon from December to July [949].

Family: Siganidae, rabbitfish

Siganus rivulatus Forsskal, 1775
Red Sea, Saudi Arabia
Lunar rhythm of spawning, 4 days after Full Moon in the months from April to August [950].

Family: Stromateidae, basket fish

Peprilus burti (gulf butterfish)
Gulf of Mexico
Lunar rhythm of catch rate, at maximum in the days of the first quarter [951].

Order: Pleuronectiformes, flat fish

Family: Scophthalmidae, turbot relatives

Rhombus (= Scophthalmus) maxinus
Linnaeus, 1758 (turbot)
Roscoff, Brittany, France
Tidal rhythm of oxygen consumption, maximum at high tide [371].

Family: Bothidae, lefteye flounders

Paralichthys dentatus Linnaeus, 1766
(summer flounder)
Schooner Creek, Little Egg Inlet, New Jersey, USA
The directon of movement of flounders depends on the level of the tides: up the tidal creek at high tide, down the tidal creek at low tide [952, 953].

Family: Pleuronectidae, plaice

Platichthys stellatus Pallas, 1787
(starry flounder)
The otolith microstructure of juvenile flounders displays growth in a tidal rhythm [954].

Pleuronectes (= Platichthys) platessa
Linnaeus, 1758 (plaice)
Roscoff, Brittany, France
Tidal rhythm of oxygen consumption, maximum at high tide [371].

North Sea, location of fish tagged with acoustic transmitters
Seasonal migration: the animals spend the summer in the southern area of the Dogger Bank, in autumn they migrate to the spawning grounds at the eastern entrance to the English Channel between the Rhine and Thames estuaries. Spawning takes place largely in January; in late February the animals are back near the Dogger Bank. Plaice use the tidal currents for their migration: they rise from the sea floor to about 2–5 m under the surface when the current flows in the required direction. As soon as the direction is reversed, they dive to the sea floor and remain there until the current which they need returns [955, compare 956 and 957 for Gadus morhua L. (cod)].

Benderloch Beach (Ardmucknish Bay), Argyll, Scotland
Juvenile animals moving along the sea floor were observed with underwater cameras only during the day, mostly at high water and shortly before sunset. The direction of movement tends to be towards the land at high tide, towards the coast at low tide. Animals were hardly observed at night. Probably nocturnal migration following the tidal currents is carried out at half height in the water [958, compare also 1074].

Coast of Netherlands
Freshly caught animals displayed a tidal rhythm for a short time (1–3 days) under constant conditions in permanent darkness which quickly turned into a circadian one under the influence of darkness or a light-dark alternation [1075, see also 959, 960].

Order: Tetraodontiformes, puffer relatives
Family: Tetraodontidae

Takijugu niphobles
Tomioka, West Kyushu, Japan
Semi-lunar spawning rhythm, at Full and New Moon [963].

Class: Amphibia, amphibians
Order: Urodela, caudate
Family: Salamandridae, salamanders and water newts

Triturus (= Diemictylus) viridescens
(red spotted newt)
W Massachusetts, USA
Lunar rhythm of oxygen consumption, at maximum 3–4 days before Full Moon [964].

Order: Anura, tail-less amphibians
Family: Ranidae, true frogs

Rana cancrivora Gravenhorst, 1829
Coastal tidal zone, Jawa, Indonesia
Lunar spawning rhythm with an annual one superimposed, maximum at New Moon and in August [965].

Rana pipiens Schreber, 1782
(leopard frog)
Syracuse, New York, USA
Experiments at Full and New Moon and in the first and third quarter: the twice daily maximums of the calcium transport activity of the duodenum at Full Moon, in the first and third quarter and at New Moon coincide with the upper and lower transit of the Moon respectively related to the meridian of the location of the study. The plasma calcium concentrations are also at maximum in that time [966].

Syracuse, New York, USA
In the period from April to June the upper transit of the Moon (related to the meridian of the location of the study) is accompanied by reduced movement activity. Also during the day in the period from October to March, the upper transit of the Moon coincides with increased movement activity. If the maximum movement is analysed for all days in the year, an oscillating pattern of behaviour emerges, the period of which is equal to the lunar tidal cycle [967].

Family: Bufonidae, true toads

Bufo americanus Holbrook, 1836
(American toad)
Lac Carré, Terrebonne County, Quebec, Canada
These toads move more actively during the period of New Moon than in other lunar phases. Movement activity is influenced by other factors such as rain and temperature [968].

Bufo biporcatus
Bali, Indonesia
Lunar spawning rhythm, maximum at Full Moon [969, 970].

Bufo fowleri
Jawa, Indonesia
Lunar spawning rhythm with an annual one superimposed, maximum at New Moon and in August [971].

Bufo melanostictus Schneider, 1799
(black-spined toad)
Jawa, Indonesia
Lunar spawning rhythm with an annual one superimposed, at maximum in the first quarter and at Full Moon, strongest spawning in the course of the

year: November/December and a subsidiary maximum in June [969, 970, 972].

Class: Reptilia, reptiles
Sub-class: Anapsida, reptiles without temporal fossa
Order: Testudines, turtles
Family: Emydidae, terrapins

Pseudemys scripta Schoepff, 1792
(lettered terrapin)
Terre Haute, Indiana, USA
The studies produced indications of a diurnal (lunar) rhythm of movement activity. Slight maximum close to nadir position and slight mininum at zenith position of Moon. Experiments in a Faraday cage indicate that the rhythm can be influenced by electrostatic fields [973].

Sub-class: Lepidosauria, squamata
Order: Squamata
Family: Gekkonidae, gecko

Ptyodactylus hasselquistii guttatus
Von Heyden, 1827 (house gecko)
Israel
Nightly activity increases and decreases respectively in parallel to the waxing and waning Moon [974].

Class: Aves, birds

Birds
Netherlands; North America
Lunar periodicity: there are numerous reports that birds die at night through artificially illuminated obstacles. Statistical analysis shows that this does not happen in nights around or at Full Moon [975].

Order: Galliformes, gallinaceous birds
Family: Phasianidae, pheasant-type

Tetrao urogallus Linnaeus, 1758
(capercaillie)
Finland
Increased reproduction always when the Full Moon coincides with a fixed spring date mainly every 3–4 years (possible sub-cycle of the Metonic cycle) [10, 11].

Lyrurus tetrix Linnaeus, 1758
(black grouse)
Finland
Increased reproduction always when the Full Moon coincides with a fixed spring date mainly every 3–4 years (possible sub-cycle of the Metonic cycle) [10, 11].

Order: Gruiformes, crane relatives
Family: Gruidae, crane

Grus grus Linnaeus, 1758 *(wintering common crane)*
Spain
The brighter the additional light from the Moon, the later these birds go to their sleeping perches after sunset [976].

Order: Charadriformes
Family: Charadriidae, plover

Vanellus vanellus Linnaeus, 1758
(lapwing)
North Downs, Hampshire, England
In the months from November to February the behaviour of lapwings (search for food mainly during the day, rest at night) is reversed in the period of Full Moon. This is not the case if the light of the Full Moons has been obscured by strong cloud cover [977, 978].

Carse of Stirling, Scotland
Before the start of the brooding season, the use of habitat by the lapwing flocks follows a pronounced lunar periodicity. With the approach of the cycle to New Moon, the percentage of birds found during the day in wheat fields (winter grain) was reduced and a larger part of the birds was encountered on grass areas [979].

Charadrius wilsonia cinnamominus Ord, 1814 (thick-billed plover)
Chacopata Lagune, N Araya Peninsula, NE Venezuela
The presence of the plover at the feeding grounds does not overall display any seasonal variation but in the months from October to March a reduced daytime share is compensated by increased nocturnal hunt for food. Here the feeding sites are mainly visited in nights when the Moon is shining brightly, irrespective of the size of the Moon's disc, i.e. of the lunar phase. The nightly hunt for food has a tendency to increase from October to January and then reduces again. The hunt for food on the tidal flats was significantly longer from October to March at nocturnal high water than at low water during the day [980].

Family: Sternidae, terns

Sterna fuscata Linnaeus, 1766 *(sooty tern)*
Ascension Island, S Atlantic
At intervals of 10 synodic months (approx. 294 days) the animals come to the island for nesting. Nesting thus takes place a little earlier each year with 4 broods over 3 years. After nesting the animals spread out widely over the Atlantic [981, 982, 983, compare also 984, 985 p.407].

Sterna sumatrana *(black-naped tern)*
Nesting colony on Eagle Island, Great Barrier Reef, Australia
The first clutches of the nesting season (September – January) and also the first following 'wave of clutches' in a season were always laid within a four-day period around New Moon in 3 years of observation. Feeding rate for the chicks was highest in the later afternoon and lowest in the early afternoon. It was also influenced by the tides and was high during spring tides (in the observation period early in the morning and late in the afternoon at Full and New Moon) and low at neap tides. Food for the chicks is clearly particularly abundant during this period. This could also explain why the laying of the eggs is related to the lunar phase. For the male must feed the female as part of his courtship immediately before the eggs are laid [986].

Family: Alcidae, auks

Ptychoramphus aleuticus Pallas, 1811 *(Cassin's auklet)*
Southeast Farallon Island, California, USA
The Cassin's Auklet nesting in small earth holes along the sea terrace of the island fell victim to marauding seagulls (*Larus occidentalis*, western gull) particularly at the time of the Full Moon irrespective of whether clouds or fog were to be observed at this time or not. The auks only visit their breeding colonies at night and stay near their holes on moonlit nights for approximately 30% less time than on dark nights. It is generally known about sea birds that their activities in their colonies is reduced on moonlit nights [987].

Order: Columbiformes, columbaceous birds
Family: Columbidae, pigeons

Columba livia Gmelin, 1789 *(homing pigeons)*
Ithaca, New York, USA
Lunar rhythm in the starting direction of flight, which varies from day to day, in orientation behaviour [988].

Order: Strigiformes, owls
Family: Strigidae

Bubo virginianus Gmelin, 1788 *(great horned owl)*
S Central Pennsylvania, USA
The owl calls were broadcast with a tape recorder and the answering calls were recorded. There was significantly more contact during the second quarter (first day after the first quarter up to and including Full Moon) than in the three remaining quarters. The owls answered more frequently on nights with little cloud than on overcast nights [989].

Strix occidentalis Xantus, 1859 *(spotted owl)*
N Arizona, USA
The calling behaviour of owls in the wild correlates with the lunar phases. Thus owls call more frequently than is to be expected statistically during the last quarter and the New Moon and less during the first quarter and at Full Moon [990].

Order: Caprinulgiformes, nightjar
Family: Caprinulgidae

Caprinulgus europaeus Linnaeus, 1758 *(nightjar)*
Plymouth, England
Lunar rhythm of evening courting song of the male: at New Moon soon after sunset (approx. 15 min), at Full Moon later (approx. 35 Min). Maturing of gonads and ovulation co-ordinated with Moon such that egg laying takes place during third quarter [991, compare 992, but also 993].

Order: Passeriformes, passerine birds
Family: Bombycillidae

Bombycilla garrulus Linnaeus, 1758 *(waxwing)*
Finland
Mass occurrence every 18 to 20 years (Metonic cycle); probably caused through coincidence of the Full Moon with a fixed spring date, which optimally triggers the reproductive cycle [9, 11].

Class: Mammalia, mammals
Sub-class: Prototheria, egg-laying mammals
Order: Monotremata, monotremes
Family: Tachyglossidae

Tachyglossus aculeatus Shaw & Nodder, 1792 *(Australian spiny anteater)* (Plate 17)
Zoo in Prague, Czech Republic
The male develops a small pouch every 28 days [994].

Sub-class: Theria
Infra-class: Metatheria

Order: Marsupialia, marsupials
Family: Macropodidae

Macropus eugenii Desmarest, 1817 *(tammor wallaby)*

Kangaroo Island, Southern Australia
The reactivation of embryonic development (= end of the diapause) is linked with the Full Moon of the summer solstice [995].

Infraclass: Eutheria
Order: Primates
Family: Cebidae, capuchines

Aotus trivirgatus griseinembra
De Boer, 1982 or
Aotus lemurinus griseinembra
Hershkovitz, 1983 (night ape)
Colombia
While the night apes are very active for the whole night at Full Moon, their activity is largely restricted to dusk and dawn at New Moon [996, 997, compare 998, 999].

Order: Chiroptera, alipeds
Family: Pteropidae

Rousettus aegyptiacus Geoffroy, 1810
(Egyptian fruit bat)
Colombia
Maximum activity with waxing Moon, minimum activity at Full Moon [996].

Family: Phyllostomidae, leaf-nosed bat

Artibeus jamaycensis Leach, 1821
(Jamaican fruit bat)
Barro Colorado Island, Canal Zone, Panama
In the nights one week before and after New Moon these animals were active from dusk to dawn. In the nights 1 week before and after Full Moon the search for food was interrupted for 1 to 7 hours at the time that the Moon was at its zenith even when the Moon was covered by thick cloud. Longer flights to seek new fruiting trees were only undertaken in the dark half of the lunar month [1000].

Artibeus lituratus Olfers, 1818
Colombia
The activity of the bat is reduced with waxing Moon in the first half, waning Moon in the second half of the night and at Full Moon during the whole night [996, compare 998].

Phyllostomus hastatus Pallas, 1767
Colombia
Maximum activity with waxing Moon, minimum activity at Full Moon [996].

Family: Vespertilionidae

Myotis yumanensis Allen, 1864
Bosque del Apache National Wildlife Refuge, New Mexico, USA
In moonlit nights flying activity is concentrated in shadow areas [1001].

Order: Rodentia, rodents
Family: Heteromyidae, kangaroo rat

Dipodomys merriami Mearns, 1890
(Merriam's kangaroo rat)
Boyd Deep Canyon Desert Research Center, Palm Desert, California, USA
At Full Moon the animals are most likely to be found in their daytime shelters, but if they do emerge they stay closer to their layers than usual. When the Moon is at half, they prefer to be outside their caves when Moon is still below the horizon. At Full Moon the suppression of nighttime activity is compensated through increased activity at sunrise and sunset [1002].

Dipodomys spectabilis Merriam, 1890
(bannertail kangaroo rat)
Portal, Arizona, USA
The maximum nocturnal activity falls in the period approx. 20 min. after sunset, and thereafter diminishes throughout the night. Activity was reduced to one third when the Moon was in the sky and shifted from open territory to territory covered by bush [1003, compare 1004].

Family: Cricetidae, burrowing animals

Mesocricetus auratus Waterhouse, *1839 (golden hamster)*
Lunar rhythm of running activity (maximum shortly before Full Moon); the pH value of the urine is minimal in this period, it is at maximum at Full Moon [1005, 1006].

Lunar rhythm of daily running activity (fluctuations) [1007].

Meriones unguiculatus Milne-Edwards, *1867 (Mongolian gerbil)*
Evanston, Illinois, USA
Under L:D 12h:12h: lundian activity rhythm in summer, autumn and winter; at maximum towards Moon rise and the setting Moon, minimal when the passes through its top and bottom transit. The activity rhythm could also be observed in permanent light with the same period [1010].

Onychomys leucogaster Wied-Neuwied, *1841 (grasshopper mouse)*
Oklahoma, USA
Time spent outside the nest is longest at New Moon [1008, compare also 1009].

Family: Microtidae, vole

Lemmus lemmus Linnaeus, 1758
(Norway lemming)
Norway
Mass reproduction probably after coincidence of the Full Moon phase with a fixed spring date every 3–4 or 10 years respectively (possible sub-cycles of half the Metonic cycle) [11, compare also 1080].

Family: Muridae, mice

Apodemus sylvaticus Linnaeus, 1758
(wood mouse)
Foveran Wood, Aberdeenshire, Scotland
In Winter, the time spent outside the nest increased with rising temperature and diminishing moonlight. Below 2–4°C the temperature tends to be the determining factor, above 2–4°C it is the moonlight [1011].

Mus musculus Linnaeus, 1758
(white mouse)
Lundian rhythm of running activity, at maximum shortly before the Moon reaches its nadir; 2 minimums shortly before the Moon rises and shortly before it reaches its zenith [97].

Movement activity was relatively greater during the days of New Moon and the first quarter than on days of Full Moon and the third quarter. The reverse applied to air pressure which was also measured [1012].

Rattus rattus Linnaeus, 1758
(house rat)
Lundian rhythm of oxygen consumption, at maximum when the Moon is at its nadir, at minimum when the Moon is at its zenith [97, 1013].

Rattus mülleri
Rattus rajah
Rattus rattus jalorensis
Rattus sabanus
Selangor, Malaysia
Lunar rhythm: more conceptions at the start of the lunar month [1014, 1015, 1016].

Family: Hystricidae, porcupines

Hystrix indica Kerr, 1792
Negev Desert, Israel
The activity phases of the porcupine lie in the winter (October – March) such that they are exposed only minimally to moonlight. This avoidance of moonlight disappears in the following period and cannot be observed in later summer (August – September) [1017].

Order: Carnivora, carnivores
Suborder: Fissipedia, fissipeds
Family: Canidae, canidae

Alopex lagopus Linnaeus, 1758
(Arctic fox)
Small island of the Kronprinsens Ejland archipelago, Disko Bugt, W Greenland
Main source of food is fish caught by the animal itself, the activity of the foxes correlates with the tides: catching fish at low tide, sleeping at high tide [1018].

Family: Felidae

Lynx canadensis Kerr, 1792
(Canadian lynx)
N North America
Population density fluctuates in a half-Metonic cycle [1024].

Suborder: Pinnipedia, seals
Family: Otariidae, eared seals

Arctocephalus galapagoensis Heller, 1904 (Galápagos fur seal)
Cabo Hammond, Fermandina, Galápagos Islands
Lunar rhythm of animals on land. At Full Moon about twice as many animals are on land as at New Moon. That applies to male and female animals and to young animals [1019].

Otaria bryonia Blainville, 1820
(maned seal)
Falkland Islands
Sections of tooth show 24 to 27 smaller lamella within the dentine annual rings. There are thus about 2 lamella for each month [1020].

Family: Phocidae, common seals

Phoca vitulina Linnaeus, 1758
(common seal)
Osteological studies on animals of various origins
The formation of concavities in close proximity on the periphery of the root area of the incisors of young common seals takes place in a monthly rhythm. The ringlike concavities always lie in parallel to the root opening and completely encircle the root area [1021, compare also 1019, 1020, 1022].

Snake Island, Strait of Georgia, British Columbia, Canada
In spring and summer, these animals show an increased tendency to stay in the water during the night at the time of the Full Moon [1023].

Order: Lagomorpha, leporines
Family: Leporidae, hares and rabbits

Lepus americanus Erxleben, 1777
(snowshoe hare)
Canada
Mass appearance every 7–12 years, on average 9.6 years = half a Metonic cycle (1849–1955). The reproductive rhythm is probably optimally triggered by the coincidence of the Full Moon with a sensitive period in spring lasting a few days. The reproduction rate is reduced depending on the extent to which the Full Moon diverges from the annually specified date [11, compare also 1079].

Order: Artiodactyla, even-toed ungulates
Family: Bovidae, horned animals

Aepyceros melampus *Lichtenstein, 1812 (impala)*
Sengwa Wildlife Research, NW Zimbabwe
The rutting period is probably influenced by the lunar cycle. The first observations of mating from the years 1973 to 1978 occurred in a period of 6 days at Full Moon [1025].

Bos prinigenius taurus *Linnaeus, 1758 (domestic cattle)*
Shinfield, Berkshire, England
Study of 57 young cows. The first oestrus occurs approx. 2 months earlier in animals which were born as the days were growing longer than in those born in the autumn. In general 4 sections in the course of the synodic cycle could be determined in which oestrus was more likely to happen, namely the 2 days before the first, second, third and fourth quarter respectively. Conception dates show a similar distribution, although this becomes less clear with increasing numbers of pregnancies of the animal [1026, compare 1027].

Bubalus arnee *Kerr, 1792*
(Indian buffalo)
India
Lunar mating rhythm, mating at New Moon, occasionally at Full Moon [1028].

Capra hircus (= aegagrus) *Linnaeus, 1758 (domestic goat, various breeds)*
Experimental farm near Kassel, Germany
The number of births appears to be higher at the time around Full Moon; but there was no statistical significance [1029].

Connochaetes taurinus *Burchell, 1823, albojubatus Thomas (brindled gnu)*
Africa
The lunar cycle of the Moon influences the time of mating and conception [1030].

Ovis ammon musinon *Pallas, 1811 (European mouflon)*
A hormonally controlled cessation of growth over several months (December – February) produces grooves around the hooves of the mouflon, the annual rings. Between 2 annual grooves is the annual spiral growth, which in turn reveals the so-called 'monthly furrows'. If these are initially 12 to 18 grooves per year at a greater distance, they reduce in number in subsequent annual spiral growth (10–12) and are closer together [1031, 1032].

Terrestrial Placentalia
Statistical analysis of 213 various species, incl. human beings
The period of pregnancy is statistically examined in a large number of terrestrial placentalia. Of the whole numbers 6 to 33 the number 30 turns our to be the one the multiplication of which best corresponds to the pregnancy periods of the animals. This figure is close to the one which the Moon requires for a synodic orbit (= lunar, 29.53 days) [193, 194].

Lunar periodicity in human beings
(Homo sapiens)
Linnaeus, 1758

Absenteeism
At Full Moon: very slight (but significant) reduction in absenteeism [1033].

Accidents, *see* **traffic accidents**

Aggression in women placed in nursing homes
The aggression rate was higher on the day of Full Moon than on other days of the lunar cycle. The second highest rate fell on the third day before Full Moon [1034, 1035].

Bacterial and viral inflammation
Medical survey in a Czech industrial city
In men and women bacterial infections and their symptoms (the start of the illness was based on the patient's report) began more frequently in the morning than in the afternoon and at Full Moon than at New Moon; in women more frequently at the time of ovulation than menstruation. The opposite applies in all cases to viral infections, i.e. they start for both sexes more frequently in the afternoon etc. [1036].

Births
New York City, USA: 500,000 live births
Close to the limit of statistical significance, it could be observed that the birth rate lies above the average before Full Moon and below the average after Full Moon [240].

New York City, USA: 120,000 births in hospitals in 1954–55, 250,000 births in municipal hospitals 1948–57
Significantly higher birth rates on the 3 days around Full Moon, least birth rate on the 3 days around New Moon [239].

New York City, 1968
An analysis of births in 1968 showed that births varied in a lunar rhythm. The time of greatest fertility was shown to be the third quarter [1037].

Freiburg, Germany: 33,000 births 1924–38; Mannhein and Stuttgart, Germany: 140,000 births 1925–38
Lunar rhythm of frequency of births, more boys are born with a waxing Moon and more girls with a waning Moon [257, 260, compare in this respect 221 p.35ff. and 259].

France: statistical study of 5,927,978 births 1968–74
There were more births between the third quarter and New Moon and fewer during the first quarter [241].

Cancer
Investigation of skin temperature in breast cancer shows: with rapidly growing, histologically undifferentiated and hot tumours circa-seven-day and circalunar rhythms are absent in comparison with the contralateral side [according to 1038 from 1039].

The circalunar temperature rhythm measured in healthy women, taken at the breast (mamma), was used for comparison with the temperature rhythm after the operation to remove the carcinoma without removal of the breast and using the healthy side as a comparison in patients before the menopause. The result was, first, that in patients with breast cancer there was hardly any circalunar temperature rhythm despite normal ovulation cycles. Second, there were increases in frequency for both breasts in the same group; thus seven-day rhythms were found, for example, which did not appear in the healthy women. Third, the amplitude of the circalunar core body temperature rhythm (oral temperature) is reduced by approx. 50% in the patient group [according to 1040 from 1039, compare also 1041].

In cancer patients the circalunar rhythm of ferritin and iron in the serum is lost [according to 1042 from 1039].

Colour vision
Lunar rhythm: greater sensitivity for green and blue (light of short wavelengths) at New Moon (more pronounced in summer); greater sensitivitry for yellow and red (light of long wavelengths) at Full Moon (more pronounced in winter) [196, 197, 198, 221].

Deaths
Lunar rhythm, broad maximum at Moon [218, 221, 1045].

Eclampsia
Lunar rhythm, accumulation of attacks on fifth day after New Moon [221, 235].

Malaria
Lunar rhythm, intermittent attacks particularly with waxing Moon, with the cycles ending at Full Moon [236].

Menstruation

University of Pennsylvania, USA: 312 women
The ovulation of women whose menstrual cycle lasts about 29.5 days, tends to fall into the dark phased of the Moon with statistical significance, i.e. into the period of the last quarter through New Moon to the first quarter. Women with irregular menstruation periods tend to ovulate in this period [246].

In women whose monthly cycle lasts 29.5 days plus/minus 1 day the start of menstruation tends to fall in Full Moon. In women with this cycle period, the probability for the start of menstruation is reduced with growing distance from the day of Full Moon [1043, 1044].

Study with femal volunteers with normal cycle. Of 826 women (16–25 years old), 28.3% menstruated at New Moon, while at other times of the lunar cycle this share only came to 8.5–12.6% [247].

Pneumonia
Lunar rhythm, more pneumonia invasions during waxing than waning Moon, accumulation of pneumonia crises in last quarter [221, 236, 237].

Poisoning

Maryland Poison Center, Baltimore, USA
A total of 22 079 telephone calls regarding cases of poisoning in a period of 13 lunar cycles were analysed by lunar phases. A large part of all calls to the Center and calls due to unintentional poisoning were made during the Full Moon period (= Full Moon +/- 2 days). A significantly larger number of unintentional poisonings occur during the Full Moon period, compared to attempted suicide and cases of drugs misuse which occur most frequently during New Moon periods (= New Moon +/- 2 days) [227].

Self-inflicted poisoning

Winnipeg General Hospital, Winnipeg, Canada, study lasting 7 lunar cycles
In women, the larger part of cases occurs during the first quarter and a smaller part during the last quarter of the lunar cycle; in men, no correlation with the lunar phases was found. In both sexes there is a relationship with the distance of the Moon from the earth: within the lunar cycle, a larger number of cases occurs in the days of the apogee and the perigee [1046].

Toxicology Information and Management Service, Hunter Valley, Australia
Of 2,215 consecutive cases between January 1987 and June 1993, at New Moon the percentage of women was 60%, at Full Moon 45% [1047].

Sleeping-waking rhythm

A psychologically normal, blind man displayed a (circadian) rhythm in his body temperature, wakefulness, performance, cortiol secretion and electrolyte discharge in his urine which was equal to that of the lunar day [207, 208, compare 217].

Traffic accidents

Asturias, Spain (1985–88)

In 1988 only it was found: the least number of accidents occurred at Full Moon, the most 2 days before Full Moon, with a second peak at New Moon. If the data is analysed in four periods according with the four quarters of the Moon, no significant differences are apparent. But in a semi-monthly analysis it emerges that there are significantly fewer traffic accidents in the period of the waning Moon (from Full Moon to New Moon) than in the period of the waxing Moon (from New Moon to Full Moon) [1048]. No relationship between phases or the distance of the Moon from the earth and traffic accidents became apparent in other studies [compare 1049, 1050, 1051, 1052].

Uricosuria

Semi-lunar and lunar rhythm; tendency for lesser uricosuria one day after Full Moon and on the fourth day after New Moon respectively [218, 219, 220, 221].

Violence

Dade County, Florida, USA

— murder and serious violence occurred more frequently at Full Moon
— psychiatric emergency wards were attended more frequently in the first lunar quarter, and particularly infrequently at New and Full Moon
— suicide rates and traffic accidents leading to fatalities also displayed lunar periodicity [225, 226].

But others dispute a connection between the lunar cycle and aggressive human bahaviour [231].

References

Outline articles and books with explanations of lunar periodicity in organisms: 28, 91, 123, 124, 147, 151, 466, 522, 823, 1053, 1054, 1055, 1056, 1057, 1058, 1059, 1060, 1061, 1062, 1063

1. Aristotle, *De Partibus Animalium,* Book 4, Chapter 5.
2. Cicero, Marcus Tullius, *De Divinatione,* Book 2, Chapter 33.
3. Plinius Secundus Cajus, *Historia naturalis* (Natural History), Book 2 (Cosmology), Chapter 41.
4. Crawshay, L.R., Possible bearing of a luminous syllid on the question of the landfall of Columbus. *Nature* 136:559f (1935).
5. Stair, J.B., Palolo, a sea worm eaten by the Samoans. *Journal of the Polynesian Society* 6:141–44 (1897).
6. Stair, J.B. & Gray, J.E., An account of Palolo, a sea worm eaten in the Navigator Islands. *Proceedings of the Zoological Society of London* 15:17f (1847).
7. Abrami, G., Correlations between lunar phases and rhythmicities in plant growth under field conditions. *Canadian Journal of Botany* 50(11):2157–66 (1972).
8. Darwin, Ch., *The formation of vegetable mould through the action of worms, with observations on their habits.* London 1881.
9. Siivonen, L., Über die Kausalzusammenhänge der Wanderungen beim Seidenschwanz Bombycilla g. garrulus (L.). *Annales Zoologici Societatis Vanamo* (Helsinki) 8:1–40 (1941).
10. Siivonen, L., On the reflection of short-term fluctuations in numbers in the reproduction of tetraonids. *Papers on Game Research* (Helsinki) 9:1–43 (1952).
11. Siivonen, L. & Koskimies, J., Population fluctuations and the lunar cycle. *Papers on Game Research* (Helsinki) 14:1–22 (1955).
12. Archibald, H.L., Is the ten-year wildlife cycle induced by a lunar cycle? *Wildlife Society Bulletin* 5(3):126–29 (1977).
13. Bulmer, M.G., A statistical analysis of the ten-year cycle in Canada. *Journal of Animal Ecology* 43(3):701–18 (1974).
14. Thompson, L.M., Effects of change in climate and weather variability on the yields of corn and soybeans. *Journal of Production Agriculture* 1(1):20–27 (1988).
15. Vines, R.G., Rainfall patterns in the Eastern USA. *Climatic Change* 6(1):79–98 (1984).
16. Stifter, A., Die Sonnenfinsternis am 8. Juli 1842. *Sämtliche Werke,* Vol.3, pp.1211ff. Wiesbaden, Emil Vollmer. (n.d.).
17. Laskar, J., Der Mond und die Stabilität des Erdklimas. *Spektrum der Wissenschaft* 9:48–55 (1993).
18. Camões, L. Vaz de (Portuguese poet, 16th century), *Os Lusiadas.* Edicao nacional. Imprensa Nacional de Lisboa 1928. (Aqui, ... onde a terra se acaba e o mar começa...).
19. Jablonski, D. & Bottjer, D.J., The origin and diversification of major groups: environmental patterns and macroevolutionary lags, in: Taylor, P.D. & Larwood, G.P. (eds.), *Major evolutionary radiations.* The Systematics Association, Special Volume 42, pp.17–57. Oxford (Clarendon) 1990.

20 Schad, W., *Der Heterochronie-Modus in der Evolution der Wirbeltierklassen und Hominiden.* Dissertation, Universität Witten-Herdecke 1992.
21 Defant, A., *Ebbe und Flut des Meeres, der Atmosphäre und der Erdfeste.* Berlin 1953.
22 Dietrich, G., Kalle, K., Krauss, W. & Siedler, G., *Allgemeine Meereskunde. Eine Einführung in die Ozeanographie.* Stuttgart 1975.
23 Sager, G., *Ebbe und Flut.* Gotha 1959.
24 Wood, F.J., *Tidal Dynamics. Coastal Flooding, and Cycles of Gravitational Force.* Dordrecht, Netherlands 1986.
25 Palmer, J.D., Clock-controlled vertical migration rhythms in intertidal organisms, in: DeCoursey, P.J. (ed.), *The Belle W. Baruch Library in Marine Sciences,* No.4. *Biological rhythms in the marine environment,* pp.239–55. Papers presented at a symposium, held April 24–27, 1974, at the Belle W. Baruch Coastal Research Center Georgetown SC. University of South Carolina Press 1976.
26 Neumann, D., Die lunare und die tägliche Schlüpfperiodik der Mücke Clunio. Steuerung und Abstimmung auf Gezeitenperiodik. *Zeitschrift für Vergleichende Physiologie* (Berlin) 53:1–61 (1966).
27 Neumann, D., Entrainment of a semilunar rhythm by simulated tidal cycles of mechanical disturbance. *Journal of Experimental Marine Biology and Ecology* 35:73–85 (1978).
28 Neumann, D., Tidal and lunar rhythms, in: Aschoff, J. (ed.), *Handbook of Behavioral Neurobiology,* Volume 4: *Biological Rhythms.* New York (Plenum) 1981.
29 Pflüger, W., Die Sanduhrsteuerung der gezeitensynchronen Schlüpfrhythmik der Mücke Clunio marinus im arktischen Mittsommer. *Oecologia* (Berlin) 11:113–50 (1973).
30 Koskinen, R., Seasonal emergence of Clunio marinus Haliday (Diptera: Chironomidae) in Western Norway. *Annales Zoologici Fennici* (Helsinki) 5:71–75 (1968).
31 Neumann, D. & Honegger, H.-W., Adaptations of the intertidal midge Clunio to arctic conditions. *Oecologica* (Berlin) 3:1–13 (1969).
32 Palmén, E. & Lindeberg, B., The marine midge, Clunio marinus Hal. (Diptera: Chironomidae) found in brackish water in the Northern Baltic. *Internationale Revue der gesamten Hydrobiologie und Hydrographie* (Leipzig) 44:383–94 (1959).
33 Caspers, H., Rhythmische Erscheinungen in der Fortpflanzung von Clunio marinus (Dipt. Chiron.) und das Problem der lunaren Periodizität bei Organismen. *Archiv für Hydrobiologie* (Stuttgart), Supp. 18:414–594 (1951).
34 Rumphius, G.E., *D'Amboinsche Rariteikamer.* Book 1. Chapter 64, Amsterdam 1705.
35 Remmert, H., Biologische Periodik. *Handbuch der Biologie 5,* pp.407ff. Wiesbaden (Akademische Verlagsgesellschaft Athenaion) 1977.
36 Caspers, H., Spawning periodicity and habitat of the palolo worm Eunice viridis (Polychaeta, Eunicidae) in the Samoan Islands. *Marine Biology* (Berlin) 79(3):229–36 (1984).
37 Collins, D., Ward, P.D. & Westermann, G.E.G., Function of cameral water in Nautilus. *Paleobiology* 6(2):168–72 (1980).
38 Denton, E.J., Gilpin-Brown, F.R.S. & Gilpin-Brown, J.B., On the Buoyancy of the Pearly Nautilus. *Journal of the Marine Biology Association of the United Kingdom* 46:723–59 (1966).
39 Ward, P., Greenwald, L. & Greenwald, O.E., Der schwebende Nautilus. *Spektrum der Wissenschaft* (Heidelberg) 12:110–18 (1980); *Scientific American* No.10 (1980).

40. Kahn, P.G.K. & Pompea, S.M., Nautiloid growth rhythms and dynamical evolution of the Earth-Moon system. *Nature* 275:606–11 (1978).
41. Hughes, W.W., Nautiloid growth rhythms and lunar dynamics. *Nature* 279:453f (1979).
42. Jones, D.S. & Thompson, I., Nautiloid growth rhythms and lunar dynamics. *Nature* 279:454f (1979).
43. Landman, N.H., Ammonoid growth rhythms. *Lethaia* 16:248 (1983).
44. Runcorn, S.K., Nautiloid growth rhythms and lunar dynamics. *Nature* 279:452f (1979).
45. Saunders, W.B. & Ward, P.D., Nautiloid growth and lunar dynamics. *Lethaia* 12:172 (1979).
46. Doguzhaeva, L., Rhythms of ammonoid shell secretion. *Lethaia* 15:385–94 (1982).
47. Cochran, J.K., Rye, D.M. & Landman, N.H., Growth rate and habitat of Nautilus pompilius inferred from radioactive and stable isotope studies. *Paleobiology* 7(4):469–80 (1981).
48. Ward, P., Greenwald, L. & Magnier, Y., The chamber formation cycle in Nautilus macromphalus. *Paleobiology* 7(4):481–93 (1981).
49. Martin, A.W., Catala-Stucki, I. & Ward, P.D., The growth rate and reproductive behavior of Nautilus macromphalus. *Neues Jahrbuch für Geologie und Paläontologie.* Abhandlungen 156(2):207–25 (1972).
50. Evans, J.W., Cockle diaries: the interpretation of tidal growth lines. *Endeavour* (New Series) 12:8–15 (1988).
51. Jones, D.S., Repeating layers in the molluscan shell are not always periodic. *Journal of Paleontology* 55(5):1076–82 (1981).
52. Kaiser, H., Cornélissen, G. & Halberg, F., Paleochronobiology circadian rhythms, gauges of adaptive Darwinian evolution; about 7-day (circaseptan) rhythms, gauges of integrative internal evolution, in: Hayes, D.K., Pauly, J.E. & Reiter, R.J. (eds.), *Progress in clinical and biological research,* Vol.341: *Chronobiology: Its role in clinical medicine, general biology, and agriculture,* part B, pp.755–62. XIX. International Conference of the International Society for Chronobiology, Bethesda, Maryland, USA, June 20–24, 1989. Chichester, England, 1990.
53. Rhoads, D.C. & Lutz, R.A. (eds.), *Skeletal growth of aquatic organisms.* New York (Plenum) 1980.
54. Brosche, P. & Sündermann, J. (eds.), *Tidal friction and the earth's rotation* I. Berlin (Springer) 1978.
55. Brosche, P. & Sündermann, J. (eds.), *Tidal friction and the earth's rotation* II. Berlin (Springer) 1983.
56. Chevallier, J.M. & Cailleux, A., The earth-moon gravitational link through the ages. *Canadian Journal of Earth Science* 9(5):479–85 (1972).
57. Hövel, W.-T., *Analyse und Anwendung von Modellen ozeanischer Gezeiten im Hinblick auf die Gezeitenreibung im Erde-Mond-System.* Dissertation, Bonn 1982.
58. Sager, G., Gezeitenreibung und Erdretardation. *Nova acta Leopoldina* NF 57, No.258:139–52 (1984).
59. Walker, J.C.G. & Zahnle, K.J., Lunar nodal tide and distance to the moon during the precambrian. *Nature* 320:600–602 (1986).
60. Pang, K. et al., Chinese had shorter days. *New Scientist* 120:33 (1988).
61. Taylor, G.J., Ursprung und Entwicklung des Mondes. *Spektrum der Wissenschaft* 9:58–65 (1994).

62 Vaas, R., Entstehung des Mondes – neue Hypothese. *Naturwissenschaftliche Rundschau* 10:409f (1987).
63 Kremer, B.P., Mikroalgen als Zellgäste. *Spektrum der Wissenschaft* 2:48–55 (1994).
64 Babcock, R.C. et al., Synchronous spawnings of 105 scleractinian coral species on the great Barrier Reef, Australia. *Marine Biology* (Berlin) 90(3):379–94 (1986).
65 Shlesinger, Y. & Loya, Y., Coral community reproductive patterns: Red Sea versus the Great Barrier Reef, Australia. *Science* 228 (4705):1333–35 (1985).
66 Bohn, G., Sur les mouvements oscillatoires des Convoluta roscoffensis. *Comptes rendus hebdomadaires des Séances et Memoires de la Société de Biologie et de ses filiales* 137:576–78 (1903).
67 Enright, J.T., The tidal rhythm of activity of a sand beach amphipod. *Zeitschrift für vergleichende Physiologie* (Berlin) 46:276–313 (1963).
68 Hsiao, S.M. & Meier, A.H., Semilunar spawning cycles of the gulf killifish Fundulus grandis in closed circulation systems. *American Zoologist* 25(4):17A (1985).
69 Hsiao, S.M. & Meier, A.H., Spawning cycles of the gulf killifish Fundulus grandis in closed circulation system. *Journal of Experimental Zoology* 240(1):105–12 (1986).
70 Lohmann, K.J. & Willows, A.O.D., Lunar modulated geomagnetic orientation by a marine mollusc. *Science* 235 (4786):331–34 (1987).
71 Bennett, M.F., Geomagnetism and circadian activity in earthworms, in: Scheving, L.E., Halberg, F. & Pauly, J.E. (eds.), *Chronobiology*, pp.700–702. Stuttgart (Thieme) 1974.
72 Brown, F.A.jr., Barnwell, F.H. & Webb, H.M., Adaptation of the magnetoreceptive mechanism of mud-snails to geomagnetic strength. *Biological Bulletin of the Marine Biological Laboratory Woods Hole* 127:221–31 (1964).
73 Brown, F.A.jr. et al., Magnetic response of an organism and its solar relationships. *Biological Bulletin of the Marine Biological Laboratory Woods Hole* 118:367–81 (1960).
74 Brown, F.A. jr., Webb, H.M. & Brett, W.J., Magnetic response of an organism and its lunar relationship. *Biological Bulletin of the Marine Biological Laboratory Woods Hole* 118:382–92 (1960).
75 Korall, H. & Martin, H., Responses of bristle field sensilla in Apis mellifica to geomagnetic and astrophysical fields. *Journal of Comparative Physiology* A 161:1–22 (1987).
76 Lindauer, M. & Martin, H., Die Schwereorientierung der Bienen unter dem Einfluß des Erdmagnetfeldes. *Zeitschrift für Vergleichende Physiologie* 60:219–43 (1968).
77 Lindauer, M., Orientierung der Tiere. *Verhandlungen der Deutschen Zoologischen Gesellschaft.* 156–83 (1976).
78 Martin, H. & Lindauer, M., Der Einfluß des Erdmagnetfeldes und die Schwereorientierung der Honigbiene (Apis mellifica). *Journal of Comparative Physiology* A 122:145–87 (1977).
79 Wiltschko, W. & Wiltschko, R., Magnetic compass of European robins. *Science* 176:62–64 (1972).
80 Brown, F.A.jr. & Scow, K.M., Magnetic induction of a circadian cycle in hamsters. *Journal of Interdisciplinary Cycle Research* 9:137–45 (1978).
81 Olcese, J., Reuss, S. & Semm, P., Geomagnetic field detection in rodents. *Life Sciences* 42(6):605–13 (1988).

82. Wever, R., Über die Beeinflussung der circadianen Periodik des Menschen durch schwache elektromagnetische Felder. *Zeitschrift für vergleichende Physiologie* 56:111–28 (1967).
83. Wever, R., Einfluß schwacher elektromagnetischer Felder auf die circadiane Periodik des Menschen. *Naturwissenschaften* 55:29–32 (1968).
84. Wever, R., Different aspects of the studies of human circadian rhythms under the influence of weak electric fields, in: Scheving, L.E., Halberg, F. & Pauly, J.E. (eds.), *Chronobiology,* pp.694–99. Stuttgart (Thieme) 1974.
85. Waterman, T.H., *Der innere Kompaß. Sinnesleistungen wandernder Tiere.* Heidelberg (Spektrum) n.d.
86. Webb, H.M., Biological clocks and the role of subtle geophysical factors, in: Tomassen, G.J.M. et al. (ed.), *Geocosmic relations; the earth and its macroenvironment,* pp.56–64. First International Congress, Amsterdam, Netherlands, April 19–22, 1989, Wageningen, Netherlands, 1990.
87. Enright, J.T., A virtuoso isopod: circalunar rhythms and their tidal fine structure. *Journal of Comparative Physiology* 77(2):141–62 (1972).
88. Palmer, J.D., The rhythmic lives of crabs. *Bioscience* 40(5):352–58 (1990).
89. Aschoff, J. (ed.), *Circadian clocks. Proceedings of the Feldafing summer school.* September 7–18, 1964. Amsterdam (North Holland Publishing Company) 1965.
90. Bünning, E., *Die physiologische Uhr – Circadiane Rhythmik und Biochronometrie.* Berlin (Springer) 1977.
91. Rensing, L., *Biologische Rhythmen und Regulation.* Stuttgart 1973.
92. Sweeney, B.M., *Rhythmic phenomena in plants.* London 1987.
93. Berthold, P., Innere Jahreskalender – Grundlage der Orientierung bei Tieren. *Biologie in unserer Zeit,* Year 9, No.1:1–8 (1979).
94. Gwinner, E., Endogene Jahresrhythmen beim Vogelzug. *Spektrum der Wissenschaft,* Juni 1986.
95. Gwinner, E., *Circannual rhythms.* Berlin (Springer) 1986.
96. Milia, A. Di & Geppetti, L., On the expansion-contraction rhythm of the sea anemone, Actinia equina L. *Experientia* 20:571f (1964).
97. Brown, F.A.jr., Response to pervasive geophysical factors and the biological clock problem. *Cold Spring Harbour Symposia on Quantitative Biology* 25:57–71 (1960).
98. Brown, F.A.jr., Extrinsic rhythmicality: a reference frame for biological rhythms under so-called constant conditions, in: Wolf, W. (Conference Editor): Rhythmic functions in the living system, pp.775–87. *Annals of the New York Academy of Sciences* 98. New York 1962.
99. Brown, F.A.jr., A unified theory for biological rhythms: rhythmic duplicity and the genesis of 'circa' periodisms, in: Aschoff, J. (ed.), Circadian clocks, pp. 231–61. *Proceedings of the Feldafing summer school.* September 7–18, 1964. Amsterdam (North Holland Publishing Company) 1965.
100. Brown, F.A.jr., A hypothesis for extrinsic timing of circadian rhythms. *Canadian Journal of Botany* 47:287–98 (1969).
101. Brown, F.A.jr., Hastings, J.W. & Palmer, J.D., *The biological clock: two views.* New York (Academic Press) 1970.
102. Brown, F.A.jr., Why is so little known about the biological clock?, in: Scheving, L.E., Halberg, F. & Pauly, J.E. (eds.), *Chronobiology,* pp.689–93. Stuttgart (Thieme) 1974.
103. Brown, F.A.jr., The biological clock phenomenon: exogen timing hypothesis. *Journal of Interdisciplinary Cycle Research* 14:137–62 (1983).

104 Hardeland, R. & Balzer, I., The cellular circadian oscillator: A fundamental biological mechanism corresponding to a geophysical periodicity. *International Journal of Biometeorology* 32:149–62 (1988).
105 Brown, F.A.jr., Studies on the physiology of Uca red chromatophores. *Biological Bulletin of the Marine Biological Laboratory Woods Hole* 98:218–26 (1950).
106 Brown, F.A.jr., An exogenous reference-clock for persistent, temperature-independent, labile biological rhythms. *Biological Bulletin of the Marine Biological Laboratory Woods Hole* 115:81–100 (1958).
107 Brown, F.A.jr. & Webb, H.M., Temperature relations endogenous daily rhythmicity in the fiddler crab Uca. *Physiological Zoology* 21:371–81 (1948).
108 Brown, F.A.jr. & Sandeen, M.I., Responses of the chromatophores of the fiddler crab, Uca, to light and temperature. *Physiological Zoology* 21:361–71 (1948).
109 Brown, F.A.jr. et al., Persistent diurnal and tidal rhythms of color change in the fiddler crab, Uca pugnax. *Journal of Experimental Zoology* 123:29–60 (1953).
110 Brown, F.A.jr. et al., Temperature-independence of the frequency of the endogenous tidal rhythm of Uca. *Physiological Zoology* 27:345–49 (1954).
111 Palmer, J.D., Contributions made to chronobiology by studies of fiddler crab rhythms. *Chronobiology International* 8(2):110–30 (1991).
112 Ward, R.R., *The living clocks.* New York 1971.
113 Balzer, I. & Hardeland, R., Influence of temperature on biological rhythms. *International Journal of Biometeorology* 32:231–41 (1988).
114 Barnwell, F.H., Comparative aspects of the chromatophoric responses to light and temperature in fiddler crabs of the genus Uca. *Biological Bulletin of the Marine Biological Laboratory Woods Hole* 134(2):221–34 (1968).
115 Hastings, J.W., Rusak, B. & Boulos, Z., Circadian rhythms: The physiology of biological timing, in: Prosser, C. L. (ed.), *Neural and Integrative Animal Physiology. Comparative Animal Physiology,* 4 ed, pp.435–546. New York 1991.
116 Hamner, K.C. et al., The Biological Clock at the South Pole. *Nature* 195:476–80 (1962).
117 Aveni, Anthony, F., *Empires of time. Calendars, clocks and cultures.* London (Tauris) 1990.
118 Alpatov, A.M., Studies of circadian rhythms in space flight: some results and prospects. *Physiologist* 34 (1 Supplement):145f (1991).
119 Fuller, C.A., Homeostasis and biological rhythms in the rat during spaceflight. *Physiologist* 28 (Supplement 6):199f (1985).
120 Mergenhagen, D. & Mergenhagen, E., The biological clock of Chlamydomonas reinhardii in space. *European Journal of Cell Biology* 43:203–7 (1987).
121 Mergenhagen, D., The circadian rhythm in Chlamydomonas reinhardii in a Zeitgeber-free environment. *Naturwissenschaften* 73:410–12 (1986).
122 Sulzmann, F.M. et al., Neurospora circadian rhythms space: a reexamination of the endogenous-exogenous question. *Science* 225:232–34 (1983).
122a Round, F.E., *The biology of algae.* London 1973 (2 ed).
123 Naylor, E., Tidal and lunar rhythms in animals and plants, in: Brady, J. (ed.), Society for experimental biology. Seminar series, Vol.14: *Biological Timekeeping,* pp.33–48. Cambridge University Press 1982.
124 Morgan, E., An appraisal of tidal activity rhythms. *Chronobiology International* 8(4):283–306 (1991).
125 Hauenschild, C., Neue experimentelle Untersuchungen zum Problem der Lunarperiodizität. *Naturwissenschaften* 43:361–63 (1956).

REFERENCES

126 Hauenschild, C., Lunar periodicity. *Cold Spring Harbor Symposion on Quantitative Biology,* Vol.25: *Biological Clocks,* pp.491–97. The Biology Laboratory, Cold Spring Harbour, New York 1960.

127 Saigusa, M., Entrainment of a semilunar rhythm by a simulated moon light cycle in the terrestrial crab Sesarma haematocheir. *Oecologia* (Berlin) 46(1):38–44 (1980).

128 Neumann, D., Tidal and lunar rhythmic adaptations of reproductive activities in invertebrate species. Conference of the European Society for Comparative Physiology and Biochemistry, Proceedings 8; Strasbourg, France, August 31 – September 2, 1986. *Comparative Physiology of Environmental Adaptations,* 3: *Adaptations to climatic changes,* pp.152–70. P. Pevet (ed.). Basel 1987.

129 Enright, J.T., Resetting a tidal clock: a phase-response curve for Excirolana, in: DeCoursey, P.J. (ed.), *The Belle W. Baruch Library in Marine Sciences,* No.4. *Biological rhythms in the marine environment,* pp.103–14. Papers presented at a symposium, held April 24–27, 1974, at the Belle W. Baruch Coastal Research Center, Georgetown, SC. University of South Carolina Press, 1976.

130 Enright, J.T., Plasticity in an isopod's clockworks: shaking shapes form and affects phase and frequency. *Journal of Comparative Physiology* 107:13–37 (1976).

131 Palmer, J.D., Are tidal rhythms ultradian, circadian or infradian, in: Hayes, D.K., Pauly, J.E. & Reiter, R.J. (eds.), Progress in clinical and biological research, Vol.341: *Chronobiology: Its role in clinical medicine, general biology, and agriculture,* part B, pp.263–70. International Conference for Chronobiology, Bethesda, Maryland, USA, June 20–24, 1989. Chichester, England, 1990.

132 Aschoff, J., Circadian rhythms: influences of internal and external factors on the period measured in constant conditions. *Zeitschrift für Tierpsychologie* 49:225–49 (1979).

133 Palmer, J.D., Comparative studies in avian persistent rhythms: spontaneous change in period length. *Comparative Biochemistry and Physiology* 12:273–82 (1964).

134 Pittendrigh, C.S. & Daan, S., A functional analysis of circadian pacemakers in nocturnal rodents. I: The stability and lability of spontaneous frequency. *Journal of Comparative Physiology* 106:223–52 (1976).

135 Pittendrigh, C.S. & Daan, S., A functional analysis of circadian pacemakers in nocturnal rodents. V: Pacemaker structure: a clock for all seasons. *Journal of Comparative Physiology* 106:333–55 (1976).

136 Turek, F.W., Earnest, D.J. & Swann, J., Splitting of circadian rhythm of activity in hamsters, in: Aschoff, J., Daan, S. & Groos, G.A. (eds.), *Vertebrate circadian systems,* pp.203–14. New York (Springer) 1982.

137 Enright, J.T., Heavy water slows biological timing processes. *Zeitschrift für vergleichende Physiologie* 72:1–16 (1971).

138 Lesauter, J. & Silver, R., Heavy water lengthens the period of free-running rhythms in lesioned hamsters bearing SCN grafts. *Physiology and Behaviour* 54(3):599–604 (1993).

139 Palmer, J.D., Are lunar-day and solar-day clocks one and the same. XIXth International Conference of the International Society for Chronobiology, *Chronobiologia* 16(2):167 (1989).

140 Palmer, J.D., Comparative studies of tidal rhythms. IX: The modifying roles of deuterium oxide and azadirachtin on circalundian rhythms. *Marine Behaviour and Physiology* 17:167–75 (1990).

141 Webb, H.M., Interactions of daily and tidal rhythms, in: DeCoursey, P.J. (ed.), *The Belle W. Baruch Library in Marine Sciences,* No.4. *Biological rhythms in the marine environment,* pp.129–35. Papers presented at a symposium, held April 24–27, 1974, at the Belle W. Baruch Coastal Research Center, Georgetown, SC. University of South Carolina Press, 1976.

142 Williams, B.G. & Naylor, E., Spontaneously induced rhythm of tidal periodicity in laboratory-reared Carcinus. *Journal of Experimental Biology* 47:229–34 (1967).

143 Reid, D.G. & Naylor, E., Different free-running periods in split components of the circatidal rhythm in the shore crab Carcinus maenas. *Marine Ecology Progress Series* 102:295–302 (1993).

144 Neumann, D., Die Kombinationen verschiedener endogener Rhythmen bei der zeitlichen Programmierung von Entwicklung und Verhalten. *Oecologica* (Berlin) 3:166–83 (1969).

145 Vielhaben, V., Zur Deutung des semilunaren Fortpflanzungszyklus von Dictyota dichotoma. *Zeitschrift für Botanik* 51:156–73 (1963).

146 Bünning, E. & Müller, D., Wie messen Organismen lunare Zyklen? *Zeitschrift für Naturforschung* 166:391–95 (1961).

147 Naylor, E., Tidally rhythmic behaviour of marine animals, in: Lavesack, M. (ed.), Symposia of the Society for Experimental Biology 39: *Psychological adaptions of marine animals,* pp.63–93. Cambridge 1985.

148 Neumann, D., Die Analyse endogener Rhythmen bei der Mücke Clunio. *Nachrichten der Akademie der Wissenschaften Göttingen,* mathematisch-naturwissenschaftliche Klasse: 113 (1967).

149 Neumann, D., Die Steuerung eines semilunaren Schlüpfrhythmus mit Hilfe eines künstlichen Gezeitenrhythmus. *Zeitschrift für vergleichende Physiologie* (Berlin) 60:63–78 (1968).

150 Neumann, D., Entrainment of a semilunar rhythm, in: De Coursey, P.J. (ed.), *The Belle W. Baruch Library in Marine Sciences,* No.4. *Biological rhythms in the marine environment,* pp.115–27. Papers presented at a symposium, held April 24–27, 1974, at the Belle W. Baruch Coastal Research Center, Georgetown, SC. University of South Carolina Press, 1976.

151 Neumann, D., Tide- und Lunarrhythmen. *Arzneimittelforschung, Drug-research* 28(II) No.10a:1842–49 (1978).

152 Neumann, D. & Heimbach, F., Circadian range of entrainment in the semilunar eclosion rhythm of the marine insect Clunio marinus. *Journal of Insect Physiology* 31(7):549–58 (1985).

153 Reid, D.G. & Naylor, E., Free-running, endogenous semilunar rhythmicity in Eurydice pulchra, a marine isopod crustacean. *Journal of the Marine Biological Association of the United Kingdom* 65(1):85–91 (1985).

154 Williams, J.A., A semilunar rhythm of locomotor activity and moult synchrony in the sandbeach amphipod Talitrus saltator, in: Naylor, E. & Hartnoll, R.G. (eds.), *Cyclic Phenomena in Marine Plants and Animals,* pp.407–14. Proceedings of the 13th Marine Biology Symposium, Isle of Man, 27 September – 4 October 1978. Oxford (Pergamon Press) 1979.

155 Williams, C.B. & Singh, B.P., Effect of moonlight on insect activity. *Nature* 4256:853 (1951).

156 Nowinszky, L. et al., The effect of the moon phases and of the intensity of polarizied moon light on the light trap catches. *Zeitschrift für Angewandte Entomologie* 88(4):337–53 (1979).

157 Semmens, E.S., Effects of moonlight on the germination of seeds. *Nature* 111:49f (1923).

158 Kolisko, L., Der Mond und das Pflanzenwachstum. *Mitteilungen des Biologischen Institutes am Goetheanum,* No.4. Stuttgart 1935.
159 Kolisko, E. & Kolisko, L., *Die Landwirtschaft der Zukunft.* Schaffhausen (Meier) 1953.
160 Spieß, H., Chronobiological investigations of crops grown under biodynamic management. I: Experiments with seeding dates to ascertain the effects of lunar rhythms on the growth of Winter Rye (Secale cereale). *Biological Agriculture and Horticulture* 7:165–78 (1990).
161 Spieß, H., Chronobiological investigations of crops grown under biodynamic management. II: Experiments with seeding dates to ascertain the effects of lunar rhythms on the growth of Little Radish (Raphanus sativus). *Biological Agriculture and Horticulture* 7:179–89 (1990).
162 Zürcher, P.E., Rythmicités dans la germination et la croissance initiale d'une essence forestière tropicale. *Schweizerische Zeitschrift für Forstwesen (Journal forestier suisse)* 143(12):951–66 (1992).
163 Suessenguth, K., Tropische Bäume, Bambuseen und Mondwechsel. *Mitteilungen der Deutschen Dendrologischen Gesellschaft* 42:97–104 (1930).
164 Sitte, P., Phylogenetische Aspekte der Zellevolution. *Biologische Rundschau* 28:1–18 (1990).
165 Chou, H.-M. et al., Rhythmic nitrogenase activity of Synechococcus sp. RF-1 established under various light-dark cycles. *Botanical Bulletin of Academia Sinica* 30:291–96 (1989).
166 Grobbelaar, N. et al., Dinitrogen-fixing endogenous rhythms in Synechococcus RF-1. FEMS (Federation of European Microbiological Societies) *Microbiology Letters* 37:173–77 (1986).
167 Huang, T.C. et al., Circadian rhythm of the prokaryote Synechococcus sp. RF-1. *Plant Physiology* 92:531–33 (1990).
168 Mitsui, A. et al., Strategy by which nitrogen-fixing unicellular cyanobacteria grow photoautotrophically. *Nature* 323:720–22 (1986).
169 Sweeney, B.M. & Borgese, M.B., A circadian rhythm in cell division in a prokaryote, the cyanobacterium Synechococcus WH7803. *Journal of Phycology* 25:183–86 (1989).
170 Walsby, A.E., Finding a time and a place. *Nature* 323:667 (1986).
171 Chen, T.H. et al., Circadian rhythm in acid uptake by Synechococcus RF-1. *Plant Physiology* 97:55–59 (1991).
172 Picophytoplankton, *Naturwissenschaftliche Rundschau,* Year 40, No.1:29 (1987).
173 Sherr, E.B., And now, small is plentiful. *Nature* 340:4 (1989).
174 Chisholm, S.W. et al., A novel free-living prochlorophyte abundant in the oceanic euphotic zone. *Nature* 334:340–43 (1988).
175 Freie Urformen von Chloroplasten, *Bild der Wissenschaft* 11:43 (1988).
176 Planktische Prochlorophyten, *Naturwissenschaftliche Rundschau,* Year 42, No.5:197 (1989).
177 Chloroplasten-Evolution: Zweischrittige Endosymbiose, *Naturwissenschaftliche Rundschau,* Year 40, No.11:445f (1987).
178 Freilebender Prokaryot mit Chlorophyll a und b, *Naturwissenschaftliche Rundschau,* Year 40, No.1:28f (1987).
179 Prochlorophyten und Chloroplasten, *Naturwissenschaftliche Rundschau,* Year 42, No.11:449f (1989).
180 Verwandtschaft von Prochlorothrix und Chloroplasten, *Naturwissenschaftliche Rundschau,* Year 43, No.1:21f (1990).

181 DeLong, E.F. et al., High abundance of Archaea in antarctic marine picoplankton. *Nature* 371:695–97 (1994).
182 Olsen, G.J., Archaea, Archaea, everywhere. *Nature* 371:657f (1994).
183 Bergh, O. et al., High abundance of viruses found in aquatic environments. *Nature* 340:467f (1989).
184 Zeitbombe im Meer, *Der Spiegel* 37:225–28 (1989).
185 Proctor, L.M. & Fuhrman, J.A., Viral mortality of marine bacteria and cyanobacteria. *Nature* 343:60–62 (1990).
186 Eichler, H., Die Bedeutung der Flechten (Lichenes) für die geowissenschaftliche Ökosystemforschung. *Heidelberger Geowissenschaftliche Abhandlungen* 6:81–93 (1986).
187 De Wilde, P.A.W.J. & Berghuis, E.M., Cyclic temperature fluctuations in a tidal mud-flat, in: Naylor, E. & Hartnoll, R.G. (eds.), *Cyclic Phenomena in Marine Plants and Animals,* pp.435–41. Proceedings of the 13th Marine Biology Symposium, Isle of Man, 27 September – 4 October 1978. Oxford (Pergamon Press) 1979.
188 Edmunds, L.N.jr., *Cellular and molecular bases of biological clocks. Models and mechanisms for circadian timekeeping.* Berlin (Springer) 1988.
189 Schweiger, H.G., Interrelationship between chloroplasts and the nucleo-cytosol compartment in Acetabularia, in: Parthier, P. & Boulter, D. (eds.), *Nucleic acids and proteins in plants* II, pp.645–62. Berlin (Springer) 1982. *(Encyclopedia of Plant Physiology,* New Series, 14b).
190 Bruce, V.G. & Bruce, N.C., Diploids of clock mutants Chlamydomonas reinhardii. *Genetics* 89:225–33 (1978).
191 Bünning, E., Über die Erblichkeit der Tagesperiodizität bei Phaseolus-Blättern. *Jahrbuch für wissenschaftliche Botanik* 77:282–320 (1932).
192 Rensing, L., Der molekulare Mechanismus der circadianen Uhr. *Biologie in unserer Zeit* 2:101–6 (1995).
193 Brown, F.M., Common 30-day multiple in gestation time of terrestrial Placentals. *Chronobiology International* 5(3):195–210 (1988).
194 Brown, F.M. & Hostrander, L., Circalunar components in mammalian gestation duration. 16th International Conference of the International Society for Chronobiology, Dublin, Ireland, July 25–29, 1982. *Chronobiologia* 10(2):113 (1983).
195 Purkinje, J.E., Beobachtungen und Versuche zur Physiologie der Sinne. No.1. *Beiträge zur Kenntnis des Sehens in subjektiver Hinsicht.* Prague 1819. No.2. *Neuere Beiträge zur Kenntnis des Sehens in subjektiver Hinsicht.* Berlin 1824.
196 Dresler, A., Über eine jahreszeitliche Schwankung der spektralen Hellempfindlichkeit. *Das Licht* 10:79 (1940).
197 Dresler, A., Die subjektive Photometrie farbiger Lichter. *Die Naturwissenschaften* 16:225–31 (1941).
198 Kohlrausch, A., Periodische Änderungen des Farbensehens. *Film und Farbe* (Berlin) 9:98–102 (1943).
199 Lang, H.J. & Birukow, G., Lunarperiodische Schwankungen der Farbempfindlichkeit beim Guppy (Lebistes reticulatus). *Nachrichten der Akademie der Wissenschaften, Göttingen,* II. Mathematisch-physikalische Klasse 19:255–62 (1964).
200 Lang, H.J., Übereinstimmungen im Verlauf lunarer Rhythmen bei verschiedenartigen biologischen Vorgängen. *Naturwissenschaften* 13:401 (1965).
201 Lang, H.J., Über lunarperiodische Schwankungen der Farbempfindlichkeit beim Guppy (Lebistes reticulatus). *Verhandlungen der Deutschen Zoologischen Gesellschaft in Kiel* (1964); *Zoologischer Anzeiger Supplementband* 28:379–86 (1965).

REFERENCES

202 Lang, H.J., Über das Licht-Rücken-Verhalten des Guppy (Lebistes reticulatus) in farbigen und farblosen Lichtern. *Zeitschrift für vergleichende Physiologie* 56: 296–340 (1967).
203 Lang, H.J., Neue Befunde über lunarperiodische Schwankungen der Farbempfindlichkeit beim Guppy (Lebistes reticulatus). *Zoologischer Anzeiger Supplementband* 32:291–98 (1969).
204 Lang, H.J., Mondphasenabhängigkeit des Farbensehens. *Umschau in Wissenschaft und Technik* 14:445f (1970).
205 Lang, H.J., Korrelation und Kausalität bei lunaren Periodizitätserscheinungen in Biologie und Geophysik. *Nachrichten der Akademie der Wissenschaften, Göttingen, Mathematisch-Physikalische Klasse* 8:30–34 (1972).
206 Lang, H.J., Lunar periodicity of color sense of fish. *Journal of Interdisciplinary Cycle Research* 8:317–21 (1977).
207 Miles, L.E.M. et al., Blind man goes lunar. Monitor in: *New Scientist* 76, No. 1080, p. 564, of 1.12 (1977).
208 Miles, L.E.M., Raynal, D.M. & Wilson, M.A., Blind man living in normal society has circadian rhythms of 24,9 hours. *Science* 198:421–23 (1977).
209 Apfelbaum, M. et al., Rythmes circadiens de l'alternance veille-sommeil pendant l'isolement souterrain de sept jeunes femmes. *Presse Medicale* 77(24): 879–82 (1969).
210 Aschoff, J. & Wever, R., The circadian system of man, in: Aschoff, J., *Handbook of Behavioral Neurobiology, Volume 4: Biological Rhythmus*, pp.311–31. New York (Plenum Press) 1981.
211 Hollwich, F. & Dieckhues, B., Changes in the circadian rhythm of blind people, in: Scheving, L.E., Halberg, F. & Pauly, J.E. (eds.), *Chronobiology*, pp.285–92. Stuttgart (Thieme) 1974.
212 Hartmann, W. & Kluge, H., Die Rhythmik der Melatoninsynthese in der Epiphyse und deren Steuerung durch Licht. *Psychiatrie, Neurologie und medizinische Psychologie* 41(4):224–29 (1989).
213 Birbaumer, N. & Schmidt, R.F., *Biologische Psychologie*. Berlin (Springer) 1990.
214 Rietveld, W.J., The suprachiasmatic nucleus and other pacemakers, in: Touitou, Y. & Haus, E. (eds.), *Biologic Rhythms in Clinical and Laboratory Medicine*, pp.55–64. Berlin (Springer) 1992.
215 Wirz-Justice, A., Neue therapeutische Ansätze aus Chronobiologie: Licht und Melatonin. Praxis: Schweizerische *Rundschau für Medizin* 79(52):1627–29 (1990).
216 Wollnik, F., Die innere Uhr der Säugetiere. *Biologie in unserer Zeit* 1:37–43 (1995).
217 Klein, T. et al., Circadian sleep regulation in the absence of light perception: chronic non-24-hour-circadian rhythm sleep disorder in a blind man with a regular 24-hour sleep-wake schedule. *Sleep* 16(4):333–43 (1993).
218 Schweig, G., *Untersuchungen über periodische Vorgänge im gesunden und kranken Organismus des Menschen*. Karlsruhe 1843.
219 Jores, A., *Gibt es einen Einfluß des Mondes auf den Menschen?* Deutsche Medizinische Wochenschrift (Leipzig) 63:12 (1937).
220 Pentz, U. von, *Untersuchungen über Harnsäureausscheidungen im menschlichen Urin während 4 Wochen. (Unter besonderer Berücksichtigung der chemischen Bestimmungsverfahren und Berücksichtigung früherer Versuche über mondperiodische Einflüsse.)* Dissertation, Hamburg 1952.
221 Heckert, H., *Lunationsrhythmen des menschlichen Organismus*. Akademische Verlagsgesellschaft, Leipzig (Gesst & Portig) 1961.

222 Müller, E.F. & Brooks, G.W., Harnsäure und Erfolgsstreben. *Bild der Wissenschaft*, No.5:402–8 (1967).
223 Lorenz, K., Haben Tiere ein subjektives Erleben?, in: *Über tierisches und menschliches Verhalten*, Vol.2. Munich (Piper) 1968.
224 Fortlage, C., *Acht psychologische Vorträge*. Ch.1: Über die Natur der Seele. Jena 1869.
225 Lieber, A.L., Lunar effect on homicides, a confirmation. *International Journal of Chronobiology* 1(4):338f (1973).
226 Lieber, A.L., Human aggression and the lunar synodic cycle. *Journal of Clinical Psychiatry* 39:385–92 (1978).
227 Oderda, G.M. & Klein-Schwartz, W., Lunar cycle and poison center calls. *Journal of Toxicology (Clinical Toxicology)* 20(5):487–95 (1983).
228 Martin, S.J., Kelly, I.W. & Saklofske, D.H., Suicide and lunar cycles: A critical review over 28 years. *Psychological Reports* 71 (3 Part 1):787–95 (1992).
229 Rotton, J. & Kelly, I.W., Much ado about the full moon: A metaanalysis of lunar-lunacy research. *Psychological Bulletin* 97(2):286–306 (1985).
230 Byrnes, G. & Kelly, I.W., Crisis calls and lunar cycles: a twenty-year review. *Psychological Reports* 71:779–85 (1992).
231 Campbell, D.E. & Beets, J.L., Lunacy and the moon. *Psychological Bulletin* 85:1123–29 (1978).
232 Culver, R., Rotton, J. & Keely, I.W., Geophysical variables and behavior. XLIX: Moon mechanisms and myths: A critical appraisal of explanations of purported lunar behavior. *Psychological Reports* 62:683–710 (1988).
233 Rotton, J., Frey, J. & Kelly, I.W., Geophysical variables and behavior. X: Detecting lunar periodicities: something old, new borrowed, and true. *Psychological Reports* 52(1):111–16 (1983).
234 Templer, D.L. & Veleber, D.M., The moon and madness: a comprehensive perspective. *Journal of Clinical Psychology* 36:865–67 (1980).
235 Bach, E. & Schluck, L., Untersuchung über den Einfluss von meteorologischen, ionosphärischen und solaren Faktoren sowie der Mondphasen auf die Auslösung von Eklampsie und Praeklampsie. *Zentralblatt für Gynäkologie* (Leipzig) 66:196 (1942).
236 Brunner, H., Über das zeitliche Auftreten der croupösen Lungenentzündung und die Beziehungen der Disposition zu atmosphärischen und kosmischen Verhältnissen. *Deutsches Archiv für klinische Medizin* 60:339 (1898).
237 Brunner, H., Über den Einfluss der Gravitation des Mondes auf Invasion und Krisis der croupösen Pneumonie. *Deutsches Archiv für klinische Medizin* 62:574 (1899).
238 Roßlenbroich, B., *Die rhythmische Organisation des Menschen. Aus der chronobiologischen Forschung*. Stuttgart (Freies Geistesleben) 1994.
239 Menacker, W. & Menaker, A., Lunar periodicity in human reproduction: a likely unit of biological time. American *Journal of Obstetrics and Gynecology* (St Louis) 77:905–17 (1959).
240 Osley, M., Summerville, D. & Borst, L.B., Natality and the moon. American *Journal of Obstetrics and Gynecology* 117(3):413–15 (1973).
241 Guillon, P. et al., Births fertility and rhythms together with the lunar cycle. A statistical study carried out on 5927978 births. *Journal de Gynecologie Obstetrique et Biologie de la Reproduction* 15(3):265–71 (1986).
242 Martens, R., Kelly, I.W. & Saklofske, D.H., Lunar phase and birthrate, a 50-year critical review. *Psychological Reports* 63(3):923–34 (1988).
243 Rudder, B. de, *Über sogenannte 'kosmische' Rhythmen beim Menschen*. Stuttgart (Georg Thieme) 1948.

REFERENCES

244 Langman, J., *Medizinische Embryologie*. Stuttgart (Thieme) 1989.
245 Schmidt, H., Mondphasen, Jahreszeiten und Geburtenrate. *Pais* 11:313f (1987).
246 Cutler, W.B., Lunar and menstrual phase locking. *American Journal of Obstetrics and Gynecology* 137(7):834–39 (1980).
247 Law, S.P., The regulation of menstrual cycle and its relationship to the moon. *Acta Obstetricia et Gynaecologica Scandinavica* 65(1):45–48 (1986).
248 Cloudsley-Thompson, J.L., *Rhythmic activity in animal physiology and behaviour. Theoretical and Experimental Biology (Monographs)*. London, New York (Academic Press) 1961.
249 Pochobradsky, J., Independence of human menstruation on lunar phases and days of the week. *American Journal of Obstetrics and Gynecology* 118(8): 1136–38 (1974).
250 Boklage, C.E., Zincone, L.H. & Kirby, C.F.jr., Annual and subannual rhythms in human conception rates: time-series analyses show annual and weekday but no monthly rhythms in daily counts for last normal menses. *Human Reproduction* 7(7):899–905 (1992).
251 Ziac, D.C., Menstrual synchrony in university women. 53rd Annual Meeting of the American Association of Physical Anthropologists, Philadelphia/PA, USA, April 11–14, 1984, *American Journal of Physical Anthropology* 63(2):237 (1984).
252 Moore, K., *Embryologie*. Stuttgart (Schattauer) 1993.
253 Conrad, K., Warum mehr Knabengeburten in Kriegs- und Krisenzeiten? *Kosmos,* Year 47, H.7:257–60 (1949).
254 Stark, D., *Embryologie*. Stuttgart 1975.
255 Knörr, K. & Knörr-Gärtner, H., Das Abortgeschehen unter genetischen Aspekten. *Der Gynäkologe* 10:3–8 (1977).
256 Bouíe, A. & Bouíe, J., Chromosomal anomalies associated with fetal malformations, in: Scrimgeour, J.B. (ed.), *Towards the prevention of fetal malformation*, pp.49–65. Edinburgh (Edinburgh University Press) 1978.
257 Bühler, W., *Über Mondwirksamkeiten in der Nativität*. Dissertation, Freiburg, Germany. Summary in *Natura,* Year 8. No. 7/8:199–243 (1940).
258 Hosemann, H., Bestehen solare und lunare Einflüsse auf die Nativität und den Menstruationszyklus? *Zeitschrift für Geburtshilfe und Gynäkologie* 133:263–85 (1950).
259 Urbantat, V., *Kosmische Rhythmik und Geburt. Diplomarbeit*, Seminar für Waldorfpädagogik in Witten-Annen, 1987.
260 Bühler, G., *Lunare Rhythmik in Geburtenhäufigkeit und Knabenüberschuß*. Dissertation, Heidelberg 1953.
261 Klages, L., Vom Wesen des Rhythmus (1923), in: Röthig, P. (eds.), *Beiträge zur Theorie und Lehre vom Rhythmus*. Stuttgart 1966.
262 Hoerner, W., *Zeit und Rhythmus. Die Ordnungsgesetze der Erde und des Menschen*. Stuttgart 1978.
263 Hensel, H., Anpassung in der Physiologie des Menschen – Erkenntnisse medizinischer Forschung. *Universitas* 27(11):1163–72 (1972).
264 Hildebrandt, G., Zur Physiologie des rhythmischen Systems. *Beiträge zu einer Erweiterung der Heilkunst* 39:8–29 (1986).
265 Schad, W., Die Zeitordnung im Menschen und ihre pädagogische Bedeutung. *Erziehungskunst* 5:393–429 (1994).
266 Amelung, W. & Hildebrandt, G. (eds.), *Balneologie und medizinische Klimatologie,* Vol.1: *Therapeutische Physiologie – Grundlagen der Kurortbehandlung*. Berlin (Springer) 1985.

267 Rensing, L., Chronobiologie des Alterns: Veränderungen der zeitlich-periodischen Ordnung. *Zeitschrift für Gerontologie* 22(2):73–78 (1989).
268 Lemmer, B., *Chronopharmakologie. Tagesrhythmen und Arzneimittelwirkung.* Stuttgart (Wissenschaftliche Verlagsgesellschaft) 1983.
269 Reinberg, A.E., Concepts in chronopharmacology. *Annual Review of Pharmacology and Toxicology* 32:51–66 (1992).
270 Halberg, F., Cornélissen, G. & Carandente, F., Chronobiology meets the need for integration in a reductionist climate of biology and medicine. *Chronobiologia* 18(2–3):93–103 (1991).
271 Haus, E. & Touitou, Y., Principles of clinical chronobiology, in: Touitou, Y. & Haus, E. (eds.), *Biologic Rhythms in Clinical and Laboratory Medicine*, pp.6–34. Berlin (Springer) 1992.
272 Hayes, D.K., Insects as novel models for research in chronobiology. *Chronobiologia* 16(4):417–20 (1989).
273 Hildebrandt, G. & Brandt-Reges, I., *Chronobiologie in der Naturheilkunde.* Heidelberg 1992.
274 Cornélissen, G. et al., Beyond circadian chronorisk: worldwide circaseptancircasemiseptan patterns of myocardial infarctions, other vascular events, and emergencies. *Chronobiologia* 20(1–2):87–115 (1993).
275 Halberg, F., More on chronomes: circaseptans and circasemiseptans in Marburg, Germany, and 47 other locations. *Chronobiologia* 20(1–2):119–22 (1993).
276 Saito, Y.Z., Gradual adjustment of circaseptan-circadian blood pressure and heart rate rhythms after a trans-9-meridian flight. *Chronobiologia* 19(1–2):67–74 (1992).
277 Wendt, H.W., Beyond circadian chronorisk. *Chronobiologia* 20(1–2):116–18 (1993).
278 Steiner, R., *Gesamtausgabe,* Vol. 201. (Lecture of April 1920). Dornach (Rudolf Steiner).
279 Pinder, W., *Rembrandts Selbstbildnisse.* Leipzig (Langewiesche) 1943; new edition, Stuttgart (Steinkopf) 1956.
280 Schubert, G.H., *Ahndungen einer allgemeinen Geschichte des Lebens*, Part 2. Vol.2. Ch.4. Leipzig 1821.
281 Schubert, G.H., *Die Geschichte der Seele.* Vol.1: §12, p.116 und §24, p.328. Hildesheim (Georg Olms) 1961.
281a Hildebrandt, G., Rhythmusprobleme der unstimmenden Therapie. *Allgemeine Therapeutik* 7(6):202–14. 1967.
282 Goethe, J.W.v., *Maximen und Reflexionen*, No.599. *Werke, Hamburger Ausgabe in 14 Bänden,* Vol.12. Munich (dtv) 1988.
283 Heraclites, in: Diels, H.: *Die Fragmente der Vorsokratiker.* Hamburg (Rowohlt) 1957.
284 Aristotle, *Physics,* 4.10–14.
285 Augustine, Aurelius, *Confessiones* Book 11.
286 Mozart, W.A., in: *Das Musikleben,* Year 1, p.18 under: Wie wird komponiert? Mainz (Melos) 1948.
287 Boisserée, S., Tagebuchnotiz vom 16.9.1815, in: Biedermann, F. von: *Goethes Gespräche.* p.352. Wiesbaden (Insel) 1949.
288 Goethe, J.W.v., *Maximen und Reflexionen*, No.93. *Werke,* Hamburger Ausgabe Vol.12. Munich (dtv) 1988.
289 Goethe, J.W.v., *Westöstlicher Divan, Buch der Sprüche*, No.12. *Werke,* Hamburger Ausgabe Vol.2. Munich (dtv) 1988.
290 Dowse, H.O. & Palmer, J.D., Entrainment of circadian activity rhythms in mice by electrostatic fields. *Nature* 222:564–66 (1969).

291 Wever, R., Influence of electric fields on some parameters of circadian rhythms, in: Menaker, M. (ed.), *Biochronometry*, pp.117–33. Washington D.C. (National Academy of Sciences) 1971.
292 Grzimeks *Tierleben.* (12 vols.) Zürich (Kindler) 1968.
293 *Lehrbuch der speziellen Zoologie, begründet von A. Kaestner.* Vol.1: Gruner, H.-E. (ed.), *Wirbellose Tiere,* Stuttgart (Fischer) 1993. Vol.2: Starck, D. (ed.), *Wirbeltiere,* Stuttgart (Fischer) 1991, 1995. For animals not treated in these volumes, see Kaestner, A., *Lehrbuch der speziellen Zoologie.* Vol.1 Stuttgart (Fischer) 1969, 1967, 1972, 1973.
294 Storch, V. & Welsch, U., *Systematische Zoologie.* Stuttgart (Fischer) 1991.
295 *Straßburger Lehrbuch der Botanik für Hochschulen.* Stuttgart (Fischer) 1991.
296 *Urania Pflanzenreich in vier Bänden*, Leipzig (Urania). 1. *Viren, Bakterien, Algen, Pilze,* 1991. 2. *Moose, Farne, Nacktsamer,* 1992. 3. *Blütenpflanzen* 1, 1993. 4. *Blütenpflanzen* 2, 1994.
Urania Tierreich in sechs Bänden, Leipzig (Urania). 1. *Wirbellose Tiere* 1, 1993. 2. *Wirbellose Tiere* 2, 1994. 3. *Insekten,* 1994. 4. *Fische, Lurche, Kriechtiere,* 1991. 5. *Vögel,* 1995. 6. *Säugetiere,* 1992.
297 Bracher, R., The light relations of Euglena limosa Gard. Part I: The influence of intensity and quality of light on phototaxy. *Journal of the Linnean Society, Botany* 51:23–42 (1938).
298 Akselman, R., Contribution to the knowledge of the family Warnowiaceae Lindemann, class Dinophyceae in the Southwestern Atlantic Ocean. *Darwiniana* (Buenos Aires) 27(1–4):9–18 (1986).
299 Balch, W.M., An apparent lunar tidal cycle of phytoplankton blooming and community succession in the Gulf of Maine, USA. *Journal of Experimental Marine Biology and Ecology* 55(1):65–78 (1981).
300 Balch, W.M., Are red tides correlated to spring-neap tidal mixing? Use of a historical record to test mechanisms responsible for dinoflagellate blooms, in: Bowman, M.J., Ventsch, C.M. & Peterson, W.T. (eds.), *Lecture notes on coastal and estuarine studies,* Vol.17: *Tidal Mixing and plankton dynamics*, pp.193–223. Berlin (Springer) 1986.
301 Fauré-Fremiet, E., The tidal rhythm of the diatom Hantzschia amphioxys. *Biological Bulletin of the Marine Biological Laboratory Woods Hole* 100(3):173–77 (1951).
302 Eaton, J.W. & Simpson, P., Vertical migrations of the intertidal dinoflagellate Amphidinium herdmaniae Kofoid u. Swezy, in: Naylor, E. & Hartnoll, R.G. (eds.), *Cyclic phenomena in marine plants and animals,* pp.339–45. Proceedings of the 13th Marine Biology Symposium, Isle of Man, 27 September – 4 October 1978. Oxford (Pergamon Press) 1979.
303 Herdman, W.A., On the occurrence of Amphidinium operculatum in vast quantity at Port Erin, Isle of Man. *Journal of the Linnean Society, Zoology* 14(212): (1911).
304 Herdman, E.C., Dinoflagellates and other organisms causing discoloration of the sand. *Transactions of the Liverpool Biological Society* 35:59–63 (1921).
305 Jorgensen, O., *Occurrence of Amphidinium operculatum at Cullercoats.* Report of the Dove Marine Laboratory, Northumberland 1918.
306 Jonsson, P.R., Tidal rhythm of cyst formation in the rock pool ciliate Strombidium oculatum Gruber (Ciliophora: Oligotrichida): A description of the functional biology and an analysis of the tidal synchronization of encystment. *Journal of Experimental Marine Biology and Ecology* 175(1):77–103 (1994).

307 Fauré-Fremiet, E., Rythme de marée d'une Chromulina psammophile. Bulletin Biologique de la France et de la Belgique (Paris) 84:207–14 (1950).
308 Palmer, J.D. & Round, F.E., Persistent, vertical-migration rhythms in benthic microflora. VI: The tidal and diurnal nature of the rhythm in the diatom Hantzschia virgata. *Biological Bulletin of the Marine Biological Laboratory Woods Hole* 132:44–55 (1967).
309 Kuckuck, P., Beiträge zur Kenntnis der Meeresalgen. 10: Neue Untersuchungen über Nemoderma Schonsboe. *Wissenschaftliche Meeresuntersuchungen, Abteilung Helgoland*, N. F. 5, No.3:119–54 (1902).
310 Hoyt, W.D., The periodic fruiting of Dictyota and its relation to the environment. *American Journal of Botany* (Lancaster) 14:592–619 (1927).
311 Hoyt, W.D., Periodicity in the production of the sexual cells of Dictyota dichotoma. *Botanical Gazette* (Chicago) 43:383–92 (1907).
312 Lewis, J.F., Periodicity in Dictyota at Naples. *Botanical Gazette* (Chicago) 50:59–64 (1910).
313 Müller, D., Untersuchungen über die lunarperiodische Entleerung der Geschlechtsorgane von Dictyota dichotoma. *Vorträge aus dem Gesamtgebiet der Botanik, herausgegeben von der Deutschen Botanischen Gesellschaft,* Neue Folge No.1:173–77 (1961).
314 Müller, D., Über jahres- und lunarperiodische Erscheinungen bei einigen Braunalgen. Botanica Marina. *Internationale Zeitschrift für die Erforschung und Auswertung von Meeresalgen* (Hamburg) 4:140–55 (1962).
315 Williams, J.L., Studies on the Dictyotaceae, III: The periodicity of the sexual cells in Dictyota dichotoma. *Annals of Botany* (London) 19:548 (1905).
316 Hofman, P.J., Featonby-Smith, B.C. & Staden, J. van, The development of Elisa and radioimmunoassay for cytokinin estimation and their application to a study of lunar periodicity in Ecklonia maxima. *Journal of Plant Physiology* 122(5): 455–66 (1986).
317 Brown, F.A.jr., Freeland, R.O. & Ralph, C.L., Persistent rhythms of O_2-consumption in potatoes, carrots and the seaweed, Fucus. *Plant Physiology* 30:280–92 (1955).
318 Brown, F.A.jr., Sandeen, M.I. & Webb, H.M., Solar and lunar rhythms of O_2-consumption in the seaweed, Fucus. *Biological Bulletin of the Marine Biological Laboratory Woods Hole* 107:306 (1954).
319 Brawley, S.H., Fertilization in natural populations of the dioecious brown alga Fucus ceranoides and the importance of the polyspermy block. *Marine Biology* (Berlin) 113(1):145–57 (1992).
320 Andersson, S., Kautsky, L. & Kalvas, A., Circadian and lunar gamete release in Fucus vesiculosus in the atidal Baltic Sea. *Marine Ecology Progress Series* 110:195–201 (1994).
321 Lausi, D. & De Cristini, P., Osservazioni sulle variazioni periodiche del continuto in acido alginico in Fucus virsoides. *Nova Thalassia* (Trieste) 3(4):1–16 (1967).
322 Tahara, M., On the periodical liberation of the oophores in Sargassum. *The Botanical Magazine* (Tokyo) 23:151 (1909).
323 Tahara, M., Oogonium liberation and the embryogeny of some fucaceous Algae. *Journal of the College of Science. Imperial University of Tokyo* 32(2):1–13 (1913).
324 Norton, T.A., Gamete expulsion and release in Sargassum muticum. *Botanica Marina* 24:465–70 (1981).
325 May, D.I. & Clayton, M.N., Oogenesis: the formation of oogonial stalks and fertilization in Sargassum vestitum, Fucales, Phaeophyta, from Southern Australia. *Phycologia* 30(3):243–56 (1991).

326 Allen, W.E., A quantitative and statistical study of the plankton of the San Joaquin River. *University of California Publications in Zoology* 22:39–45, 115, 124, 131–34, 198–200, 213–33, 263–64 (1920).
327 Christie, A.O. & Evans, L.V., Periodicity in the liberation of gamets and zoospores of Enteromorpha intestinalis Link. *Nature* 193:193–94 (1962).
328 Smith, G.M., On the reproduction of some Pacific Coast Species of Ulva. *American Journal of Botany* 34:80–87 (1947).
329 Hollenberg, G.J., A study of Halicystis ovalis. II: Periodicity in the formation of gametes. *American Journal of Botany* 23:1–3 (1936).
330 Kuckuck, P., Über den Bau und die Fortpflanzung von Halicystis Areschoug und Valonia Ginnani. *Botanische Zeitung* (Leipzig) 65:139–85 (1907).
331 Drew, E.A. & Abel, K.M., Studies on Halimeda, II: Reproduction, particularly the seasonality of gametangia formation in a number of species from the Great Barrier Reef, Province Australia. *Coral Reefs* 6(3–4):207–18 (1988).
332 Rounds, H.D., A semilunar periodicity of neurotransmitter-like substances from plants. *Physiologia Plantarum* 54(4):495–99 (1982).
333 Rounds, H.D., A lunar rhythm in the occurrence of blood-borne factors in cockroaches mice and men. *Comparative Biochemistry and Physiology,* C 50(2): 193–98 (1975).
334 Rounds, H.D., Semilunar cyclicity of neurotransmitter-like materials from Coleus blumei and Phaseolus vulgaris. 112th Annual Meeting of the Kansas Academy of Science, Hays, Kansas, USA, March 28, 1980, *Transactions of the Kansas Academy of Science* 83(3):151 (1980).
335 Rounds, H.D., Semilunar cyclicity of neurotransmitter-like substances in the central nervous system of Periplaneta americana. *Comparative Biochemistry and Physiology,* C 69(2):293–300 (1981).
336 Rounds, H.D., Lunar and seasonal variation in cardiac response to acetylcholine and noradrenaline. *Comparative Biochemistry and Physiology,* C 74(2):373–76 (1983).
337 Rounds, H.D., Pulses of cholinergic receptor escape from atropine blocking reflect major natural periodicities. *Comparative Biochemistry and Physiology,* C 75(1):115–20 (1983).
338 Rounds, H.D., A cholinergic receptor 12 hour interval timer displaying circa periodicities. *Comparative Biochemistry and Physiology,* C 77(1):59–64 (1984).
339 Spieß, H., Haben lunare Rhythmen Bedeutung für den ökologischen Landbau?, in: Zerger, U. (ed.), *Forschung im ökologischen Landbau. Beiträge zur 2. Wissenschaftstagung im ökologischen Landbau*, pp.397–403. Bad Dürkheim 1993 (= Stiftung im ökologischen Landbau, Sonderausgabe 42.)
340 Spieß, H., *Chronobiologische Untersuchungen mit besonderer Berücksichtigung lunarer Rhythmen im biologisch-dynamischen Pflanzenbau*. Darmstadt (Institut für Biologisch-Dynamische Forschung, Schriftenreihe Vol.3) 1994.
341 Graviou, E., Analogies between rhythms in plant material, in atmospheric pressure, and solar lunar periodicities. *International Journal of Biometeorology* 22(2):103–11 (1978).
342 Graviou, E., Variabilité non aléatoire d'un matériel végétal dans ses échanges gazeux et sa croissance. *Bulletin Mensuel da la Société linnéenne de Lyon* 46(4): 108–12 (1977).
343 Graviou, E., A complex rhythm problem: the possibility of a lunar modulation of plant function. *Journal of Interdisciplinary Cycle Research* 9(4):243–68 (1978).

344 Gaertner, T. von & Braunroth, E., Über den Einfluß des Mondlichtes auf den Blühtermin der Lang- und Kurztagspflanzen. *Beihefte zum Botanischen Zentralblatt* 53:554–63 (1935).

345 Brown, F.A.jr., Lunar correlated variations in water uptake by bean seeds. *Biological Bulletin of the Marine Biological Laboratory Woods Hole* 145(2):265–78 (1973).

346 Spruyt, E., Verbelen, J.P. & De Greff, J.A., Expression of circaseptan and circannual rhythmicity in the inhibition of dry stored bean seeds. *Plant Physiology* 84(3):707–10 (1987).

347 Millet, B., *Analyse des rythmes de croissance de la fíeve (Vicia faba L.)*. Thíese Ph.D. présenté à la Faculté des Sciences de l'Université de Besançon 1970.

348 Brown, F.A.jr., Freeland, R.O. & Bennett, M. F., Persistent rhythms of O_2-consumption in carrots and potatoes. *Biological Bulletin of the Marine Biological Laboratory Woods Hole* 107:305f (1954).

349 Panzram, H., Gibt es einen Einfluß des Mondes auf die Witterung? *Naturwissenschaftliche Rundschau*, Year 26, No.1:28f (1973).

350 Maw, M.G., Periodicities in the influences of air ions on the growth of garden cress Lepidium sativum L. *Canadian Journal of Plant Science* (Ottawa) 47: 499–505 (1967).

351 Brown, F.A.jr., Seasonal unimodal lunar day oscillations in potato oxygen consumption. *Journal of Interdisciplinary Cycle Research* 10(3):165–72 (1979).

352 Cox, P.A., Bestäubung im Wasser. *Spektrum der Wissenschaft*, December 1993.

353 Bijma, J., Erez, J. & Hemleben, C., Lunar and semilunar reproductive cycles in some spinose planktonic foraminifers. *Journal of Foraminiferal Research* 20(2): 117–27 (1990).

354 Bijma, J., Hemleben, C. & Wellnitz, K., Lunar-influenced carbonate flux of the planktic foraminifer Globigerinoides sacculifer (Brady) from the central Red Sea. Deep-Sea Research, Part I. *Oceanographic Research Papers*. 41(3):511–30 (1994).

355 Ray, H. & Chakraverty, M., Lunar periodicity in the conjugation of Conchophthirius lamellidens Gosh. *Nature* 134:663f (1934).

356 Fauré-Fremiet, E., Le rythme de marée du Strombidium oculatum Gruber. *Bulletin Biologique de la France et de la Belgique* (Paris) 82:3–23 (1948).

357 Fauré-Fremiet, E., The ecology of some infusorian communities of intertidal pools. *Journal of Animal Ecology* 17:127–30 (1948).

358 Hoppe, W.F. & Reichert, M.J.M., Predictable annual mass release of gemetes by the coral reef sponge Neofibularia nolitangere (Porifera: Demospongiae). *Marine Biology* (Berlin) 94:277–85 (1987).

359 Hoppe, W.F., Reproductive patterns in three species of large coral reef sponges. *Coral Reefs* 7:45–50 (1988).

360 Siewing, R., *Lehrbuch der Zoologie* (3 ed.), p.142. Stuttgart (Fischer) 1985.

361 Fromont, J. & Bergquist, P.R., Reproductive biology of three sponge species of the genus Xestospongia (Porifera: Demospongiae: Petrosida) from the Great Barrier Reef. *Coral Reefs* 13:119–26 (1994).

362 Sarano, F., Synchronized spawning in Indonesian sponges. *Coral Reefs* 10:166 (1991).

363 Heyward, A. et al., Sexual reproduction of corals in Okinawa, Japan. *Galaxea* 6(2):331–43 (1987).

364 Brock, M.A., Differential sensitivity to temperature steps in the circannual rhythm of hydranth longevity in the marine cnidarian Campanularia flexuosa. *Comparative Biochemistry and Physiology* A 64(3):381–90 (1979).

365 Elmhirst, R., Lunar periodicity in Obelia. *Nature* 116:358–459 (1925).
366 Harrison, P.L. et al., Mass spawning in tropical reef corals. Science 223:1186–89 (1984).
367 Bull, G., Distribution and abundance of coral plankton. Coral Reefs 4:197–200 (1986).
368 Babcock, R.C., Willis, B.L. & Simpson, C.J., Mass spawning of corals on a high latitude coral reef. Coral Reefs 13:161–69 (1994).
369 Bohn, G., La persistance du rythme des marées chez l'Actinia equina. *Comptes rendus hebdomadaires des Séances et Memoires de la Société de Biologie et de ses filiales* 61:661–63 (1906).
370 Bohn, G. & Piéron, H., Les rythmes des marées et le phénomíene de l'anticipation réflexe. *Comptes hebdomadaires des Séances et Memoires de la Société de Biologie et de ses filiales* 61:660f (1906).
371 Gompel, M., Recherches sur la consommation d'oxygíene de quelques animaux aquatique littoraux. *Comptes rendus hebdomadaires des Séances de l'Académie des Sciences* 205:816–18 (1937).
372 Piéron, H., La réaction aux marées par anticipation réflexe chez Actinia equina. Comptes rendus hebdomadaires des *Séances et Memoires de la Société de Biologie et de ses filiales* 61:658–60 (1906).
373 Piéron, H., La rythmicité chez Actinia equina L. *Comptes rendus hebdomadaires des Séances et Memoires de la Société de Biologie et de ses filiales* 65:726–28 (1908).
374 Piéron, H., Sur les facteurs des mouvements d'ascension et de descente chez les Convoluta. *Comptes rendus hebdomadaires des Séances et Memoires de la Société de Biologie et de ses filiales* 64/65:673–75 (1908).
375 Piéron, H., *De l'actinie à l'homme. L'anticipation et mémoire bases de l'evolution psychique.* Presse Universitaire de France, Paris 1958.
376 Kojis, B.L., Sexual reproduction in Acropora (Isopora) (Coelenterata: Scleractinia). I: A. cuneata and A. palifera on Heron Island Reef, Great Barrier Reef. *Marine Biology* (Berlin) 91(3):291–309 (1986).
377 Szmant-Froelich, A., Riggs, L. & Reutter, M., Sexual reproduction in Caribbean reef corals. *American Zoologist* 23(4):961 (1993).
378 Szmant-Froelich, A., Reutter, M. & Riggs, L., Sexual reproduction of Favia fragum, lunar patterns of gametogenesis, embryogenesis and planulation in Puerto Rico. *Bulletin of Marine Science of the Gulf and Caribbean* 37(3):880–92 (1985).
379 Soong, K., Sexual reproductive patterns of shallow-water reef corals in Panama. *Bulletin of Marine Science of the Gulf and Caribbean* 49(3):832–46 (1991).
380 Abe, N., Post-Larval development of the coral Fungia actiniformis var. palawensis Döderlein. *The Palao Tropical Biological Station Studies* 1:73–93 (1937).
381 Abe, N., Ecological survey of Iwayama Bay, Palao. Kurze Bemerkung über Fungia actiniformis var. palawensis. T*he Palao Tropical Biological Station Studies* 2:217–324 (1937).
382 Babcock, R.C., Reproduction and distribution of two species of Goniastrea (Scleractinia) from the Great Barrier Reef Province. *Coral Reefs* 2:187–95 (1984).
383 Szmant, A.M., Sexual reproduction by the Caribbean reef corals Montastrea annularis and M. cavernosa. *Marine Ecology Progress Series* 74:13–25 (1991).
384 Yeemin, T., Nojima, S. & Kikuchi, T., Sexual Reproduction of the scleractinian coral Montastrea valenciennesi from a high-latitude coral community Southwest Japan. *Publications. Amakusa Marine Biological Laboratory. Kyushu University* 10(2):105–22 (1990).

385 Wyers, S.C., Barnes, H.S. & Smith, S.R., Spawning of hermatypic corals in Bermuda. A pilot study. Fifth International Conference on Coelenterate Biology, Southampton, England, July 10–14, 1989. *Hydrobiologia* 216/217:109–16 (1991).

386 Gittings, S.R. et al., Mass spawning and reproductive viability of reef corals al the East Flower Garden Bank, Northwest Gulf of Mexico. *Bulletin of Marine Science of the Gulf and Caribbean* 51(3):420–28 (1992).

387 Johnson, K.G., Synchronous planulation of Manicina areolata (Scleractinia) with lunar periodicity. *Marine Ecology Progress Series* 87:265–73 (1992).

388 Glynn, P.W. et al., Reef coral reproduction in the Eastern Pacific, Costa Rica, Panama and Galapagos Island, Ecuador. I: Pocilloporidae. *Marine Biology* (Berlin) 109(3):355–68 (1991).

389 Richmond, R.H. & Jokiel, P.L., Lunar periodicity in larva release in the reef coral Pocillopora damicornis at Enewetak, Marshall Islands and Hawaii, USA. *Bulletin of Marine Science of the Gulf and Caribbean* 34(2):280–87 (1984).

390 Jokiel, P.L., Ito, R.Y. & Liu, P.M., Night irradiance and synchronization of lunar release of planula larvae in the reef coral Pocillopora damicornis. *Marine Biology* (Berlin) 88(2):167–74 (1985).

391 Stoddart, J.A. & Black, R., Cycles of gametogenesis and planulation in the coral Pocillopora damicornis. *Marine Ecology Progress Series* 23(2):153–64 (1985).

392 Stimson, J.S., Location quantity and rate of change in quantity of lipids in tissue of hawaiian hermatypic corals. *Bulletin ot Marine Science of the Gulf and Caribbean* 41(3):889–904 (1987).

393 Marshall, S.M. & Stephenson, T.A., The breeding of reef animals. Part II: The corals. *Scientific Report on the Great Barrier Reef Expedition* (British Museum, London) 3(8):219 -245 (1933).

394 Kinzie, R.A.III, Spawning in the reef corals Pocillopora verrucosa and Pocillopora eydouxi at Sesoko Island, Okinawa. *Galaxea* 11:93– 105 (1993).

395 Coffroth, M.A., Cyclical formation of mucous sheet by 3 coral species. *American Zoologist* 23(4):960 (1983).

396 Coffroth, M.A., Mucous sheet formation on poritid corals. An evaluation of coral mucous as a nutrient source on reefs. *Marine Biology* (Berlin) 105(1):39–50 (1990).

397 Coffroth, M.A., Cyclical mucous sheet formation on poritid corals in the San Blas Islands, Panama. *Marine Biology* (Berlin) 109(1):35–40 (1991).

398 Chornesky, E.A. & Peters, E.C., Sexual reproduction and colony growth in the scleractinian coral Porites astreoides. *Biological Bulletin of the Marine Biological Laboratory Woods Hole* 172:161–77 (1987).

399 Buddemeier, R.W., Environmental controls over annual and lunar monthly cycles in hermatypic coral calcification, in: Cameron, A.M. et al. (eds.), *Proceedings of the second International Symposium on coral reefs*, Vol.2. June 22 -July 2, 1973. The Great Barrier Reef Committee, Brisbane, Australia, 1974.

400 Buddemeier, R.W. & Kinzie, R.A., The chronometric reliability of contemporary corals, in: Rosenberg, G.D. & Runcorn, S.K. (eds.), *Growth rhythms and the history of the earth's rotation*, pp.135–47. London (John Wiley & Sons) 1975.

401 Barnes, D.J. & Lough, J.M., The nature of skeletal density banding in scleractinian corales: fine banding and seasonal patterns. *Journal of Experimental Marine Biology and Ecology* 126(2):119–34 (1989).

402 Glynn, P.W. et al., Reef coral reproduction in the Eastern Pacific, Costa Rica, Panama and Galapagos Islands, Ecuador. II: Poritidae. *Marine Biology* (Berlin) 118(2):191–208 (1994).

REFERENCES

403 Ryland, J.S. & Babcock, R.C., Annual cycle of gametogenesis and spawning in a tropical zoanthid Protopalythoa sp. Fifth International Conference on Coelenterate Biology, Southampton, England, July 10–14, 1989. *Hydrobiologia* 216–17:117–24 (1991).

404 Flügel, W., 'Rugae' und Wachstumszonen bei Korallen. *Paläontologische Zeitschrift* 51:117–30 (1977).

405 Babcock, R.C., Reproduction and development of the Blue Coral Heliopora coerulea, Alcyonaria, Coenothecalia. *Marine Biology* (Berlin) 104(3):475–82 (1990).

406 Benayahu, Y., Reproductive cycle and developmental processes during embryogenesis of Clavularia hamra, Cnidaria, Octocorallia. *Acta Zoologica* (Stockholm) 70(1):29–36 (1989).

407 Bohn, G., Périodicité vitale des animaux soumis aux oscillations du niveau des hautes mers. *Comptes rendus hebdomadaires des Séances et Memoires de la Société de Biologie et de ses filiales* 139:610–11 (1904).

408 Gamble, G.W. & Keeble, F., The bionomics of Convoluta roscoffensis with special reference to its green cells. *Proceedings of the Royal Society of London* 72:93–98 (1903).

409 Gamble, G.W. & Keeble, F., The bionomics of Convoluta roscoffensis with special reference to its green cells. *Quaterly Journal of Microscopical Science* 47:363–431 (1904).

410 Keeble, F., *Plant-animals*. Cambridge 1910.

411 Kremer, B.P., Ein Strudelwurm spielt Pflanze. *Kosmos* (Stuttgart) 73:838–40 (1977).

412 Brown, F.A.jr., An orientational response to weak gamma radiation. *Biological Bulletin of the Marine Biological Laboratory Woods Hole* 125:206–26 (1963).

413 Brown, F.A.jr. & Park, Y.H., A persistent monthly variation of planarians to light and its annual modulation. *International Journal of Chronobiology* 3(1):57–62 (1975).

414 Goodenough, J.E., The monthly orientation rhythm of planarians is not generated by the interaction of solar day and lunar day rhythms. *Journal of Interdisciplinary Cycle Research* 11(2):117–24 (1980).

415 Goodenough, J.E., The lack of effect of deuterium oxide on the period and phase of the monthly orientation rhythm in planarians. *International Journal of Chronobiology* 5(3):465–76 (1978).

416 May, E. & Birukow, G., Lunar-periodische Schwankungen des Lichtpräferendums bei der Bachplanarie. *Die Naturwissenschaften* 53:182 (1966).

417 Stephenson, A., The breeding of reef animals, Part II: Invertebrates other than coral. *Scientific Report on the Great Barrier Reef Expedition* (London) 3(9):247–72 (1934).

418 Yoshioka, E., Annual Reproductive cycle of the chiton Acanthopleura japonica. *Marine Biology* (Berlin) 96(3):371–74 (1987).

419 Yoshioka, E., Phase shift of semilunar spawning periodicity of the chiton Acanthopleura japonica Lischke by artificial regimes of light and tide. Journal of Experimental *Marine Biology and Ecology* 129(2):133–40 (1989).

420 Crozier, W.Y., An observation on the 'cluster-formation' of the sperms of Chiton. *American Naturalist* 56 (1922).

421 Grave, B.H., An analysis of the spawning habits and spawning stimuli of Chaetopleura apiculata (Say). *Biological Bulletin of the Marine Biological Laboratory Woods Hole* 42:234–56 (1922).

422 Zann, L.P., Interactions of the circadian and circatidal rhythms of the littoral gastropod Melanerita atramantosa Reeve. *Journal of Experimental Marine Biology and Ecology* 11:249–61 (1973).

423 Zann, L.P., Relationships between intertidal zonation and circatidal rhythmicity in littoral gastrapods. *Marine Biology* (Berlin) 18:243–50 (1973).

424 Corpuz, G.C., The role of gametogenic and spawning synchrony in the reproductive success of Cellana exarata (Gastropoda, Patellidae). *American Zoologist* 22(4):905 (1982).

425 Funke, W., Heimfindevermögen und Ortstreue bei Patella L. (Gastropoda: Prosobranchia). *Oecologia* 2:19–142 (1968).

426 Della Santina, P., Naylor, E. & Chelazzi, G., Long term field actography to assess the timing of foraging excursions in the limpet Patella vulgata L. *Journal of Experimental Marine Biology and Ecology* 178(2):193–203 (1994).

427 Della Santina, P. & Naylor, E., Endogenous circatidal rhythm of the homing behaviour in the limpet Patella vulgata L. *Journal of Molluscan Studies* 59:87–91 (1993).

428 Chelazzi, G., Della Santina, P. & Santini, G., Rhythmical timing and spatial scattering of foraging in a homer limpet (Patella rustica). *Behavioral Ecology* 5:288–92 (1994).

429 Little, C., Partridge, J.C. & Teagle, L., Foraging activity of limpets in normal and abnormal tidal regimes. *Journal of the Marine Biological Association of the United Kingdom* 71:537–54 (1991).

430 Hartnoll, R.G. & Wright, J.R., Foraging movements and homing in the limpet Patella vulgata L. *Animal Behaviour* 25:806–10 (1977).

431 Bell, L.J., Reproduction and larval development of the West Indian Topshell Cittarium pica, Trochidae, in the Bahamas. *Bulletin of Marine Science of the Gulf and Caribbean* 51(2):250–66 (1992).

432 Heslinga, G.A. & Hillmann, A., Hatchery culture of the commercial top snail Trochus niloticus in Palau, Caroline Islands, West Pacific Ocean. *Aquaculture* 22(1–2):35–44 (1981).

433 Hahn, K.O., The reproductive cycle of the tropical top shell, Trochus niloticus, in French Polynesia. *Invertebrate Reproduction and Development* 24(2):143–55 (1993).

434 Lenderking, R.E., Some recent observations on the biology of Littorina angulifera Lam. of Biscayne and the Virginia Keys, Florida. *Bulletin of Marine Science of the Gulf and Caribbean* 3:273–96 (1954).

435 Sandeen, M.I., Stephens, G.C. & Brown, F.A.jr., Persistent, daily and tidal rhythms of oxygen consumption in two species of marine snails. *Physiological Zoology* 27:350–56 (1954).

436 Grahame, J., Spawning in Littorina littorea (L.) (Gastropoda: Prosobranchiata). *Journal of Experimental Marine Biology and Ecology* 18(2):185–96 (1975).

437 Mileikovsky, S.A., Lunar periodicity in spawning of littoral and upper sub-littoral invertebrates of the White Sea and other areas. *Doklady Akademii Nauk SSSR* 123:1006–9 (1958).

438 Smith, J.E. & Newell, G.E., The dynamics of the zonation of the common periwinkle (Littorina littorea (L.)) on a stony beach. *Journal of Animal Ecology* 24:35–56 (1955).

439 Lysaght, A.M., The biology and Trematode parasites of the gastropod Littorina neritoides (L.) on the Plymouth breakwater. *Journal of the Marine Biological Association of the United Kingdom* 25:41–80 (1941).

440 Haseman, J.D., The rhythmical movements of Littorina littorea synchroneus with ocean tides. *Biological Bulletin of the Marine Biological Laboratory Woods Hole* 21:113–21 (1911).

REFERENCES

441 Thamdrup, H., Beiträge zur Ökologie der Wattfauna. *Meddeleser fra Kommissionen for Havunder sgeleser.* Ser. Fiskeri (Copenhagen) 10:2 (1935).
442 Ross, B. & Berry, A.J., Annual and lunar reproductive cycles in Littorina saxatilis Olivi and differences between breeding in the marine Firth of Forth and the Forth estuary. *Journal of Molluscan Studies* 57(3):347–58 (1991).
443 Berry, A.J., Semilunar and lunar spawning periodicity in some tropical littorinid gastropods. *Journal of Molluscan Studies* 52(2):144–49 (1986).
444 Abe, N., Ecological observations on Melarhaphe (Littorinopsis) scabra L., mainly on its locomotion. *Ecological Review* 2(1):132–36 (1936) (Japanese).
445 Abe, N., Ecological observations on Melarhaphe (Littorinopsis) scabra L. inhabiting on mangrove-trees. *The Palao Tropical Biological Station Studies* 6:(1939).
446 Brown, F.A.jr., Webb, H.M. & Brett, W.J., Exogenous timing of solar and lunar periodism in metabolism of the mud snail Ilyanassa (=Nassarius) obsoleta, in laboratory constant conditions. *Journal of Medicine and Science* 8:233–42 (1959).
447 Stephens, G.C., Sandeen, M.I. & Webb, H.M., A persistent tidal rhythm of activity in the mud snail, Nassa obsoleta. *Anatomical Records* (Philadelphia) 117:635 (1953).
448 Abe, N., On the locomotion of a limpet-like Pulmonata, Siphonaria atra Quey et Gaimard. *The Venus* 5:206–13 (1935) (Japanese with English summary).
449 Abe, N., Ecological observations on a limpet, Siphonaria atra Quey et Gaimard with a note on homing behaviour of limpets. *The Palao Tropical Biological Station Studies* 6: (1939).
450 Creese, R.G., Reproductive cycles and fecundities of 2 species of siphonaria, mollusca, pulmonata in Southeastern Australia. *Australian Journal of Marine and Freshwater Research* 31(1):37–48 (1980).
451 Abe, N., The homing, spawning and other habits of a limpet, Siphonaria japonica Donovan. *Science Reports of The Thoku Imperial University, Biology, Sendai, Japan,* 15:59–95 (1940).
452 Price, C.H., Physical factors and neurosecretion in the control of reproduction in Melampus, Mollusca, Pulmonta. *Journal of Experimental Zoology* 207(2): 269–82 (1979).
453 Price, C.H., Water relations and physiological ecology of the salt marsh snail Melampus bidentatus. *Journal of Experimental Marine Biology and Ecology* 45:51–68 (1980).
454 Sturmwasser, F., The demonstration and manipulation of a circadian rhythm in a single neuron, in: Aschoff, J. (ed.), *Circadian clocks,* pp.442–62. Proceedings of the Feldafing summer school. September 7–18, 1964. Amsterdam (North Holland Publishing Company) 1965.
455 Battle, H.J., Rhythmic sexual maturity and spawning of certain bivalve mollusks. *Contributions to Canadian Biology and Fisheries* (Ottawa) 7:257–67 (1932).
456 Battle, J., *The moon and when to eat mussels.* Progress Reports of The Atlantic Coast Stations. Atlantic Biological Station, St Andrews, N.B., and Atlantic Fisheries Experimental Station, Halifax, N.S. 3:5–6 (1932).
457 Rao, K.P., Tidal rhythmicity of rate of water propulsion in Mytilus and its modifiability by transplantation. *Biological Bulletin of the Marine Biological Laboratory Woods Hole* 106:353–59 (1954).
458 Rao, K.P., Rate of water propulsion in Mytilus californianus as a function of latitude. *Biological Bulletin of the Marine Biological Laboratory Woods Hole* 104:171–81 (1953).
459 Aschoff, J., *Handbook of Behavioral Neurobiology,* Vol.4: *Biological Rhythms.* New York (Plenum Press) 1981.

460 Tang, S.-F., The breeding of the escallop (Pecten maximus L.) with a note on the growth rate. *Proceedings and Transactions of the Liverpool Biological Society* 54:9–28 (1941).
461 Parsons, G.J. et al., Intra-annual and long-term patterns in the reproductive cycle of giant scallops Placopecten magellanicus (Bivalvia: Pectinidae) from Passamaquoddy Bay, New Brunswick, Canada. *Marine Ecology Progress Series* 80:203–14 (1992).
462 Amirthalingam, C., On lunar periodicity in reproduction of Pecten operculairs near Plymouth in 1927/28. *Journal of the Marine Biological Association of the United Kingdom* (Plymouth) 15:605–41 (1928).
463 Brown, F.A.jr., Persistent activity rhythms in the oyster. *American Journal of Physiology* 178:510–14 (1954).
464 Loosanoff, V.L. & Nomejko, Ch.A., Spawning and setting of the American Oyster, Ostrea virginica, in relation to lunar phases. *Ecology Offical Publication of the Ecological Society of America* (Durham) 32:113f (1951).
465 Loosanoff, V.L. & Nomejko, Ch.A., Existence of physiologically different races of oysters, Crassostrea virginica. *Biological Bulletin of the Marine Biological Laboratory Woods Hole* 101(2):151–56 (1951).
466 Korringa, P., Relations between the moon and periodicity in the breeding of marine animals. *Ecological Monographs* 17:343–81 (1947).
467 Loosanoff, V.L. & Nomejko, Ch.A., Feeding of oysters in relation to tidal stages and to periods of light and darkness. *Biological Bulletin of the Marine Biological Laboratory Woods Hole* 90:244–64 (1946).
468 Orton, J.H., On lunar periodicity in spawning of normally grown Falmouth oysters in 1925 with comparison of spawning capacity of normally grown and dumpy oysters. *Journal of the Marine Biological Association of the United Kingdom* (Plymouth) 14:199f (1926).
469 House, M.R. & Farrow, G.E., Daily Growth Banding in the shell of the cockle Cardium edule. *Nature* 219:1384–86 (1968).
470 Richardson, C.A., Crisp, D.J. & Runham, N.W., Tidally deposited growth bands in the shell of the common cockle Cerastoderma edule L. Proceedings of the 6th European Malacological Congress, Amsterdam, August 15–20, 1977. *Malacologia* 18(1–2):277–90 (1979).
471 Richardson, C.A., Crisp, D.J. & Runham, N.W., An endogenous rhythm in shell deposition in Cerastoderma edule. *Journal of the Marine Biological Association of the United Kingdom* 60:991–1004 (1980).
472 Richardson, C.A. et al., The use of tidal growth bands in the shell of Cerastoderma edule to measure seasonal growth rates under cool temperature and sub-arctic conditions. *Journal of the Marine Biological Association of the United Kingdom* 60:977–89 (1980).
473 Richardson, C.A., Crisp, D.J. & Runham, N.W., Factors influencing shell deposition during a tidal cycle in the intertidal bivalve Cerastoderma edule. *Journal of the Marine Biological Association of the United Kingdom* 61:465–76 (1981).
474 Richardson, C.A., Crisp, D.J. & Runham, N.W., Factors influencing shell growth in Cerastoderma edule. *Proceedings of the Royal Society of London B* 210 (1181):513–32 (1980).
475 Lonne, O.J. & Gray, J.S., Influence of tides on microgrowth bands in Cerastoderma edule from Norway. *Marine Ecology Progress Series* 42:1–7 (1988).
476 Lutz, R.A. & Rhoads, D.C., Anaerobiosis and a theory growth line formation. Micro- and ultrastructural growth patterns within the molluscan shell reflect periodic respiratory changes. *Science* 198:1222–27 (1977).

477 Bourget, E. & Brock, V., Short-term shell growth in bivalves. Individual, regional and age-related variation in the rhythm of deposition of Cerastoderma edule equals Cardium edule. *Marine Biology* (Berlin) 106(1):103–8 (1990).
478 Armonies, W., Migratory rhythms of drifting juvenile molluscs in tidal waters of the Wadden Sea. *Marine Ecology Progress Series* 83:197–206 (1992).
479 Braley, R.D., Reproduction in the giant clams Tridacna gigas and T. derasa in situ on the North-Central Great Barrier Reef, Australia, and Papua New Guinea. *Coral Reefs* 3:221–27 (1984).
480 Gompel, M., Recherches sur la consommation d'oxygíene de quelques animaux aquatiques litteraux. Rythme des oxydations et rythme des marées. *Annales de Physiologie* 14:914–31 (1938).
481 Bennett, M.F., The rhythmic activity of the quahog, Venus mercenaria, and its modification by light. *Biological Bulletin of the Marine Biological Laboratory Woods Hole* 107:174–91 (1954).
482 Palincsar, J.S., *Periodism in amount of spontaneous activity in the quahog Venus mercenaria.* Dissertation Northwestern University, Evanston/Illinois, USA, 1958.
483 Grave, B.H., An analysis of the spawning habits and spawning stimuli of Cumingia tellinoides. *Biological Bulletin of the Marine Biological Laboratory Woods Hole* 52:418–35 (1927).
484 Donn, T.E.jr. et al., Distribution and abundance of Donax serra roding, Bivalvia, Donacidae, as related to beach morphology, I: Semilunar migrations. *Journal of Experimental Marine Biology and Ecology* 102(2–3):121–32 (1986).
485 McLachlan, A., Wooldridge, T. & Horst, G. van der, Tidal movements of the macro fauna on an exposed sandy beach in South Africa. *Journal of Zoology* (London) 187:433–42 (1979).
486 Roch, F., Die Teredinen des Mittelmeeres. Thalassa. *Revue mensuelle des Valeurs Mediterranéennes* 4(3), Marseille 1940.
487 Fage, L. & Legendre, R., Essaimage et rythme lunaire d'un Phyllodocien (Eulalia punctifera Grube). *Comptes rendus hebdomadaires des Séances de l'Académie des Sciences* 182:721–23 (1926).
488 Fage, L. & Legendre, R., Pôches planctonique à la lumíere effectuées à Banyuls-sur-mer et Concarneau, I: Annélides Polychíetes. *Archives de Zoologie Expérimentale et Générale* 67:23–222 (1927).
489 Herpin, R., Recherches biologique sur la reproduction et le développement de quelques Annélides Polychíetes. *Bulletin de la Société des Sciences Naturelles de l'Ouest de la France,* Série 4 (1925), Vol.5:1–250 (1926).
490 Galloway, T.W. & Welch, P.S., Studies on a phosphorescent Bermudan annelid, Odontosyllis enopla Verril. *Transactions of the American Microscopical Society* 30:13–35 (1911).
491 Huntsman, A.G., Odontosyllies at Bermuda and lunar periodicity. *Journal of the Fisheries Research Board, Canada* 7:363–69 (1948).
492 Markert, R.E., Markert, B.J. & Vertrees, N.J., Lunar periodicity in spawning and luminescence in Odontosyllis enopla. *Ecology* 42:414f (1961).
493 Lummel, L.A.E. van, Over lichtende wormpjes in de baai van Batavia. *De Tropische Natuur* (Weltefreden) 21:85–87 (1932).
494 Fraser, C.M., The swarming of Odontosyllis. *Transactions of the Royal Society of Canada, Mémoires de la Société royale du Canada,* Ser.4, Vol.9:43–49 (1915).
495 Potts, F.A., The swarming of Odontosyllis. *Proceeding of the Cambridge Philosophical Society* 17:193–200 (1913).

496 Tsuji, F.I. & Hill, E., Repetitive cycles of bioluminescence and spawning in the polychaete Odontosyllis phosphorea. *Biological Bulletin of the Marine Biological Laboratory Woods Hole* 165(2):444–49 (1983).
497 Herpin, R., L'Essaimage et développement d'un Eunicien et d'un Syllidien. *Comptes rendus hebdomadaires des Séances de l'Académie des Sciences* 179:-1209ff (1924).
498 Franke, H.D., Zur Determination der zeitlichen Verteilung von Fortpflanzungsprozessen in Laborkulturen des Polychaeten Typosyllis prolifera. *Helgoländer Wissenschaftliche Meeresuntersuchungen* 34(1):61–84 (1980).
499 Franke, H.D., Endocrine mechanisms mediating light-temperature effects on male reproductive activity in Typosyllis prolifera (Polychaeta: Syllidae). *Wilhelm Roux's Archives of Developmental Biology* 192(2):95–102 (1983).
500 Franke, H.D., Endocrine control of reproductive periodicity in male Typosyllis prolifera (Polychaeta: Syllidae). *International Journal of Invertebrate Reproduction* 6(4):229–38 (1983).
501 Franke, H.D., A clock-like mechanism timing lunar-rhythmic reproduction in Typosyllis prolifera, Polychaeta. *Journal of Comparative Physiology A* 156(4):553–62 (1985).
502 Franke, H.D., Resetting a circalunar reproduction rhythm with artificial moonlight signals, phase-response curve and moon-off effect. *Journal of Comparative Physiology A* 159(4):569–76 (1986).
503 Franke, H.D., The role of light and endogenous factors in the timing of the reproductive cycle of Typosyllis prolifera and some other Polychaetes. *American Zoologist* 26:433–45 (1986).
504 Fage, L. & Legendre, R., Les danses nuptiales de quelques Néréidiens. *Comptes rendus hebdomadaires des Séances de l'Académie des Sciences* 177:1150–52 (1923).
505 Fage, L. & Legendre, R., Rythmes lunaires de quelques Néréidiens. *Comptes rendus hebdomadaires des Séances de l'Académie des Sciences* 177:982–85 (1923).
506 Izuka, A., On the breeding habit and development of Nereis japonica new species. *Annotationes Zoologicae Japonenses* 6:295–305 (1908).
507 Durchon, M., L'essaimage d'une polychíete en eau calme et saumâtre Nereis succinea Leuck. *Archives de Zoologie Expérimentale et Générale,* Paris, Vol.854 (1948).
508 Lillie, F.R. & Just, E.E., Breeding habits of the heteronereis form of Nereis limbata at Whitstable (Mass.). *Biological Bulletin of the Marine Biological Laboratory Woods Hole* 24:147–60 (1913).
509 Hardege, J.D. et al., Induction of swarming of Nereis succinea. *Marine Biology* (Berlin) 104(2):291–95 (1990).
510 Gallien, L., Observations sur l'essaimage de Nereis (Eunereis) longissima Johnston sur la cóote Boulonnaise. *Bulletin de la Société Zoologique de France* 61:407–11 (1936).
511 Herpin, R., Etudes sur les essaimages des Annélides Polychíetes. *Bulletin Biologique de la France et de la Belgique* (Paris) 62:308–77 (1928).
512 Herpin, R., Etudes sur les essaimages des Annélides, Polychíetes (Note complémentaire). *Bulletin Biologique de la France et de la Belgique* (Paris) 63:85–94 (1929).
513 Hardege, J.D. et al., Environmental control of reproduction in Perinereis nuntia var. brevicirrus. *Journal of the Marine Biological Association of the United Kingdom* 74:903–18 (1994).

514 Fong, P.P., Lunar control of epitokal swarming inpolychaete Platynereis bicanaliculata Baird from Central California. *Bulletin of Marine Science of the Gulf and Caribbean* 52(3):911–24 (1993).
515 Enright, J.T., Endogenous tidal and lunar rhythms. *Proceedings of the XVI. International Congress of Zoology* 4:355–59 (1963).
516 Fage, L., Migrations verticales périodique des animaux benthiques litteraux. *Rapports et procíes-verbaux Conseil permanent international pour l'exploration de la mer* (Copenhagen) 85:60–69 (1933).
517 Georgévitch, J., Note sur la biologie de Platynereis dumerilii. *Godisnjak Oceanogr. Inst. Kralj. Jugoslavijc (Annuaire de la Institution Oceanographique Royaume Yugosl.)* 1:141–48. French summary: 149–52 (1938).
518 Gravier, Ch. & Dantan, J.L., Sur quelques résultats à obtenus au cours des pôches nocturnes (à la lumíere) dans la Baie d'Alger. *Comptes rendus hebdomadaires des Séances de l'Academie des Sciences* 186:1327–30 (1928).
519 Gravier, Ch. & Dantan, J.L., Pôches nocturnes à la lumíere dans la Baie d'Alger. *Annales de l'Institut Océanographique* 5:1–185 (1928).
520 Hauenschild, C., Photoperiodizität als Ursache des von der Mondphase abhängigen Metamorphose-Rhythmus bei dem Polychaeten Platynereis dumerilii. *Zeitschrift für Naturforschung* 106:658–62 (1955).
521 Hempelmann, Fr., Zur Naturgeschichte von Nereis dumerilii Aud. et Edw. *Zoologica* (Stuttgart) 25(62):1–135 (1911).
522 Korringa, P., Lunar periodicity. In: *Treatise on Marine Ecology and Palaeoecology*. Committee on Marine Ecology and Paleoecology, New York. (= Geological Society of America. Memoir 67, Vol.1 (Reprint): 917–34 (1957)).
523 Legendre, R., La lune et les ôtres marins. *Revue Scientifique (Revue Rose)* (Paris) 63:225–36 (1925).
524 Ranzi, S., Richerche sulla biologia sessuale degli Anellidi. *Publicazione della Stazione Zoologica Napoli* (Naples) 11:271–92 (1931).
525 Ranzi, S., Maturita sessuale degli Anellidi e fasi lunari. *Bollettino della Societa Italiana di Biologica Esperimentale* (Naples) 6:18 (1931).
526 Just, E.E., Breeding habits of the Heteronereis form of Platynereis megalops at Woods Hole (Mass.). *Biological Bulletin of the Marine Biological Laboratory Woods Hole* 27:201–12 (1914).
527 Aiyar, R. G. & Panikkar, N. K., Observations on the swarming habits and lunar periodicity of Platynereis sp. from the Madras harbor. *Proceedings of the Indian Academy of Science* 5(B):245–60 (1937).
528 Gravier, Ch., Sur l'évolution de la forme épigame du Palolo japonais, Ceratocephale osawai Izuka. *Compte rendu des Séances du Congríes International de Zoologie.* Monaco (IX. Kongress, 1913):223ff (1914).
529 Izuka, A., Observations on the Japanese Palolo, Ceratocephale osawai, new species. *Journal of the College of Science. Imperial University of Tokyo* 17(11):1–37 (1903).
530 Osawa, K., Über die japanischen Palolo. *Verhandlungen des V. Internationalen Zoologenkongresses,* Jena: 751–55 (1902).
531 Clark, L.B., Observations on the swarming of the Atlantic Palolo. *Yearbook of The Carnegie Institution of Washington* 36:89–90 (1937).
532 Clark, L.B., Observations on the Atlantic Palolo. *Yearbook of The Carnegie Institution of Washington* 37:87f (1938).
533 Mayer, A.G., An Atlantic 'Palolo' Staurocephalus gregaricus. *Bulletin of the Museum of Comparative Zoology at Harvard College, Cambridge*, 36:1–14 (1900).

534 Mayer, A.G., The Atlantic Palolo. *Brooklyn Museum Bulletin. Brooklyn Institute of Arts and Services* 1(3):93–103 (1902).
535 Mayer, A.G., The annual breeding-swarm of the Atlantic Palolo. *Papers from the Tortugas Laboratory, Carnegie Institution of Washington* 1:107–12 (1908).
536 Treadwell, A.L., Annellids of Tortugas. *Yearbook of the Carnegie Institution of Washington* 8:150 (1909).
537 Treadwell, A.L., Researches upon Annelids at Tortugas. *Yearbook of the Carnegie Institution of Washington* 13:220–22 (1914).
538 Burrows, W., Periodic spawning of Palolo-worms in Pacific waters. *Nature* 155:47f (1945).
539 Caspers, H., Beobachtungen über Lebensraum und Schwärmperiodizität des Palolowurmes, Eunice viridis (Polychaeta, Eunicidae). I: Die tages- und mondzeitliche Konstanz des Schwärmens, Bibliographie. *Internationale Revue der gesamten Hydrobiologie* (Berlin) 46(2):175–83 (1961).
540 Collin, A., Bemerkungen über den eßbaren Palolowurm, Lysidice viridis Gray. Anhang zu A. Krämer: *Über den Bau der Korallenriffe und die Planktonverteilung an den samoanischen Küsten*, pp.164–74. Kiel (Lipsius & Fischer) 1897.
541 Corney, B.G., Abstracts of a paper on the periodicity of the swarming Palolo (Eunice viridis Gray). *Journal of the Torquay Natural History Society* 3:126–30 (1922).
542 Friedländer, B., Über den sogenannten Palolowurm. *Biologisches Zentralblatt* 18:337–57 (1898).
543 Friedländer, B., Nochmals der Palolo und die Frage nach unbekannten kosmischen Einflüssen auf physiologische Vorgänge. *Biologisches Zentralblatt* 19:241–69 (1899).
544 Friedländer, B., Verbesserungen und Zusätze zu meinen Notizen über den Palolo. *Biologisches Zentralblatt* 19:553–57 (1899).
545 Friedländer, B., Über noch wenig bekannte kosmische Einflüsse auf physiologische Vorgänge. *Verhandlungen der Physiologischen Gesellschaft Berlin vom 10. März 1899 und Archiv für Anatomie und Physiologie*: 570–74 (1899).
546 Friedländer, B., Herrn Alfred Goldborough Mayers's Entdeckung eines Atlantischen Palolo und deren Bedeutung für die Frage nach unbekannten kosmischen Einflüssen auf biologische Vorgänge. Zugleich eine Beleuchtung der darwinistischen Betrachtungsweise. *Biologisches Zentralblatt* 21:312–17 and 352–66 (1901).
547 Friedländer, B., Zur Geschichte der Palolofrage. *Zoologischer Anzeiger* 27:716–22 (1904).
548 Gravier, Ch., Sur le 'Palolo' des Nouvelles-Hébrides (D'apríes les renseignements fournis par le P. Suas, missionaire à Aoba (Ile des Lépreux)). *Bulletin du Muséum National d'Histoire Naturelle* 30:472–74 (1924).
549 Hauenschild, C., Fischer, A. & Hofman, D.K., Untersuchungen am pazifischen Palolowurm Eunice viridis (Polychaeta) in Samoa. *Helgoländer Wissenschaftliche Meeresuntersuchungen* 18:254–95 (1968).
550 Horst, R., Wawo and Palolo worms. *Nature* 69:582 (1904).
551 Krämer, A., Palolountersuchungen. *Biologisches Zentralblatt* 19:15–30 (1899).
552 Krämer, A., Palolountersuchungen im Oktober und November 1898 in Samoa. *Biologisches Zentralblatt* 19:237–39 (1899).
553 Krämer, A., *Die Samoainseln. Entwurf einer Monographie, 2 Bände*. Stuttgart 1899, 1903.

554 MacDonald, J.D., On the external anatomy and natural history of the Genus of Annelida named Palolo by the Samoans and Tonguese, and Mbalolo by the Fijians. *Transactions of the Linnean Society of London* 22 (No.16):237–39 (1858).
555 MacIntosh, W.C., On the Pacific, Atlantic and Japanese Palolo. Notes from the Gathy Marine Laboratory of St Andrews. *Annals and Magazine of Natural History* 15(7):33–36 (1905).
556 Powell, Th., Remarks on the structure and habits of the coral-reef annelid, Palolo viridis. *Journal of the Linnean Society, Zoology* 16:393–96 (1882).
557 Whitmee, S.J., On the habits of the Palolo viridis. *Proceedings of the Zoological Society of London*: 496–502 (1875).
558 Woodworth, W., Preliminary report on the 'Palolo" worm of Samoa, Eunice viridis Gray. *American Naturalist* 37:875–81 (1903).
559 Woodworth, W., Vorläufiger Bericht über den Palolowurm. In: Krämer, A.: *Die Samoa-Inseln*, Vol.2:399–403, Stuttgart 1903.
560 Woodworth, W., The Palolo worm, Eunice viridis Gray. *Bulletin of the Museum of Comparative Zoology at Harvard College* 51:3–21 (1907).
561 Horst, R., Over de 'Wawo' van Rumphius (Lysidice oele, new species). *Rumphius Gedenkboek Kolon. Mus.* Haarlem: 105–8 (1902).
562 Okada, K., The gametogenesis, the breeding habits and the early development of Arenicola cristata Stimpson, a tubicolous Polychaete. *Science Reports of the Tohoku Imperial University, Serie 4, Biology, Sendai, Japan*, 16:99–145 (1941).
563 Scott, J.W., Some egg-laying habits of Amphitrite ornata Verril. *Biological Bulletin of the Marine Biological Laboratory Woods Hole* 17:327–40 (1909).
564 Schroeder, P.C. & Hermans, C.O., Annelida: Polychaeta, in: Giese, A.C. & Pearse, J.S. (eds.), *Reproduction of marine invertebrates*. Vol.3, p. 96. New York (Academic Press) 1975.
565 Garbarini, P., Rythme d'émission des larves chez Spirorbis borealis Daudin. *Comptes rendus des Séances et Memoires de la Société de Biologie et de ces Filiales* 112:1204f (1933).
566 Knight-Jones, E.W., Gregariousness and some other aspects of the settling behaviour of Spirorbis. *Journal of the Marine Biological Association of the United Kingdom* 30:201–22 (1951).
567 Bennett, M.F., Presistent seasonal variations in the diurnal cycle of earthworms. Zeitschrift für vergleichende *Physiologie* 60:34–40 (1968).
568 Ralph, C.L., Persistent rhythms of activity and O_2-consumption in the Earthworm. *Physiological Zoology* 30:41–55 (1957).
569 Ross-Farhang, S., Gadient, J.L. & Wasserman, G.S., Electrophysiological and behavioral measures of seasonal and monthly variations in limulus ventral eye sensitivity. *Journal of Interdisciplinary Cycle Research* 21(2):97–118 (1990).
570 Wasserman, G.S., Unconditioned response to light in Limulus: Mediation by lateral, median, and ventral eye loci. *Vision Research* 13:95–105 (1973).
571 Wasserman, G.S. & Patton, D.G., Limulus visual threshold obtained from light-elicited unconditioned tail movements. *Journal of Comparative and Physiological Psychology* 73(1):111–16 (1970).
572 Rudloe, A., The breeding behavior and patterns of movement of horseshoe crabs, Limulus polyphemus, in the vicinity of breeding beaches in Apalachee Bay, Florida. *Estuaries* 3:177–83 (1980).
573 Cohen, J.A. & Brockmann, H.J., Breeding activity and mate selection in the horseshoe crab, Limulus polyphemus. *Bulletin of Marine Science of the Gulf and Caribbean* 33:274–81 (1983).

574 Moser, P.W., Schwertschwänze – Besucher aus anderer Zeit. *Geo* 10:138–52 (1991).
575 Rudloe, A., Variation in the expression of lunar and tidal behavioral rhythms in the horsehoe crab, Limulus polyphemus. *Bulletin of Marine Science of the Gulf and Caribbean* 36(2):388–95 (1985).
576 Rudloe, A., Locomotor and light responses of larvae of the horseshoe crab, Limulus polyphemus (L.). *Biological Bulletin of the Marine Biological Laboratory Woods Hole* 157:494–505 (1979).
577 Rudloe, A., Aspects of the biology of juvenile horseshoe crabs, Limulus polyphemus. *Bulletin of Marine Science of the Gulf and Caribbean* 31:125–33 (1981).
578 Foster, W.A. et al., Short-term changes in activity rhythms in an intertidal arthropod (Acarina: Bdella interrupta Evans). *Oecologia* 38:291–301 (1979).
579 d'Albertis, L.M., *New Guinea; what I did and what I saw*. Vol.1, London 1880.
580 Oudemans, A.C., Die bis jetzt bekannten Larven von Thrombidiidae und Erythraeidae. *Zoologische Jahrbücher* (Jena), Suppl. 14, No.1:1–230 (1913).
581 Mikulecky, M. & Zemek, R., Does the moon influence the predatory activity of mites? *Experientia* 48(5):530–32 (1992).
582 Gliwicz, M.Z., A lunar cycle in zooplankton. *Ecology* 67(4):883–97 (1986).
583 Fryer, G., Lunar cycles in lake plankton. *Nature* 322:306 (1986).
584 Jacoby, C.A. & Greenwood, J.G., Emergent zooplankton in Moreton Bay, Queensland, Australia: seasonal, lunar, and diel patterns in emergence and distribution with respect to substrata. *Marine Ecology Progress Series* 51:131–54 (1989).
585 Jacoby, C.A. & Greenwood, J.G., Spatial, temporal, and behavioral patterns in emergence of zooplankton in the lagoon of Heron Reef, Great Barrier Reef, Australia. *Marine Biology* (Berlin) 97(3):309–28 (1988).
586 Patra, A.K., Nayak, L.D. & Patnaik, E., Lunar rhythm in the planktonic biomass of the river Mahanadi at Sambalpur, India. *Tropical Ecology* 27(1):40–48 (1986).
587 Alldredge, A.L. & King, J.M., Effects of moonlight on the vertical migration patterns of demersal zooplankton. *Journal of Experimental Biology and Ecology* 44(2–3):133–56 (1980).
588 Dittel, A.I. & Epifanio, C.E., Seasonal and tidal abundance of crab larvae in a tropical mangrove system, Gulf of Nicoya, Costa Rica. *Marine Ecology Progress Series* 65(1):25–34 (1990).
589 Hough, A.R. & Naylor, E., Endogenous rhythms of circatidal swimming activity in the estuarine copepod Eurytemora affinis (Poppe). *Journal of Experimental Marine Biology and Ecology* 161(1):27–32 (1992).
590 Hough, A.R. & Naylor, E., Field studies on retention of the planktonic copepod Eurytemora affinis in a mixed estuary. *Marine Ecology Progress Series* 76:115–22 (1991).
591 Reaka, M.L., Lunar and tidal periodicity of molting and reproduction in stomatopod crustacea: a selfish herd hypothesis. *Biological Bulletin of the Marine Biological Laboratory Woods Hole* 150:468–90 (1976).
592 Hough, A.R. & Naylor, E., Distribution and position maintenance behaviour of the estuarine mysid Neomysis integer. *Journal of the Marine Biological Association of the United Kingdom* 72(4):869–76 (1992).
593 Dieleman, J., Swimming rhythms, migration and breeding cycles in the estuarine amphipods Gammarus chevreuxi and Gammarus zaddachi, in: Naylor, E. & Hartnoll, R.G. (eds.), *Cyclic Phenomena in Marine Plants and Animals*, pp.415–22. Proceedings of the 13th Marine Biology Symposium, Isle of Man, 27 September – 4 October 1978. Oxford (Pergamon Press) 1979.

REFERENCES

594 Hough, A.R. & Naylor, E., Biological and physical aspects of migration in the estuarine amphipod Gammarus zaddachi. *Marine Biology* (Berlin) 112(3): 437–43 (1992).

595 Fincham, A.A., Rhythmic swimming and rheotropism in the amphipod Marinogammarus marinus. *Journal of Experimental Marine Biology and Ecology* 8:19–26 (1972).

596 Wildish, F.J., Locomotor activity rhythms in some littoral Orchestia. *Journal of the Marine Biological Association of the United Kingdom* 50:241–52 (1970).

597 Enright, J.T., Lunar orientation of Orchestoidea corniculata Stout (Amphipoda). *Biological Bulletin of the Marine Biological Laboratory Woods Hole* 120: 148–56 (1961).

598 Hoffmann, K., Clock-mechanisms in celestial orientation of animals, in: Aschoff, J. (ed.), *Circadian clocks*, pp.426–41. Proceedings of the Feldafing summer school. September 7–18, 1964. Amsterdam (North Holland Publishing Company) 1965.

599 Papi, F. & Pardi, L., Ricerche sull'orientamento di Talitrus saltator (Montagu) (Crustacea, Amphipoda). *Zeitschrift für vergleichende Physiologie* 35:490–518 (1953).

600 Pardi, L. & Papi, F., Ricerche sull'orientamento di Talitrus saltator (Montagu) (Crustacea, Amphipoda). I: L'orientamento durante il giorno in una popolazione del litorale Tirrenico. *Zeitschrift für vergleichende Physiologie* 35:459–89 (1953).

601 Papi, F. & Pardi, L., On the lunar orientation of sandhoppers (Amphipoda, Talitridae). *Biological Bulletin of the Marine Biological Laboratory Woods Hole* 124:97–105 (1963).

602 Edwards, J.M. & Naylor, E., Endogenous circadian changes in orientational behaviour of Talitrus saltator. *Journal of the Marine Biological Association of the United Kingdom* 67:17–26 (1987).

603 Weeks, J.M. & Moore, P.G., The effect of synchronous molting on body copper and zinc concentrations in four species of talitrid amphipods, crustacea. *Journal of the Marine Biological Association of the United Kingdom* 71(2):481–88 (1991).

604 Benson, J.A. & Lewis, R.D., An analysis of the activity rhythm of the sand beach amphipod, Talorchestia quoyana. *Journal of Comparative Physiology* 105:339–52 (1976).

605 Forward, R.B.jr., Phototaxis of a sand-beach amphipod physiology and tidal rhythms. *Journal of Comparative Physiology* 135:243–50 (1980).

606 Fincham, A.A., Rhythmic behaviour of the intertidal amphipod Bathyporeia pelagica. *Journal of the Marine Biological Association of the United Kingdom* 50:1057–68 (1970).

607 Preece, G.S., The swimming rhythm of Bathyporeia pilosa (Crustacea: Amphipoda). *Journal of the Marine Biological Association of the United Kingdom* 51:777–91 (1971).

608 Fish, J.D., Development, hatching and brood size in Bathyporeia pilosa and Bathyporeia pelagica (Crustacea, Amphipoda). *Journal of the Marine Biological Association of the United Kingdom* 55(2):357–68 (1975).

609 Fish, J.D. & Mills, A., The reproductive biology of Corophium volutator and Corophium arenarium (Crustacea: Amphipoda). *Journal of the Marine Biological Association of the United Kingdom* 59(2):355–68 (1979).

610 Morgan, E., The activity rhythm of the amphipod Corophium volutator (Pallas) and its possible relationship to changes in hydrostatic pressure associated with the tides. *Journal of Animal Ecology* 34:731–46 (1965).

611 Holmström, W.F. & Morgan, E., Variation in the naturally occurring rhythm of the estuarine amphipod Corophium volutator (Pallas). *Journal of the Marine Biological Association of the United Kingdom* 63(4):833–50 (1983).

612 Holmström, W.F. & Morgan, E., Laboratory entrainment of the rhythmic swimming activity of Corophium volutator (Pallas) to cycles of temperature and periodic inundation. *Journal of the Marine Biological Association of the United Kingdom* 63:861–70 (1983).

613 Holmström, W.F. & Morgan, E., The effects of low temperature pulses in rephasing the endogenous activity rhythm of Corophium volutator (Pallas). *Journal of the Marine Biological Association of the United Kingdom* 63:851–60 (1983).

614 Omori, K., Tanaka, M. & Kikuchi, T., Seasonal changes of short term reproductive cycle in Corophium volutator (Crustacea: Amphipoda): Semilunar or lunar cycle? *Publications. Amakusa Marine Biological Laboratory. Kyushu University*. 6(2):105–8 (1982).

615 Hardy, A.C. & Gunther, E.R., The plankton of the South Georgia whaling grounds and adjacent waters, 1926/27. *Discovery Reports* (Cambridge) 11:270–72 (1935).

616 De Ruyck, A.M.C., McLachlan, A. & Donn, T.E.jr., The activity of three intertidal sand beach isopods, Flabellifera: Cirolanidae. *Journal of Experimental Marine Biology and Ecology* 146(2):163–80 (1991).

617 Alheit, J. & Naylor, E., Behavioral basis of intertidal zonation in Eurydice pulchra Leach. *Journal of Experimental Marine Biology and Ecology* 23(2):135–44 (1976).

618 Fish, J.D. & Fish, S., The swimming rhythm of Eurydice pulchra Leach and a possible explanation of intertidal migration. *Journal of Experimental Marine Biology and Ecology* 8:195–200 (1972).

619 Hastings, M.H. & Naylor, E., Ontogeny of an endogenous rhythm in Eurydice pulchra. *Journal of Experimental Marine Biology and Ecology* 46(2–3):137–45 (1980).

620 Hastings, M.H., Semilunar variations of endogenous circatidal rhythms of activity and respiration in the isopod Eurydice pulchra. *Marine Ecology Progress Series* 4(1):85–90 (1981).

621 Hastings, M.H., The entraining effect of turbulence on the circatidal activity rhythm and its semilunar modulation in Eurydice pulchra. *Journal of the Marine Biology Association of the United Kingdom* 61:151–60 (1981).

622 Jones, D.A. & Naylor, E., The swimming rhythm of the sand beach isopod Eurydice pulchra. *Journal of Experimental Marine Biology and Ecology* 4:188–99 (1970).

623 Reid, D.G. & Naylor, E., An entrainment model for semilunar rhythmic swimming behavior in the marine isopod Eurydice pulchra. *Journal of Experimental Marine Biology and Ecology* 100(1–3):25–35 (1986).

624 Warman, C.G., Reid, D.G. & Naylor, E., Circatidal variability in the behavioural responses of a sandbeach isopod Eurydice pulchra (Leach) to orientational cues. *Journal of Experimental Marine Biology and Ecology* 168(1):59–70 (1993).

625 Enright, J.T., Entrainment of a tidal rhythm. *Science* 147:864–67 (1965).

626 Klapow, L.A., Natural and artificial rephasing of a tidal rhythm. *Journal of Comparative Physiology* 79:233–58 (1972).

627 Klapow, L.A., Lunar and tidal rhythms of an intertidal crustacean, in: De Coursey, P.J. (ed.), *The Belle W. Baruch Library in Marine Sciences*, No.4. *Biological rhythms in the marine environment*, pp.215–24. Papers presented at a symposium, held April 24–27, 1974, at the Belle W. Baruch Coastal Research Center, Georgetown, SC. University of South Carolina Press 1976.

628 Heusner, A.A. & Enright, J.T., Long-term activity recordings in small aquatic animals. *Science* 154:532f (1966).
629 Klapow, L.A., Fortnightly molting and reproductive cycles in the sand-beach isopod Excirolana chiltoni. *Biological Bulletin of the Marine Biological Laboratory Woods Hole* 143:568–91 (1972).
630 Wieser, W., Adaptions of two intertidal isopods. I: Respiration and feeding in Naesa bidentata (Adams) (Sphaeromatidae). *Journal of the Marine Biological Association of the United Kingdom* 42:665–82 (1962).
631 Marsh, B.A. & Branch, G.M., Circadian and circatidal rhythms of oxygen consumption in the sandy beach isopod Tylos granulatus. *Journal of Experimental Marine Biology and Ecology* 37:77–90 (1979).
632 Paula, J., Rhythms of larval release of decapod crustaceans in the mira estuary Portugal. *Marine Biology* (Berlin) 100(3):309–12 (1989).
633 Forward, R.B.jr., Larval release rhythms of decapod crustaceans. An overview. Second International Symposium on Indo-Pacific Marine Biology of the Western Society of Naturalists, Guam, Mariana Islands, June 23–28, 1986. *Bulletin of Marine Science of the Gulf and Caribbean* 41(2):165–76 (1987).
634 Forward, R.B.jr., Lohmann, K. & Cronin, T.W., Rhythms in larval release by an estuarine crab (Rhithropanopeus harrisii). *Biological Bulletin of the Marine Biological Laboratory Woods Hole* 163(2):287–300 (1982).
635 Forward, R.B.jr., Douglass, J.K. & Kenney, B.E., Entrainment of the larval release rhythm of the crab Rhithropanopeus harrisii (Brachyura: Xanthidae) by cycles in salinity change. *Marine Biology* (Berlin) 90:537–44 (1986).
636 Cronin, T.W. & Forward, R.B.jr., Tidal vertical migration: an endogenous rhythm in estuarine crab larvae. *Science* 205:1020–22 (1979).
637 Cronin, T.W. & Forward, R.B.jr., Vertical migration rhythms of newly hatched larvae of the estuarine crab, Rhithropanopeus harrisii. *Biological Bulletin of the Marine Biological Laboratory Woods Hole* 165:139–53 (1983).
638 Gunaga, V., Neelakantan, K. & Neelakantan, B., Effect of tidal and lunar phases on the penaeid prawn seed abundance. *Fishery Technology* 28(1):88f (1991).
639 Vance, D.J., Activity patterns of juvenile penaeid prawns in response to artificial tidal and day-night cycles: a comparison of three species. *Marine Ecology Progress Series* 87:215–26 (1992).
640 Vance, D.J. & Staples, D.J., Catchability and sampling of three species of juvenile penaeid prawns in the Embley River, Gulf of Carpentaria, Australia. *Marine Ecology Progress Series* 87:201–13 (1992).
641 Wheeler, J.F.G. & Brown, F.A., The periodic swarming of Anchistoides antiguensis Schmitt (Crustacea: Decapoda) at Bermuda. *Journal of the Linnean Society, Zoology* 39:413–28 (1936).
642 Wheeler, J.F.G., Further observations on lunar periodicity. *Journal of the Linnean Society, Zoology* 40:325–45 (1937).
643 Hughes, D.A., Factors controlling emergence of pink shrimp (Penaeus duorarum) from the substrate. *Biological Bulletin of the Marine Biological Laboratory Woods Hole* 134:48–59 (1968).
644 Wickham, D.A., *Observations on the patterns of persistent activity in juvenile pink shrimp, Penaeus duorarum.* Master of Science Thesis, University of Miami, 1966.
645 Wickham, D.A., Observations on the activity patterns juveniles of the pink shrimp, Penaeus duorarum. *Bulletin of Marine Science of the Gulf and Caribbean* 17:769–86 (1967).

646 Fuss, C.M.jr., Observations in the burrowing behaviour of the shrimp, Penaeus duorarum (Burkenroad). *Bulletin of Marine Science of the Gulf and Caribbean* 14:62–73 (1964).
647 Hughes, D.A., On the endogenous control of tide-associated displacements of pink shrimp, Penaeus duorarum Burkenroad. *Biological Bulletin of the Marine Biological Laboratory Woods Hole* 142:271–80 (1972).
648 Bishop, J.M. & Herrnkind, W.F., Burying and molting of pink shrimp Penaeus duorarum, Crustacea, Penaeidae under selected photoperiods of white light and UV light. *Biological Bulletin of the Marine Biological Laboratory Woods Hole* 150(2):163–82 (1976).
649 Natarajan, P., Persistent locomotor rhythmicity in the prawns Penaeus indicus and Penaeus monodon. Marine Biology (Berlin) 101(3):339–46 (1989).
650 Natarajan, P., External synchronizers of tidal activity rhythms in the prawns Penaeus indicus and Penaeus monodon. *Marine Biology* (Berlin) 101:347–54 (1989).
651 Staples, D.J. & Vance, D.J., Effects of changes in catchability on sampling of juvenile and adolescent banana prawns, Penaeus merguiensis. *Australian Journal of Marine and Freshwater Research* 30(4):511–20 (1979).
652 Nascimento, I.A. et al., Reproduction of ablated and unablated Penaeus schmitti in captivity using diets consisting of fresh-frozen natural and dried formulated feeds. *Aquaculture* 99(3–4):387–98 (1991).
653 Vance, D.J., Heales, D.S. & Loneragan, N.R., Seasonal, diel and tidal variation in beam-trawl catches of juvenile grooved tiger prawns, Penaeus semisulcatus (Decapoda: Penaeidae), in the Embley River, North-eastern Gulf of Carpentaria, Australia. *Australian Journal of Marine and Freshwater Research* 45:35–42 (1994).
654 Price, A.R.G., Temporal variations in abundance of penaeid shrimp Larvae and oceanographic conditions of Ras Tanura, Western Arabien Gulf, Saudi-Arabia. *Estuarine and Coastal Marine Science* 9:451–66 (1979).
655 Holtschmit, K.H. & Romero, J.M., Maturation and spawning of blue shrimp Penaeus stylirostris Stimpson under hypersaline conditions. *Journal of the World Aquaculture Society* 22(1):45–50 (1991).
656 Griffith, D.R.W. & Wigglesworth, J.M., Growth rhythms in the shrimp Penaeus vannamei and Penaeus schmitti. *Marine Biology* (Berlin) 115(2):295–99 (1993).
657 Goswami, S.C. & Goswami, U., Lunar, diel and tidal variability in penaeid prawn larval abundance in the Mandovi Estuary, Goa. *Indian Journal of Marine Sciences* 21:21–25 (1992).
658 Rodriguez, G. & Naylor, E., Behavioural rhythms in littoral prawns. *Journal of the Marine Biological Association of the United Kingdom* 52:81–95 (1972).
659 Kneib, R.T., Seasonal abundance, distribution and growth of postlarval and juvenile grass shrimp (Palaemonetes pugio) in a Georgia, USA, salt marsh. *Marine Biology* (Berlin) 96(2):215–24 (1987).
660 Al-Adhub, A.H.Y. & Naylor, E., Emergence rhythms and tidal migrations in the brown shrimp Crangon crangon (L.). *Journal of the Marine Biological Association of the United Kingdom* 55:801–10 (1975).
661 Skewes, T.D., Pitcher, C.R. & Trendall, J.T., Changes in the size structure, sex ratio and molting activity of a population of ornate rock lobsters, Panulirus ornatus, caused by an annual maturation molt and migration. *Bulletin of Marine Science of the Gulf and Caribbean* 54(1):38–48 (1994).
662 Drzewina, A., Les variations périodiques du signe du phototropisme chez les Pagures misanthropes. *Comptes rendus hebdomadaires des Séances de l'Académie des Sciences* 145:1208f (1907).

663 De Wilde, P.A.W.J, Migrations and egg deposition in Coenobita clypeatus. *Netherlands Journal of Zoology* (London) 19(2):283 (1969).
664 Chandrashekaran, M.K., Persistent tidal and diurnal rhythms of locomotory activity and oxygen consumption in Emerita asiatica (M. Edw.). *Zeitschrift für Vergleichende Physiologie* 50:137–50 (1965).
665 Aldrich, J.C. & McMullan, P.M., Observations on nonlocomotory manifestations of biological rhythms and excitement in the oxygen consumption rates of crabs. *Comparative Biochemistry and Physiology A* 62(3):707–10 (1979).
666 Chatterton, T.D. & Williams, B.G., Activity patterns of the New Zealand cancrid Cancer novaezelandiae (Jacquinot) in the field and laboratory. *Journal of Experimental Marine Biology and Ecology* 178(2):261–74 (1994).
667 Fingerman, M., Persistent, daily, and tidal rhythms of color change in Callinectes sapidus. *Biological Bulletin of the Marine Biological Laboratory Woods Hole* 109:255–64 (1955).
668 Ryer, C.H., Montfrans, J. van & Orth, R.J., Utilization of seagrass meadow and tidal marsh creek by blue crabs Callinectes sapidus. II: Spatial and temporal patterns of molting. Blue Crab Conference, Virginia Beach, USA, May 15–17, 1988. *Bulletin of Marine Science of the Gulf and Caribbean* 46(1):95–104 (1990).
669 Arudpragasam, K.D. & Naylor, E., Gill ventilation and the role of reversed respiratory currents in Carcinus maenas (L.). *Journal of Experimental Biology* 41:299–307 (1964).
670 Arudpragasam, K.D. & Naylor, E., Gill ventilation volumes, oxygen consumption and respiratory rhythms in Carcinus maenas (L.). *Journal of Experimental Biology* 41:309–21 (1964).
671 Blume, J., Bünning, E. & Müller, D., Periodenanalyse von Aktivitätsrhythmen bei Carcinus maenas. *Biologisches Zentralblatt* 81:569–73 (1962).
672 Bünning, E., *Zeitmessung bei Pflanzen und Tieren mit tagesperiodischen Schwingungen. Aufnahme und Verarbeitung von Nachrichten durch Organismen*, pp.126–37, Stuttgart 1961.
673 Naylor, E., Tidal and diurnal rhythms of locomotory activity in Carcinus maenas (L.). *Journal of Experimental Biology* 35:602–10 (1958).
674 Naylor, E., Temperature relationships of the locomotor rhythm of Carcinus. *Journal of Experimental Biology* 40:669–79 (1963).
675 Naylor, E., Locomotory rhythms in Carcinus maenas (L.) from non-tidal conditions. *Journal of Experimental Biology* 37:481–88 (1960).
676 Naylor, E., Spontaneous locomotor rhythm in Mediterranean Carcinus. *Publicazioni della Stazione Zoologica Napoli* 32:58–63 (1961).
677 Williams, B.G. & Naylor, E., Synchronisation of the locomotor tidal rhythm of Carcinus. *Journal of Experimental Biology* 51:715–25 (1969).
678 McGaw, I.J. & Naylor, E., Distribution and rhythmic locomotor patterns of estuarine and open-shore populations of Carcinus maenas. *Journal of the Marine Biological Association of the United Kingdom* 72 (3):599–609 (1992).
679 Naylor, E. & Williams, B.G., Effects of eyestalk removal on rhythmic locomotor activity in Carcinus. *Journal of Experimental Biology* 49:107–16 (1968).
680 Taylor, A.C. & Naylor, E., Entrainment of the locomotor rhythm of Carcinus by cycles of salinity change. *Journal of the Marine Biological Association of the United Kingdom* 57:273–77 (1977).
681 Warman, C.G., Reid, D.G. & Naylor, E., Variation in the tidal migratory behaviour and rhythmic light responsiveness in the shore crab Carcinus maenas. *Journal of the Marine Biological Association of the United Kingdom* 73 (2):355–64 (1993).

682 Depledge, M.H., On the tidal, diurnal and seasonal modulation of endogenous cardiac activity rhythms in the shore crab (Carcinus maenas L.). *Oebalia* 18 N.S.: 53–67 (1992).

683 Queiroga, H., Costlow, J.D. & Moreira, M.H., Larval abundance patterns of Carcinus maenas (Decapoda, Brachyura) in Canal de Mira (Ria de Aveiro, Portugal). *Marine Ecology Progress Series* 111:63–72 (1994).

684 Chatterji, A. et al., Effect of lunar periodicity on the abundance of crabs from the Goa coast. *Indian Journal of Marine Sciences* 23:180f (1994).

685 Christy, J.H., Timing of larval release by intertidal crabs on an exposed shore. *Bulletin of Marine Science of the Gulf and Caribbean* 39:176–91 (1986).

686 Kosuge, T., Murai, M. & Poovachiranon, S., Breeding cycle and mating behaviour of the tropical ocypodid Ilyoplax gangetica (Kemp 1919) (Crustacea, Brachyura). *Tropical Zoology* 7:25–34 (1994).

687 Kosuge, T., Poovachiranon, S. & Murai, M., Male courtship cycles in three species of tropical Ilyoplax crabs (Decapoda, Brachyura, Ocypodidae). *Hydrobiologia* 285:93–100 (1994).

688 Tankersley, R.A. & Forward, R.B.jr., Endogenous swimming rhythms in two estuarine crab megalopae: Implications for flood tide transport. *American Zoologist* 33(5):118A (1993).

689 Tankersley, R.A. & Forward, R.B.jr., Endogenous swimming rhythms in estuarine crab megalopae: Implications for flood-tide transport. *Marine Biology* (Berlin) 118(3):415–23 (1994).

690 De Vries, M.C. et al., Abundance of estuarine crab larvae is associated with tidal hydrologic variables. *Marine Biology* (Berlin) 118(3):403–13 (1994).

691 Kosuge, T., Molting and breeding cycles of the rock-dwelling ocypodid crab Macrophthalmus boteltobagoe (Sakai, 1939) (Decapoda, Brachyura). *Crustaceana* (Leiden) 64(1):56–65 (1993).

692 Henmi, Y., The description of wandering behavior and its occurrence varying in different tidal areas in Macrophthalmus japonicus (Crustacea: Ocypodiae). *Journal of Experimental Marine Biology and Ecology* 84(3):211–24 (1984).

693 Brooke, M. de L., Size as a factor influencing the ownership of copulation burrows by the ghost crab (Ocypode ceratophthalmus). *Zeitschrift für Tierpsychologie* 55:63–78 (1981).

694 Barnwell, F.H., The role of rhythmic systems in the adaptation of fiddler crabs to the intertidal zone. *American Zoologist* 8:569–83 (1968).

695 Crane, J., Aspects of social behavior in fiddler crabs, with special reference to Uca maracoani (Latreille). *Zoologica* (Stuttgart) 43:113–30 (1958).

696 Zucker, N., Monthly reproductive cycles in three sympatric hood-building tropical fiddler crabs (genus Uca). *Biological Bulletin of the Marine Biological Laboratory Woods Hole* 155(2):410–24 (1978).

697 Zucker, N., Behavioral rhythms in the fiddler crab Uca terpsichores, in: DeCoursey, P.J. (ed.), *The Belle W. Baruch Library in Marine Sciences*, No.4. Biological rhythms in the marine environment, pp.145–59. Papers presented at a symposium, held April 24–27, 1974, at the Belle W. Baruch Coastal Research Center, Georgetown, SC. University of South Carolina Press 1976.

698 Honegger, H.-W., Rhythmic motor activity responses of the California fiddler crab Uca crenulata. *Marine Biology* (Berlin) 18:18–31 (1973).

699 Severinghaus, L.L. & Lin, H.C., The reproductive behavior and mate choice of the fiddler crab Uca lactea lactea in Mid-Taiwan. *Behaviour* 113(3–4):292–308 (1990).

700 Crane, J., Crabs of the genus Uca from the west coast of Central America. *Zoologica* (Stuttgart) 26:145–208 (1941).

701 Barnwell, F.H., Observations on daily and tidal rhythms in some fiddler crabs from equatorial Brazil. *Biological Bulletin of the Marine Biological Laboratory Woods Hole* 125:399–415 (1963).
702 Barnwell, F.H., Daily and tidal patterns of activity in individual fiddler crab (genus Uca) from the Woods Hole Region. *Biological Bulletin of the Marine Biological Laboratory Woods Hole* 130:1–17 (1966).
703 Fingerman, M., Lowe, M.E. & Mobberly, W.C.jr., Environmental factors involved in setting the phases of tidal rhythms in the fiddler crab Uca puligator and Uca minax. *Limnology and Oceanography* (Baltimore) 3:271–82 (1958).
704 Christy, J.H., Adaptive significance of semilunar cycles of larval release in fiddler crabs genus Uca. Test of an hypothesis. *Biological Bulletin of the Marine Biological Laboratory Woods Hole* 163(2):251–63 (1982).
705 Bergin, M.E., Hatching rhythms in Uca pugilator (Decapoda: Brachyura). *Marine Biology* (Berlin) 63:151–58 (1981).
706 DeCoursey, P.J., Egg-hatching rhythms in three species of fiddler crabs, in: Naylor, E. & Hartnoll, R.G. (eds.), *Cyclic Phenomena in Marine Plants and Animals*, pp.399–406. Proceedings of the 13th Marine Biology Symposium, Isle of Man, September 27 – October 4, 1978. Oxford (Pergamon Press) 1979.
707 Bennett, M.F., Brown, F.A.jr. & Webb, H.M., Fluctuation in the form of the daily rhythm of O_2-consumption in fiddler crabs. *Biological Bulletin of the Marine Biological Laboratory Woods Hole* 107:304f (1954).
708 Fingerman, M., Phase differences in the tidal rhythms of color change in two species of fiddler crabs. *Biological Bulletin of the Marine Biological Laboratory Woods Hole* 110:274–90 (1956).
709 Christy, J.H., Adaptive significance of reproductive cycles in the fiddler crab, Uca pugilator: a hypothesis. *Science* 199:453–55 (1978).
710 Bennett, M.F., Shriner, J. & Brown, F.A.jr., Persistent tidal cycles of spontaneous motor activity in the fiddler crab, Uca pugnax. *Biological Bulletin of the Marine Biological Laboratory Woods Hole* 112:267–75 (1957).
711 Brown, F.A.jr., Sandeen, M.I. & Ralph, C.L., The primary lunar rhythm of O_2-consumption in the fiddler crab. *Biological Bulletin of the Marine Biological Laboratory Woods Hole* 107:306 (1954).
712 Bennett, M.F., The phasing of the cycle of motor activity in the fiddler crab, Uca pugnax. *Zeitschrift für Vergleichende Physiologie* 47:431–37 (1963).
713 Hines, M.N., Use of automized legs in determining the phases of the tidal rhythm in Uca pugnax. *Biological Bulletin of the Marine Biological Laboratory Woods Hole* 105:375f (1953).
714 Hines, M.N., A tidal rhythm in behavior of melanophores in automized legs of Uca pugnax. *Biological Bulletin of the Marine Biological Laboratory Woods Hole* 107:386–96 (1953).
715 Webb, H.M. & Brown, F.A.jr., Interaction of diurnal and tidal rhythms in the fiddler crab, Uca pugnax. *Biological Bulletin of the Marine Biological Laboratory Woods Hole* 129:582–91 (1965).
716 Greenspan, B.N., Semimonthly reproductive cycles in male and female fiddler crabs, Uca pugnax. *Animal Behaviour* 30:1084–92 (1982).
717 Dowse, H.B. & Palmer, J.D., Evidence for ultradian rhythmicity in an intertidal crab, in: Hayes, D.K., Pauly, J.E. & Reiter, R.J. (eds.), *Progress in clinical and biological research*, Vol.341: *Chronobiology: Its role in clinical medicine, general biology, and agriculture,* part B, pp.691–97. XIX. International Conference of the International Society for Chronobiology, Bethesda, Maryland, USA, June 20–24, 1989. Chichester, England, 1990.

718 Altevogt, R., Ökologische und ethologische Studien an Europas einziger Winkerkrabbe Uca tangeri. *Zeitschrift für Morphologie und Ökologie der Tiere* 48:123–46 (1959).
719 Hagen, H.-O. von, Freilandstudien zur Sexual- und Fortpflanzungsbiologie von Uca tangeri in Andalusien. *Zeitschrift für Morphologie und Ökologie der Tiere* 51:611–725 (1962).
720 Barnwell, F.H. & Martini, J., Daily and tidal rhythms of activity in the Andalusian fiddler crab, Uca tangeri. *American Zoologist* 33(5):91A (1993).
721 Wolfrath, B., Observations on the behaviour of the European fiddler crab Uca tangeri. *Marine Ecology Progress Series* 100:111–18 (1993).
722 Marsden, I.D. & Dewa, R.S., Diel and tidal activity patterns of the smooth shore crab Cyclograpsus lavauxi (Milne Edwards 1853). *Journal of the Royal Society of New Zealand* 24:429–38 (1994).
723 Naylor, E. & Williams, B.G., Phase-responsiveness of the circatidal locomotor activity rhythm of Hemigrapsus edwardsi (Hilgendorf) to simulated high tide. *Journal of the Marine Biological Asscociation of the United Kingdom* 64:81–90 (1984).
724 Masayuki, S., Adaptive significance of a semilunar rhythm in the terrestrial crab Sesarma. *Biological Bulletin of the Marine Biological Laboratory Woods Hole* 160:311–21 (1981).
725 Saigusa, M. & Hidaka, T., Semilunar rhythm in the zoea release activity of the land crab Sesarma. *Oecologia* (Berlin) 37(2):163–76 (1978).
726 Saigusa, M. & Hidaka, T., Induction of a semilunar rhythm under an artificial moon light cycle in the terrestrial crab Sesarma haematocheir. *Zoological Magazine* (Tokyo) 89(2):166–70 (1980).
727 Saigusa, M., Adaptive significance of semilunar rhythm in the terrestrial crab Sesarma. *Biological Bulletin of the Marine Biological Laboratory Woods Hole* 160:311–21 (1981).
728 Saigusa, M., Larval release rhythm coinciding with solar day and tidal cycles in the terrestrial crab Sesarma – harmony with the semilunar timing and its adaptive significance. *Biological Bulletin of the Marine Biological Laboratory Woods Hole* 162(3):371–86 (1982).
729 Saigusa, M., Adaptation and evolution of a biological timing system. *Zoological Science* (Tokyo) 1(6):998 (1984).
730 Saigusa, M., The circatidal rhythm of larval release in the incubating crab Sesarma. *Journal of Comparative Physiology A* 159:21–31 (1986).
731 Saigusa, M., Entrainment of tidal and semilunar rhythms by artificial moonlight cycles. *Biological Bulletin of the Marine Biological Laboratory Woods Hole* 174:126–38 (1988).
732 Palmer, J.D., Daily and tidal components in the persistent rhythmic activity of the crab Sesarma. *Nature* 215:64–66 (1967).
733 Zimmermann, T.L. & Felder, D.L., Reproductive cycling in a newly recognized gulf coast species of Sesarma, Decapoda, Brachyura. *American Zoologist* 30(4):137A (1990).
734 Zimmermann, T.L. & Felder, D.L., Reproductive ecology of an intertidal brachyuran crab Sesarma sp. (nr. reticulatum) from the Gulf of Mexico. *Biological Bulletin of the Marine Biological Laboratory Woods Hole* 181(3):387–401 (1991).
735 Francé, R.H., *Lebenswunder der Tierwelt – eine Tierkunde für Jedermann*, pp.48f, Berlin 1940.
736 Saha, S.K. & Mukhopadhyaya, M.C., Semilunar rhythm in the mating activity of the grassland millipede Orthomorpha coarctata, Polydesmida, Paradoxosomatidae. *Revue d'Ecologie et de Biologie du Sol* 18(4):521–30 (1981).

REFERENCES

737 Bowden, J. & Church, B.M., The influence of moon light on catches of insects in light traps in Africa, Part 2: The effect of moon phase on light trap catches. *Bulletin of Entomological Research* 63(1):129–42 (1973).

738 Brown, E.S. & Taylor, L.R., Lunar cycles in the distribution and abundance of airborne insects in the equatorial highlands of East Africa. *Journal of Animal Ecology* 40(3):767–79 (1971).

739 Brack, V.jr. & Laval, R.K., Food habits of the indiana bat myotis-sodalis in Missouri USA. *Journal of Mammalogy* 66(2):308–15 (1985).

740 Siddorn, J.W. & Brown, E.S., A robinson light trap modified for segregating samples at predetermined time intervals with notes on the effect of moon light on the periodicity of catches of insects. *Journal of Applied Ecology* 8(1):69–75 (1971).

741 Doiron, N. & De Oliveira, D., Seasonal activity of Nocturnal insects in a swamp white oak grove in the Haut Richelieu Valley, Quebec, Canada. *Phytoprotection* 59(1):3–11 (1978).

742 Hora, S.L., Lunar periodicity in the reproduction of insects. Journal and *Proceedings of the Royal Asiatic Society of Bengal,* N.S. (Calcutta) 23:339–41 (1927).

743 Müller, A., 1: Observations on the habits of Oligoneuria rhenana Imhoff. *The Entomologist's Monthly Magazine* Vol.1, London 1864/1865. 2: Further notes on Oligoneuria rhenana Imhoff. *The Entomologist's Monthly Magazine* Vol.2, London 1865/1866.

744 Corbet, P.S., Lunar periodicity of aquatic insects in Lake Victoria. *Nature* 182:330f (1958).

745 Hartland-Rowe, R., Lunar rhythm in the emergence of an ephemeropteran. *Nature* 176:657 (1955).

746 Hartland-Rowe, R., The biology of a tropical mayfly Povilla adusta Navas (Ephemeroptera: Polymitarcidae) with special reference to the lunar rhythm of emergence. *Revue de Zoologie et de Botanique Africaines* (Brussels) 58:185–201 (1958).

747 Tjønneland, A., The flight activity of many flies as expressed in some East African species. *Universitet i Bergen, Arbok, Mat.-Naturv.* (Bergen, Norway), Serie 1:1–88 (1960).

748 Corbet, S.A., Sellick, R.D. & Willoughby, N.G., Notes on the biology of the mayfly Povilla adusta in West Africa. *Journal of Zoology* (London) 172:491–502 (1974).

749 Duviard, D., Flight activity of Belostomatidae in central Ivory-Coast. *Oecologia* (Berlin) 15(4):321–28 (1974).

750 Mukhopadhyay, S., Lunation included variation in catchment areas of light-traps to monitor rice green leafhoppers Nephotettix spp. in West Bengal. *Indian Journal of Agricultural Science* (London) 61(5):337–40 (1991).

751 Bowden, J. et al., Analysis of light-trap catches of Nephotettix spp. Hemiptera, Cicadellidae, in West Bengal India. *Indian Journal of Agricultural Science* 58(2):125–30 (1988).

752 Youthed, G.J. & Moran, V.C., The lunar-day activity rhythm of myrmeleontid larvae. *Journal of Insect Physiology* 15:1259–71 (1969).

753 Pandian, R.S. & Chandrashekaran, M.K., Rhythms in the biting behaviour of a mosquito Armigeres subalbatus. *Oecologia* (Berlin) 47(1):89–95 (1980).

754 Kerfoot, W.B., The lunar periodicity of Specodogastra texana, nocturnal bee (Hymenoptera: Halictidae). *Animal Behaviour* 15:479–86 (1967).

755 Kerfoot, W.B., Lunar periodicity in a north american bee. *Umschau in Wissenschaft und Technik* 68(14):444 (1968).

756 Oehmke, M., *Lunarperiodische und tagesrhythmische Flugaktivität der Bienen.* Dissertation, Frankfurt 1971.

757 Oehmke, M., Lunar periodicity in flight activity of honeybees. *Journal of Interdisciplinary Cycle Research* 4(4):319–35 (1973).

758 Mohssine, E.H., Bounias, M. & Cornuet, J.M., Lunar phase influence on the glycemia of worker honeybees. *Chronobiologia* 17(3):201–7 (1990).

759 Basset, Y., The seasonality of arboreal arthropods foraging within an Australian rainforest tree. *Ecological Entomology* 16(3):265–78 (1991).

760 Taylor, R.A.J., Time series analysis of numbers of Lepidoptera caught at light traps in East Africa and the effect of moonlight on trap efficiency. *Bulletin of Entomological Research* 76(4):593–606 (1986).

761 Stradling, D.J., Legg, C.J. & Bennett, F.D., Observations of the Sphingidae, Lepidoptera, of Trinidad. *Bulletin of Entomological Research* 73(2):201–32 (1983).

762 Persson, B., Distribution of catch in relation to emergence of adults in some noctuid pest species in South Coastal Queensland, Australia. *Australian Journal of Zoology* 25(1):95–102 (1977).

763 Nag, A. & Nath, P., Effect of moon light and lunar periodicity on the light trap catches of cutworm Agrotis ipsilon Hufn. moths. *Journal of Applied Entomology* 111(4):358–60 (1991).

764 Foley, D.H., Barnes, A. & Bryan, J.H., Anopheles annulipes Walker (Diptera: Culicidae) at Griffith, New South Wales. 4: Phenology of two sibling species. *Journal of the Australian Entomological Society* 31:91–96 (1992).

765 Chadee, D.D., Effects of moonlight on the landing activity of Anopheles bellator Dyar and Knab (Diptera: Culicidae) in Trinidad, West Indies. *Bulletin of the Society for Vector Ecology* 17(2):119–24 (1992).

766 Bhatt, R.M. et al., Biting rhythms of malaria vector Anopheles culicifacies in Kheda District Gujarat. *Indian Journal of Malariology* 28(2):91–97 (1991).

767 Charlwood, J.D. et al., The influence of moonlight and gonothrophic age on the biting activity of Anopheles farauti (Diptera: Culicidae) from Papua New Guinea. *Journal of Medical Entomology* 23(2):132–35 (1985).

768 Chadee, D.D. & Tikasingh, E.S., Diel biting activity of Culex caudelli in Trinidad, West Indies. *Medical and Veterinary Entomology* 3(3):231–37 (1989).

769 Birley, M.H. & Charlwood, J.D., The effect of moonlight and other factors on the oviposition cycle of Malaria vectors in Madang, Papua New Guinea. *Annals of Tropical Medicine and Parasitology* 83(4):415–22 (1989).

770 Davies, J.B., Moonlight and biting activity of Culex (Melanoconion) portesi Senevet & Abonnenc and C. (M.) taeniopus D. & K. (Diptera, Culicidae) in Trinidad forests. *Bulletin of Entomological Research* 65:81–96 (1975).

771 Lillie, T.H., Kline, D.L. & Hall, D.W., Host seeking activity of Culicoides spp., Diptera, Ceratopogonidae, near Yankeetown, Florida, USA. *Journal. American Mosquito Control Association* 4(4):484–93 (1988).

772 Lillie, T.H., Kline, D.L. & Hall, D.W., Diel and seasonal activity of Culicoides spp., Diptera, Ceratopogonidae, near Yankeetown, Florida, USA, monitores with a vehicle-mounted insect trap. *Journal of Medical Entomology* 24(4):503–11 (1987).

773 Tokunaga, M. & Esaki, T., A new midge from the Palau islands with its biological notes. *Mushi* (Fukuaka, Japan) 9(1):(1936).

774 Lutz, A., Beiträge zur Kenntnis der blutsaugenden Ceratopogoninen Brasiliens. *Memorias do Instituto Oswaldo Cruz* (Rio de Janeiro) Vol.5(1): (1913).

775 Provost, M.W., Mating and male swarming in Psorophora mosquitoes. *Proceedings of the Xth International Congress of Entomology, Montreal*, 2:553–61 (1958).

776 MacDonald, W.W., Observations on the biology of Chaoborids and Chironomids in Lake Victoria and on the feeding habits of the 'Elephant-snout fish' (Mormyrus kannume Forsk). *Journal of Animal Ecology* 25:36–53 (1956).
777 Mitchell, S.A., Observations on the distribution emergence and behavior of Central African Chaoboridae. *Journal. Limnological Society of Southern Africa* 14(2):102–7 (1988).
778 Fryer, G., Lunar rhythms of emergence, differential behavior of the sexes, and other phenomenons in the african midge, Chironomus brevibucca (Kieff). *Bulletin of Entomological Research* 50:1–8 (1959).
779 Carpenter, G.H., Clunio marinus Haliday, a marine Chironomid. *Entomologist's Monthly Magazine* Series I, Vol.30, London 1984.
780 Caspers, H., Über Lunar-Periodizität bei marinen Chironomiden. *Verhandlungen der Deutschen Zoologischen Gesellschaft, Supplement* 12:148–57 (1939).
781 Chevrel, R., Sur un diptfere marin du genre Clunio Haliday. *Archives de Zoologie expérimentale et générale* 3(2):583–98 (1894).
782 Endraß, U., Physiologische Anpassungen eines marinen Insekts. I: Die zeitliche Steuerung der Entwicklung. *Marine Biology* (Berlin) 34:361–68 (1976).
783 Heimbach, F., Sympatric species, Clunio marinus Hal. and Clunio balticus n. sp. (Diptera: Chironomidae), isolated by differences in diel emergence time. *Oecologia* (Berlin) 32(2):195–202 (1978).
784 Neumann, D., Photoperiodische Steuerung der 15-tägigen lunaren Metamorphose-Periodik von Clunio-Populationen (Diptera: Chironomidae). *Zeitschrift für Naturforschung* (Tübingen) 20b(8):818f (1965).
785 Neumann, D., Die intraspezifische Variabilität der lunaren und täglichen Schlüpfzeiten von Clunio marinus (Diptera: Chironomidae). *Zoologischer Anzeiger, Supplementband* (Leipzig) 29:223–33 (1966).
786 Neumann, D., Genetic adaptation in emergence time of Clunio populations to different tidal conditions. *Helgoländer Wissenschaftliche Meeresuntersuchungen* 15(1–4):163–71 (1967).
787 Neumann, D., The temporal programming of development in the intertidal chironomid Clunio marinus (Diptera: Chironomidae). *The Canadian Entomologist* 103(3):315–18 (1971).
788 Neumann, D., Neue Ergebnisse über die Kontrolle der lunaren Schwärmzeiten der Clunio-Populationen. *Chironomus* 1 (10–11):88 (1972).
789 Neumann, D., Mechanismen für die zeitliche Anpassung von Verhaltensund Entwicklungsleistungen an den Gezeitenzyklus. *Verhandlungen der Deutschen Zoologischen Gesellschaft*: 9–28 (1976).
790 Neumann, D. & Heimbach, F., Time cues for semilunar reproduction rhythms in European populations of Clunio marinus. II: The influence of tidal temperature cycles. *Biological Bulletin of the Marine Biological Laboratory Woods Hole* 166(3):509–24 (1984).
791 Neumann, D., Photoperiodic influences of the moon on behavioral and developmental performances of organisms. 10th International Biometeorological Congress, Tokyo, July 26–30, 1984, Part 1 (Abstracts). *International Journal of Biometeorology* 29 (Suppl. 1):11 (1985). Und: Part 2 (Lectures and Reports), pp.165–77.
792 Neumann, D. & Heimbach, F., Time cues for semilunar reproduction rhythms in European populations of Clunio marinus. I: The influence of tidal cycles of mechanical disturbance. In: Naylor, E. & Hartnoll, R.G. (eds.), *Cyclic Phenomena in Marine Plants and Animals*, pp.423–33. Proceedings of the 13th Marine Biology Symposium, Isle of Man, 27 September – 4 October 1978. Oxford (Pergamon Press) 1979.

793 Neumann, D., Diel eclosion rhythm of a sublittoral population of the marine insect Pontomyia pacifica. *Marine Biology* (Berlin) 90(3):461–65 (1986).
794 Neumann, D., Temperature compensation of circasemilunar timing in the intertidal insect Clunio. *Journal of Comparative Physiology A* 163(5):671–76 (1988).
795 Neumann, D., Circadian components of semilunar and lunar timing mechanisms. Symposium on Biological Clocks and Environmental Time, held in Honor of Professor Dr. Jürgen Aschoff, Munich, West Germany, January 25–26, 1988, *Journal of Biological Rhythms* 4(2):285–94 (1989).
796 Thienemann, A., *Haffmücken und andere Salzwasser-Chironomiden*. Kieler Meeresforschung Vol.1, Kiel 1936.
797 Oka, H., Morphologie und Ökologie von Clunio pacificus Edwards. *Zoologische Jahrbücher, Abteilung für Systematik, Ökologie und Geographie der Tiere* (Jena) 59:253–80 (1930).
798 Oka, H. & Hashimoto, H., Lunare Periodizität in der Fortpflanzung einer pazifischen Art von Clunio (Diptera: Chironomidae). *Biologisches Zentralblatt* 78:545–59 (1959).
799 Vanderplank, F.L., Activity of Glossina pallidipes and the lunar cycle (Diptera). *Proceedings of the Royal Entomological Society London* (A) 16:61–64 (1941).
800 Walker, T.J., A live trap for monitoring Euphasiopteryx and tests with Euphasiopteryx ochracea, Diptera, Tachinidae. *Florida Entomologist* 72(2):314–19 (1989).
801 Kobayashi, N., Spawning periodicity of sea urchins at Seto IV: Hemicentrotus pulcherrimus. *Publications of Seto Marine Biological Laboratory* 35(6):335–45 (1992).
802 Lessios, H.A., Presence and absence of monthly reproductive rhythms among eight caribbean echinoids off the coast of Panama. *Journal of Experimental Marine Biology and Ecology* 153(1):27–48 (1991).
803 Pearse, J.S., A monthly reproductive rhythm in the diadematid sea-urchin Centrostephanus coronatus Verrill. *Journal of Experimental Marine Biology and Ecology* 8(2):167–86 (1972).
804 Pearse, J.S., Lunar reproductive rhythms in sea-urchins. A review. *Journal of Interdisciplinary Cycle Research* 6(1):47–52 (1975).
805 Kennedy, B. & Pearse, J.S., Lunar synchronization of the monthly reproductive rhythm in the sea-urchin Centrostephanus coronatus. *Journal of Experimental Marine Biology and Ecology* 17(3):323–32 (1975).
806 Carson, R.L., *The edge of the sea.*
807 Pearse, J.S., Synchrony of monthly gametogenic cycles and possible reproductive isolation among Indo-Pacific populations of Diadema. Third International Symposium on Marine Biogeography and Evolution in the Pacific, Hongkong, June 26 – July 3, 1988. *Bulletin of Marine Science of the Gulf and Caribbean* 47(1):259 (1990).
808 Truschel, M., Marchase, R. & McClay, Dr., Cortical granule protein synthesis in Diadema antillarum. A sea-urchin with a lunar cycle of oogenesis. 23rd Annual Meeting of the American Society for Cell Biology, San Antonio, Texas, USA, November 29 – December 3, 1983. *Journal of Cell Biology* 97(5 Part 2):20A (1983).
809 Iliffe, T.M. & Pearse, J.S., Annual and lunar reproductive rhythms of the sea-urchin Diadema antillarum in Bermuda. *International Journal of Invertebrate Reproduction* 5(3):139–48 (1982).
810 Fox, H.M., Lunar periodicity in living organisms. *Cairo Scientific Journal* 11:45 (1922).

811 Fox, H.M., Lunar periodicity in living organisms. *Science Progress* 17:273 (1922).
812 Fox, H.M., Lunar periodicity in reproduction. *Nature* 109:273 (1922).
813 Fox, H.M., Lunar periodicity in reproduction. *Proceedings of the Royal Society,* Ser. B 95:523–50 (1923).
814 Fox, H.M., Lunar periodicity in reproduction. *Nature* 130:23 (1932).
815 Moore, H.B. et al., The biology of Lytechinus variegatus. *Bulletin of Marine Science of the Gulf and Caribbean* 13:23–53 (1963).
816 Tennent, D.H., Variation in Echinoid Plutei. *Journal of Experimental Zoology* 9:657–714 (1910).
817 Dotan, A., Reproduction of the slate pencil sea-urchin Heterocentrotus mammillatus L. in the Northern Red Sea. *Australian Journal of Marine and Freshwater Research* 41(4):457–66 (1990).
818 Babcock, R.C. et al., Predictable and unpredictable spawning events: in situ behavioural data from free-spawning coral reef invertebrates. *Invertebrate Reproduction and Development* 22:213–27 (1992).
819 Kubota, T., Semilunar spawning rhythm, hermaphroditism and sex conversion during one breeding season in a sea-cucumber Polycheira rufescens. 56th Annual Meeting of the Zoological Society of Japan, October 10–12, 1985. *Zoological Science* (Tokyo) 2(6):942 (1985).
820 Brown, C.A. & Gruber, S.H., Age assessment of the lemon shark Negaprion brevirostris using tetracycline validated vertebral centra. *Copeia* (3):747–53 (1988).
821 Thresher, R.E., Reproduction in Reef Fishes. Ascot (T.F.H. Publications) 1984.
822 Jens, G., Über den lunaren Rhythmus der Blankaalwanderung. *Archiv für Fischereiwissenschaft* (Hamburg) 4:94–110 (1952/53).
823 Gibson, R.N., Tidal and lunar rhythms in fish, in: Thorpe, J.E. (ed.), *Rhythmic activity of fishes*, pp.201–13. London (Academic Press) 1978.
824 Knöpp, H., Die Laichwanderung des Aales (Anguilla vulgaris) und ihre Beziehungen zu Mondstand und Wetter. *Zoologischer Anzeiger* (Leipzig) 149 (7–8):160–77 (1952).
825 Lowe, R.H., The influence of light and other factors on the seabird migration of the silver eel (Anguilla anguilla L.). *Journal of Animal Ecology* 21:275–309 (1952).
826 Haraldstad, O., Volestad, L.A. & Jonsson, B., Descent of European Silver Eels Anguilla anguilla in a norwegian watercourse. *Journal of Fish Biology* (London) 26(1):37–42 (1985).
827 Creutzberg, F., The role of tidal streams in the navigation of migrating elvers (Anguilla vulgaris Turt.). *Ergebnisse der Biologie* 26:118–27 (1963).
828 Wippelhauser, G.S. & McCleave, J.D., Rhythmic activity of migrating juvenile American eels Anguilla rostrata. *Journal of the Marine Biological Association of the United Kingdom* 68:81–91 (1988).
829 Todd, P.R., Timing and periodicity of migrating New-Zealand fresh water Eels Anguilla spp. *New Zealand Journal of Marine and Freshwater Research* 15(3):225–36 (1981).
830 Hain, J.H.W., The behavior of migratory eels, Anguilla rostrata, in response to current, salinity and lunar period. *Helgoländer Wissenschaftliche Meeresuntersuchungen* 27(2):211–33 (1975).
831 Winn, H.E., Richkus, W.A. & Winn, L.K., Sexual dimorphism and natural movements of the american eel Anguilla rostrata in Rhode-Island USA streams and estuaries. *Helgoländer Wissenschaftliche Meeresuntersuchungen* 27(2):156–66 (1975).

832 O'Leary, J.A. & Kynard, B., Behavior, length, and sex ratio of seaward-migrating juvenile American Shad Alosa sapidissima and Blueback Herring Alosa aestivalis in the Connecticut River, Massachusetts, USA. *Transactions of the American Fisheries Society* 115(4):529–36 (1986).

833 Hay, D.E., Tidal influence on spawning time of Pacific herring Clupea harengus pallasi. *Canadian Journal of Fisheries and Aquatic Sciences* 47(12):2390–2401 (1990).

834 Erdmann, Der Einfluß des Mondes auf die ostenglische Heringsfischerei. *Der Fischmarkt* (Cuxhaven, Germany) 12:315–17 (1934).

835 Savage, R.E. & Hodgson, W.L., Lunar influence on the East Anglican herring fishery. *Journal du Conseil Permanent International pour l'Exploration de la Mer* 9(2), Copenhagen 1934.

836 Savage, R.E. & Hardy, Phytoplankton and the herring, Part I: 1921 to 1932. *Fishery Investigations of the Ministery of Agriculture, Food and Fisheries, London, Ser. 2*, Vol.14(2), (1935).

837 Schnakenbeck, Einflüsse auf die Erträge der Heringsfischerei. *Der Fischmarkt* (Cuxhaven, Germany) 4:93–101 (1935).

838 Thorrold, S.R., Estimating some early life history parameters in a tropical clupeid Herklotsichthys castelnaui from daily growth increments in otoliths. *US National Marine Fisheries Service. Fishery Bulletin* 87(1):73–84 (1989).

839 Kavaliers, M., Endogenous lunar rhythm in the behavioral thermoregulation of a teleost fish, the white sucker, Catastomus commersoni. *Journal of Interdisciplinary Cycle Research* 13(1):23–27 (1982).

840 Thorpe, J.E. et al., Movement rhythms in juvenile Atlantic salmon Salmo salar L. *Journal of Fish Biology* 33(6):931–40 (1988).

841 Boeuf, G. & Prunet, P., Measurements of gill sodium potassium atpase activity and plasma thy roid hormones during smoltification in Atlantic salmon salmosalar. Workshop on Salmonid Smoltification 2, Stirling, Scotland. *Aquaculture* 45(1–4):111–20 (1985).

842 Nishioka, R.S. et al., Attempts to intensify the Thyroxine surge in coho and king salmon by chemical stimulation. Workshop on Salmonid Smoltification 2, Stirling. Scotland. *Aquaculture* 45(1–4):215–26 (1985).

843 Youngson, A.F. et al., The autumn and spring emigrations of juvenile Atlantic salmon Salmo salar from the girnock burn, Aberdeenshire, Scotland, UK, environmental release of migration. *Journal of Fish Biology* 23(6):625–40 (1983).

844 Farbridge, K.J. & Leatherland, J.F., Lunar periodicity of growth cycles in rainbow trout Salmo gairdneri Richardson. *Journal of Interdisciplinary Cycle Research* 18(3):169–78 (1987).

845 Quartier, A.A., Influence of the moon on net fishing yields in Lake Neuchâtel, Switzerland. *Schweizerische Zeitschrift für Hydrologie* 37(2):220–24 (1975).

846 Winans, G.A. & Nishioka, R.S., A multivariate description of change in body shape of coho salmon Oncorhynchus kisutch during smoltification. *Aquaculture* 66(3–4):235f (1987).

847 Grau, E.G. et al., Lunar phasing of the thyroxine surge preparatory to seaward migration of salmonid fish. *Science* 211:607–9 (1981).

848 Grau, E.G. et al., Factors determining the occurrence of the surge in thyroid activity in salmon during smoltification. Symposium on Salmonid Smoltification, La Jolla/CA, USA. *Aquaculture* 28(1–2):49–57 (1982).

849 Grau, E.G., Environmental influences on thyroid function in teleost fish. Symposium on Comparative Endocrinology of the Thyroid Presented at the Annual Meeting of the American Society of Zoologists, Nashville/TN, USA. *American Zoologist* 28(2):329–36 (1988).

850 Farbridge, K.J. & Leatherland, J.F., Lunar cycles of coho salmon Oncorhynchus kisutch. I: Growth and feeding. *Journal of Experimental Biology* 129(0):165–78 (1987).
851 Farbridge, K.J. & Leatherland, J.F., Lunar cycles of coho salmom Oncorhynchus kisutch. II: Scale amino acid uptake nucleic acids metabolic reserves and plasma thyroid hormones. *Journal of Experimental Biology* 129(0):179–90 (1987).
852 Sweeting, R.M., Wagner, G.F. & McKeown, B.A., Changes in plasma glucose, amino acid nitrogen and growth hormone during smoltification and seawater adaption in coho salmon, Oncorhynchus kisutch. Workshop on Salmonid Smoltification 2, Stirling, Scotland, July 3–6, 1984. *Aquaculture* 45:185–97 (1985).
853 Yamauchi, K. et al., Physiological and behavioral changes occurring during smoltification in the masu salmon Oncorhynchus masou. Workshop on Salmonid Smoltification 2, Stirling, Scotland, July 3–6, 1984. *Aquaculture* 45(1–4): 227–36 (1985).
854 Farbridge, K.J. & Leatherland, J.F., Biweekly patterns of change in food consumption, plasma thyroid hormone and growth hormone levels and in-vitro hepatic monodeiodination of T4 in trout Oncorhynchus mykiss. *Journal of Interdisciplinary Cycle Research* 22(3):237–48 (1991).
855 Hopkins, C.L. & Sadler, W.A., Rhythmic changes in plasma thyroxine concentrations in Hatchery reared chinook salmon Oncorhynchus tshawytscha. *New Zealand Journal of Marine and Freshwater Research* 21(1):31–34 (1987).
856 Hopkins, C.L., A relationship between adult recoveries of chinook salmon Oncorhynchus tshawytscha and lunar phase at time of their release from a hatchery on the Rakaia River New Zealand. *Aquaculture* 101(3–4):305–15 (1992).
857 Witkowski, A. & Kowalewski, M., Migration and structure of spawning population of european grayling Thymallus thymallus L. in the dunajec basin, Poland. *Archiv für Hydrobiologie* 112(2):279–98 (1988).
858 Hefford, A.E., *Report on fisheries for the year ended 31st March, 1930*. New Zealand Marine Dept., Wellington: 32 (1931).
859 Hefford, A.E., *Report on fisheries for the year ended 31st March, 1930*. New Zealand Marine Dept., Wellington: 20 (1931).
860 Battle, H.J., Spawning periodicity and embryonic death rate of Enchelyopus cimbrius L. in Passamaquoddy Bay. *Contributions to Canadian Biology and Fisheries* (Ottawa) 5:363–80 (1930).
861 Muller, K., Lunar periodicity of seaward migrating juvenile burbot. A short communication. *Aquilo Serie Zoologica* 22(0):147–48 (1983).
862 Gunn, J.S. et al., Timing and location of spawning of blue grenadier Macruronus novaezelandiae, Teleostei, Merlucciidae, in Australian coastal waters. *Australian Journal of Marine and Freshwater Research* 40(1):97–112 (1989).
863 Greeley, M.S.jr., Spawning by Fundulus pulvereus and Adinia xenica, Cyprinodontidae, along the Alabama USA Gulf Coast is associated with the semilunar tidal cycles. *Copeia* 1984(3):797–800 (1984).
864 Hsiao, S.M. & Meier, A.H., Semilunar ovarian activity of the gulf killifish Fundulus grandis under controlled laboratory conditions. *Copeia* 1988(1):188–95 (1988).
865 Hsiao, S.M. & Meier, A.H., Comparison of semilunar cycles of spawning activity in Fundulus grandis and F. heteroclitus held under constant laboratory conditions. *Journal of Experimental Zoology* 252:213–18 (1989).
866 Miller, C.A., Wilson, J.M. & Meier, A.H., Induction of semilunar rhythms of reproductive indices in Fundulus grandis. *American Zoologist* 21(4):995 (1981).

867 Hsiao, S.M. & Meier, A.H., Ovarian activity during the semilunar spawning cycle of the gulf Killifish Fundulus grandis. *American Zoologist* 27(4):71A (1987).

868 Waas, B.P. & Strawn, K., Seasonal and lunar cycles in gonadosomatic indices and spawning readiness of Fundulus grandis. *Contributions in Marine Science* 26:127–41 (1983).

869 Emata, A.C., Meier, A.H. & Hsiao, S.M., Daily variations in plasma hormone concentrations during the semilunar spawning cycle of the gulf killifish Fundulus grandis. *Journal of Experimental Zoology* 259(3):343–54 (1991).

870 Greeley, M.S.jr. & MacGregor, R. III, Annual and semilunar reproductive cycles of the gulf killifish, Fundulus grandis, on the Alabama Gulf Coast. *Copeia* 1983(3):711–18 (1983).

871 Greeley, M.S.jr., MacGregor, R. III & Marion, K.R., Changes in the ovary of the gulf killifish Fundulus grandis Baird & Girard during seasonal and semilunar spawning cycles. *Journal of Fish Biology* 33(1):97–108 (1988).

872 Greeley, M.S.jr., MacGregor, R. III & Marion, K.R., Variation in plasma estrogens and androgens during the seasonal and semilunar spawning cycles of female gulf killifish Fundulus grandis Baird & Girard. *Journal of Fish Biology* 33(3):419–30 (1988).

873 Cochran, R.C. et al., Serum hormone levels associated with spawning activity in the mummichog Fundulus heteroclitus. *General and Comparative Endocrinology* 70(2):345–54 (1988).

874 Bradford, C.S. & Taylor, M.H., Cortisol and estradiol during the semilunar spawning cycle of Fundulus heteroclitus. *American Zoologist* 22(4):948 (1982).

875 Bradford, C.S. & Taylor, M.H., Semilunar changes in estradiol and cortisol coincident with gonadal maturation and spawning in the killifish Fundulus heteroclitus. *General and Comparative Endocrinology* 66(1):71–78 (1987).

876 Taylor, M.H. et al., Dynamics of the in-vitro response of Fundulus heteroclitus oocytes to 17a,20b-Dihydroxy-4-pregnen-3-one. *American Zoologist* 26(4):25A (1986).

877 Kneib, R.T., Size specific patterns in the reproductive cycle of the killifish Fundulus heteroclitus, Pisces, Fundulidae, from Sapelo Island, Georgia, USA. *Copeia* 1986(2):342–51 (1986).

878 Taylor, M.H., Lunar synchronization of fish reproduction. Symposium on rhythmicity in fishes held in conjunction with the 113th meeting of the American Fisheries Society, Milwaukee, WI, USA, August 18–20, 1983. *Transactions of the American Fisheries Society* 113(4):484–93 (1984).

879 Taylor, M.H., Di Michele, L. & Leach, G.J., Lunar spawning cycle in Fundulus heteroclitus. *American Zoologist* 17(4):900 (1977).

880 Taylor, M.H. & Di Michele, L., Photoperiod influence on the semilunar spawning cycle of Fundulus heteroclitus. *American Zoologist* 18(3):668 (1978).

881 Taylor, M.H. et al., Lunar spawning cycle in the mummichog Fundulus heteroclitus (Pisces: Cyprinodontidae). *Copeia* 1979(2):291–97 (1979).

882 Taylor, M.H. & Di Michele, L., Ovarian changes during the lunar spawning cycle of Fundulus heteroclitus. *Copeia* 1980(1):118–25 (1980).

883 Hines, A.H., Osgood, K.E. & Miklas, J.J., Semilunar reproductive cycles in Fundulus heteroclitus (Pisces: Cyprinodontidae) in an area without lunar tidal cycles. *National Marine Fisheries Service. Fishery Bulletin* 83(3):467–72 (1985).

884 Greeley, M.S.jr., Tide-controlled reproduction in longnose killifish Fundulus similis. *American Zoologist* 22:870 (1982).

885 Greeley, M.S.jr., Marion, K.R. & MacGregor, R. III, Semilunar spawning cycles of Fundulus similis, Cyprinodontidae. *Environmental Biology of Fishes* 17(2):125–32 (1986).
886 Walker, B.W., A guide to the grunion. *State of California, Fish and Game Commission Bulletin* (San Francisco) 38:409–20 (1952).
887 Thomson, D.A. & Muench, K.A., Influence of tides and waves on the spawning behavior of the gulf of California grunion Leuresthes sardina. *Bulletin of the Southern California Academy of Sciences* 75(2):198–203 (1976).
888 Clark, F.N., The life history of Leuresthes tenuis, an atherine fish with tide controlled spawning habits. *State of California, Fish and Game Commission Bulletin* (San Francisco) 10:1–51 (1925).
889 Fingerman, M., Lunar rhythmicity in marine organisms. *American Naturalist* 91:67–78 (1957).
890 Idyll, C.P., Grunion. The fish that spawns on land. *National Geographic* 135(5):714–23 (1969).
891 Sherrill, M.T. & Middaugh, D.P., Spawning periodicity of the inland silverside, Menidia beryllina, (Pisces: Atherinidae) in the laboratory: relationship to lunar cycles. *Copeia* 2:522–28 (1993).
892 Middaugh, D.P., Reproductive ecology and spawning periodicity of the Atlantic Silverside Menidia menidia (Pisces: Atherinidae). *Copeia* 4:766–76 (1981).
893 Conover, D.O. & Kynard, B.E., Field and laboratory observations of spawning periodicity and behavior of a northern population of the Atlantic Silverside Menidia menidia, Pisces, Atherinidae. *Environmental Biology of Fishes* 11(3):161–72 (1984).
894 Conover, D.O., Field and laboratory assessment of patterns in fecundity of a multiple spawning fish: the Atlantic Silverside Menidia menidia. *US National Marine Fisheries Service. Fishery Bulletin* 83(3):331–42 (1985).
895 Middaugh, D.P. & Hemmer, M.J., Reproductive ecology of the tidewater Silverside Menidia peninsulae (Pisces: Atherinidae) from Santa Rosa Island, Florida, USA. *Copeia* 3:727–32 (1987).
896 Middaugh, D.P. & Hemmer, M.J., Spawning of the tidewater silverside Menidia peninsulae (Goode and Bean), in response to tidal and lighting schedules in the laboratory. *Estuaries* 7:139–48 (1984).
897 Kruchinin, O.N., Kuznetsov, Y.U.A. & Sorokin, M.A., Diurnal rhythm of activity in some far-eastern fish species. *Voprosy Ikhtiologii* 21:134–40 (1981).
898 Toledo, J.D., Marte, C.L. & Castillo, A.R., Spontaneous maturation and spawning of sea bass Lates calcarifer in floating net cages. *Journal of Applied Ichthyology* 7(4):217–22 (1991).
899 Garcia, L.M.B., Lunar synchronization of spawning in sea bass, Lates calcarifer (Bloch): effect of luteinizing hormone-releasing hormone analogue (LHRHa) treatment. *Journal of Fish Biology* 40(3):359–70 (1992).
900 Donaldson, T.J., Pair spawning of Cephalopholis boenack (Serranidae). *Japanese Journal of Ichthyology* 35(4):497–500 (1989).
901 Colin, P.L., Shapiro, D.Y. & Weiler, D., Aspects of the reproduction of two groupers, Epinephelus guttatus and Epinephelus striatus in the West Indies. *Bulletin of Marine Science of the Gulf and Caribbean* 40(2):220–30 (1987).
902 Johannes, R.E., Reproductive strategies of coastal marin fishes in the tropics. *Environmental Biology of Fishes* 3:65 -84 (1978).
903 Randall, J.E. & Brock, V.E., Observations on the ecology of epinepheline and lutjanid fishes of the Society Islands, with emphasis on food habits. *Transactions of the American Fisheries Society* 89:9–16 (1960).

904 Shapiro, D.Y., Sadovy, Y. & McGehee, M.A., Size, composition, and spatial structure of the annual spawning aggregation of the red hind, Epinephelus (Pisces: Serranidae). *Copeia* 2:399–406 (1993).
905 Keener, P. et al., Ingress of postlarval gag Mycteroperca microlepis, Pisces, Serranidae, through a South Carolina, USA, barrier island inlet. *Bulletin of Marine Science of the Gulf and Caribbean* 42(3):376–96 (1988).
906 Doherty, P.J. et al., Monitoring the replenishment of coral trout (Pisces: Serranidae) populations. *Bulletin of Marine Science of the Gulf and Caribbean* 54(1):343–55 (1994).
907 McFarland, W.N. et al., Recruitment patterns in young french grunts Haemulon flavolineatum (Family Haemuldiae), at St Croix, Virgin Islands, USA. US National Marine Fisheries Service. *Fishery Bulletin* 83(3):413–26 (1985).
908 Worthmann, H.O., A comparative study of the growth of the postlarval and juvenile pescadas Plagioscion squamoissimus and Plagioscion monti in a white water lake of the Central Amazon Brazil. *Amazoniana* 7(4):465–77 (1983).
909 Lobel, P.S., Diel, lunar, and seasonal periodicity in the reproductive behavior of the pomacanthid fish, Centropyge potteri, and some other reef fishes in Hawaii. *Pacific Science* (Honolulu) 32(2):193–207 (1978).
910 Thresher, R.E., Courtship and spawning in the emperor angelfish Pomacanthus imperator, with comments on reproduction by other pomacanthid fishes. *Marine Biology* (Berlin) 70(2):149–56 (1982).
911 Robertson, D.R., Patterns of lunar settlement and early recruitment in Caribbean reef fishes at Panama. *Marine Biology* (Berlin) 114(4):527–37 (1992).
912 Lasker, H.R., Prey preferences and browsing pressure of the butterflyfish Chaetodon capistratus on caribbean gorgonians. *Marine Ecology Progress Series* 21(3):213–20 (1985).
913 Rossiter, A., Lunar spawning synchroneity in a freshwater fish. *Naturwissenschaften* 78(4):182–84 (1991).
914 Schwanck, E., Lunar periodicity in the spawning of Tilapia mariae in the Ethiop River, Nigeria. *Journal of Fish Biology* 30(5):533–38 (1987).
915 Nakai, K. et al., Lunar synchronization of spawning in cichlid fishes of the tribe Lamprologini in Lake Tanganyika. *Journal of Fish Biology* 37(4):589–98 (1990).
916 Foster, S.A., Diel and lunar patterns of reproduction in the caribbean and pacific sergeant major damselfishes Abudefduf saxatilis and Abudefduf troschelii. *Marine Biology* (Berlin) 95(3):333–44(1987).
917 Fricke, H.W., Öko-Ethologie des monogamen Anemonenfisches Amphiprion bicinctus (Freiwasseruntersuchung aus dem Roten Meer). *Zeitschrift für Tierpsychologie* 36:429–512 (1978).
918 Ochi, H., Temporal patterns of breeding and larval settlement in a temperate population of the tropical anemonefish, Amphiprion clarkii. *Japanese Journal of Ichthyology* 32(2):248–57 (1985).
919 Ross, R.M., Reproductive behavior of the Anemonefish Amphiprion melanopus on Guam. *Copeia* 1:103–7 (1978).
920 Booth, D.J. & Beretta, G.A., Seasonal recruitment, habitat associations, and survival of pomacentrid reef fish in the US Virgin Islands. *Coral Reefs* 13(2):81–89 (1994).
921 Williams, A.H., Ecology of three-spotted damselfish social organization, age structure, and population stability. *Journal of Experimental Marine Biology and Ecology* 34(3):197–214 (1978).
922 Doherty, P.J., Diel, lunar, and seasonal rhythms in the reproduction of 2 tropical damselfishes Pomacentrus flavicauda and Pomacentrus wardi. *Marine Biology* (Berlin) 75(2–3):215–24 (1983).

923 Moyer, J.T., Reproductive behavior of the damselfish Pomacentrus nagasakiensis at Miyake-Jima, Japan. *Japanese Journal of Ichthyology* 22(3):151–63 (1975).

924 Petersen, C.W., The occurrence and dynamics of clutch loss and filial cannibalism in two caribbean damselfishes. *Journal of Experimental Marine Biology and Ecology* 135(2):117–34 (1990).

925 Petersen, C.W. & Hess, H.C., The adaptive significance of spawning synchronization in the caribbean damselfish Stegastes dorsopunicans Poey. *Journal of Experimental Marine Biology and Ecology* 151(2):155–68 (1991).

926 Robertson, D.R., Green, D.G. & Victor, B.C., Relations in the lunar periodicities of larval production and settlement by a caribbean reef fish. Second International Symposium on Indo-Pacific Marine Biology of the Western Society of Naturalists. *Bulletin of Marine Science of the Gulf and Caribbean* 41(2):641 (1987).

927 Robertson, D.R., Green, D.G. & Victor, B.C., Temporal coupling of production and recruitment of larvae of a caribbean reef fish. *Ecology* 69(2):370–81 (1988).

928 Robertson, D.R., Petersen, C.W. & Brawn, J.D., Lunar reproductive cycles of benthic-brooding reef fishes: reflections of larval biology or adult biology? *Ecological Monographs* 60(3):311–30 (1990).

929 Ahmed, S. & Sheshappa, D.S., Preliminary observations on the lunar and tidal influences on the catches on mullets (Family: Mugilidae) by Gillnet in the Estuary of Mangalore, Karnataka. *Fishery Technology* 29(2):95–98 (1992).

930 Hunt von Herbing, I. & Hunte, W., Spawning and recruitment of the bluehead wrasse Thalassoma bifasciatum in Barbados, West Indies. *Marine Ecology Progress Series* 72(1–2):49–58 (1991).

931 Victor, B.C., Larval settlement and juvenile mortality in a recruitment-limited coral reef fish population. *Ecological Monographs* 56:145–60 (1986).

932 Ross, R.M., Annual, semilunar, and diel reproductive rhythms in the Hawaiian, USA, labrid Thalassoma duperrey. *Marine Biology* 72(3):311–18 (1983).

933 Hoffman, K.S. & Grau, E.G., Daytime changes in oocyte development with relation to the tide for the Hawaiian saddleback wrasse, Thalassoma duperrey. *Journal of Fish Biology* 34(4):529–46 (1989).

934 Gibson, R.N., Rhythmic activity in littoral fish. *Nature* 207:544–45 (1965).

935 Gibson, R.N., Experiments on the tidal rhythm of Blennius pholis. *Journal of the Marine Biological Association of the United Kingdom* (Plymouth) 47:97–111 (1967).

936 Gibson, R.N., Factors affecting the rhythmic activity of Blennius pholis L. (Teleostei). *Animal Behaviour* 19:336–43 (1971).

937 Horn, M.H. & Gibson, R.N., *Fische der Gezeitenzone, in: Biologie der Meere. Mit einer Einführung von Gotthilf Hempel*, pp.132–38. Verständliche Forschung. Spektrum (Akademischer) 1991.

938 Morgan, E. & Cordiner, S., Entrainment of a circatidal rhythm in the rock-pool blenny Lipophrys pholis by simulated wave action. *Animal Behaviour* 47(3):663–69 (1994).

939 Northcott, S.J., Gibson, R.N. & Morgan, E., The presistence and modulation of endogenous circatidal rhythmicity in Lipophrys pholis, Teleostei. *Journal of the Marine Biological Association of the United Kingdom* 70(4):815–28 (1990).

940 Gibson, R.N., The tidal rhythm of activity of Coryphoblennius galerita (L.) (Teleostei: Blenniidae). *Animal Behaviour* 18:539–43 (1970).

941 Marraro, Ch. & Nursall, J.R., The reproductive periodicity and behaviour of Ophioblennius atlanticus, Pisces, Blenniidae, at Barbados. *Canadian Journal of Zoology* 61(2):317–25 (1983).

942 Dufour, V. & Galzin, R., Colonization patterns of reef fish larvae to the lagoon at Moorea Island, French Polynesia. *Marine Ecology Progress Series* 102:143–52 (1993).
943 Sawara, Y. & Azuma, N., Tidal rhythm and predator-prey relationship in estuarine fishes. *Ergebnisse der Limnologie* 35:145–59 (1992).
944 Sawara, Y., Differences in the activity rhythms of juvenile gobiid fish, Chasmichthys gulosus, from different tidal localities. *Japanese Journal of Ichthyology* 39(3):201–9 (1992).
945 Sponaugle, S. & Cowen, R.K., Larval durations and recruitment patterns of two Caribbean gobies (Gobiidae): contrasting early life histories in demersal spawners. *Marine Biology* (Berlin) 120:133–43 (1994).
946 Thorrold, S.R. et al., Temporal patterns in the larval supply of summer-recruiting reef fishes to Lee Stocking Island, Bahamas. *Marine Ecology Progress Series* 112:75–86 (1994).
947 Voss, J., Rhythmik verschiedener Verhaltensweisen eines Thunmakrelenbestandes, untersucht am Beispiel der Fischerei vor NW-Afrika. Wissenschaftliche Zeitschrift der Humboldt-Universität zu Berlin, *Mathematisch-Naturwissenschaftliche Reihe* 26(4):417–19 (1977).
948 Robertson, D.R., The spawning behavior and spawning cycles of surgeonfishes (Acanthuridae) from the Indo Pacific. *Environmental Biology of Fishes* 9(3–4):193–224 (1983).
949 Grzimek, B., Doktorfische. Hochzeit bei Vollmond. *Grzimeks Tierleben*, Vol.5: 207f.
950 Lichatowich, T. et al., The spawning cycle fry appearance and mass collection techniques for fry of Siganus rivulatus in the Red Sea. *Aquaculture* 40(3): 269–72 (1984).
951 Render, J.H. & Allen, R.L., The relationship between lunar phase and Gulf Butterfish, Peprilus burti, catch rate. *US National Marine Fisheries Service. Fishery Bulletin* 85(4):817–19 (1987).
952 Rountree, R.A. & Able, K.W., Foraging habits, growth, and temporal patterns of saltmarsh creek habitat use by young-of-year summer flounder in New Jersey. *Transactions of the American Fisheries Society* 121:765–76 (1992).
953 Szedlmayer, S.T. & Able, K.W., Ultrasonic telemetry of age-0 summer flounder, Paralichthys dentatus, movements in a Southern New Jersey Estuary. *Copeia* 3:728–36 (1993).
954 Campana, S.E., Lunar cycles of otolith growth in the juvenile starry flounder Platichthys stellatus. *Marine Biology* (Berlin) 80(3):239–46 (1984).
955 *Neue Züricher Zeitung* of June 1, 1994: 'Forschung und Technik': 'Migration der Scholle in der Gezeitenströmung".
956 Metcalfe, J.D., Arnold, G.P. & Webb, P.W., The energetics of migration by selective tidal stream transport: an analysis for plaice tracked in the southern North Sea. *Journal of the Marine Biology Association of the United Kingdom* 70: 149–62 (1990).
957 Arnold, G.P. et al., Movements of cod (Gadus morhua L.) in relation to the tidal streams in the southern North Sea. *ICES Journal of Marine Science* 51:207–32 (1994).
958 Burrows, M.T. et al., Temporal patterns of movement in juvenile flatfishes and their predators: underwater television observations. *Journal of Experimental Marine Biology and Ecology* 177(2):251–68 (1994).

REFERENCES

959 Gibson, R.N., Comparative studies on the rhythms of juvenile flatfish, in: DeCoursey, P.J. (ed.), *The Belle W. Baruch Library in Marine Sciences*, No.4: *Biological rhythms in the marine environment*, pp.199–213. Papers presented at a symposium, held April 24–27, 1974, at the Belle W. Baruch Coastal Research Center, Georgetown, SC. University of South Carolina Press 1976.

960 Gibson, R.N., Blaxter, J.H.S. & De Groot, S.J., Developmental changes in the activity rhythms of the plaice (Pleuronectes platessa L.), in: Thorpe, J.E. (ed.), *Rhythmic activity of fishes*, pp.169–86. London (Academic Press) 1978.

961 Otubusin, S.O., Effects of lunar periods and some other parameters on fish catch in lake Kainji Nigeria. *Fisheries Research* (Amsterdam) 8(3):233–46 (1990).

962 Milton, D.A. & Blaber S.J.M., Maturation, spawning, seasonality, and proximate spawning stimuli of six species of Tuna baitfish in the Solomon Islands. *National Marine Fisheries Service. Fishery Bulletin* 89(2):221–37 (1991).

963 Yamahira, K., Combined effects of tidal and diurnal cycles on spawning of the puffer, Takijugu niphobles (Tetraodontidae). *Environmental Biology of Fishes* 40:255–61 (1994).

964 Brown, F.A.jr. et al., Evidence for an exogenous contribution to presistent diurnal and lunar rhythmicity under so-called constant conditions. *Biological Bulletin of the Marine Biological Laboratory Woods Hole* 109:238–54 (1955).

965 Church, G., The effects of seasonal and lunar changes on the breeding pattern of the edible Javanese frog, Rana cancrivora Gravenhorst. *Treubia* (Batavia/Jakarta, Indonesia) 25:215–33 (1960).

966 Robertson, D.R., Diurnal and lunar periodicity of intestinal calcium transport and plasma calcium in the frog Rana pipiens. *Comparative Biochemistry and Physiology A* 54(2):225–31 (1976).

967 Robertson, D.R., The light dark cycle and a nonlinear analysis of lunar perturbations and barometric pressure associated with the annual locomotor activity of the frog Rana pipiens. *Biological Bulletin of the Marine Biological Laboratory Woods Hole* 154(2):302–21 (1978).

968 Fitzgerald, G.J. & Bider, J.R., Influence of moon phase and weather factors on locomotor activity in Bufo americanus. *Oikos* 25(3):338–40 (1974).

969 Church, G., Annual and lunar periodicity in the sexual cycle of the Javanese Toad, Bufo melanostictus Schneider. *Zoologica* 45:181–88 (1960).

970 Church, G., Seasonal and lunar variation in the numbers of mating toads in Bandung (Java). *Herpetologica* (Chicago) 17(2):122–26 (1961).

971 Ferguson, D., Observations on movements and behaviour of Bufo fowleri in residential areas. *Herpetologica* (Chicago)16:112–14 (1960).

972 Church, G. et al., The reproductive cycle of the Javanese toad, Bufo melanostictus. *Proceedings of the First Indonesian Science Congress, Section C, Biology, Paris*, I-IV:5–37 (1958).

973 Brett, W.J., Persistent rhythms of locomotion activity in the turtle Pseudemys scripta. *Comparative Biochemistry and Physiology A* 40(4):925–34 (1971).

974 Frankenberg, E. & Werner, Y.L., Effect of lunar cycle on daily activity rhythm in a gekkonid lizard, Ptyodactylus. *Israel Journal of Zoology* 28(4):224–28 (1979).

975 Verheijen, F.J., Bird kills at lighted man made structures not on nights close to a full moon. *American Birds* 35(3):251–54 (1981).

976 Alonso, J.A., Alonso, J.C. & Veiga, J.P., The influence of moonlight on the timing of roosting flights in common cranes Grus grus. *Ornis Scandinavica* 16(4):314–18 (1985).

977 Milson, T.P., Diurnal behavior of lapwings Vanellus vanellus in relation to moon phase during winter. *Bird Study* 31(2):117–20 (1984).

978 Milson, T.P., Rochard, J.B.A. & Poole, S.J., Activity patterns of lapwings Vanellus vanellus in relation to the lunar cycle. *Ornis Scandinavica* 21(2):147–56 (1990).

979 Galbraith, H., Arrival and habitat use by lapwings Vanellus vanellus in the early breeding season. *Ibis* 131(3):377–88 (1989).

980 Thibault, M. & McNeil, R., Day/night variation in habitat use by Wilson's Plovers in Northeastern Venezuela. *Wilson Bulletin* 106(2):299–10 (1994).

981 Chapin, J.P., The calendar of Wideawake Fair, Sterna fuscata. *Auk* 71(1):1–15 (1954).

982 Chapin, J.P. & Wing, L.W., The Wideawake calendar 1953 to 1958. Sterna fuscata. *Auk* 76(2):153–58 (1959).

983 Kipp, F.A., Mondperiodischer Fortpflanzungszyklus der Ruß-Seeschwalbe Sterna fuscata. *Naturwissenschaftliche Rundschau* 10:400 (1960).

984 Enright, J.T., Relationships between circannual rhythms and endogenous lunar and tidal rhythms. *International Journal of Chronobiology* 2(2):118f (1974).

985 Remmert, H., Biologische Periodik. *Handbuch der Biologie* 5:341–411. Wiesbaden (Akademische Verlagsgesellschaft Athenaion) 1977.

986 Smith, G.C., Factors influencing egg laying and feeding in black-naped terns Sterna sumatrana. *Emu* 90(2):88–96 (1990).

987 Nelson, D.A., Gull predation on Cassin's auklet varies with the lunar cycle. *Auk* 106(3):495–97 (1989).

988 Larkin, T. & Keeton, W.T., An apparent lunar rhythm in the day-to-day variations in initial bearings of homing pigeons, in: Schmidt-König, K. & Keeton, T.W. (eds.), *Animal Migration, Navigation, and Homing*, pp.92–106. Berlin (Springer) 1978.

989 Morrell, T.E., Yahner, R.H. & Harkness, W.L., Factors affecting detection of great horned owls by using broadcast vocalizations. *Wildlife Society, Bulletin* 19(4):481–88 (1991).

990 Ganey, J.L., Calling behavior of spotted owls in Northern Arizona, USA. *Condor* 92(2):485–90 (1990).

991 Wynne-Edwards, V.C., On the waking time of the nightjar (Caprimulgus e. europaeus). *Journal of Experimental Biology* (Edinburgh, Cambridge) 7:241–47 (1930).

992 Mills, A.M., The influence of moonlight on the behavior of goatsuckers (Caprimulgidae). *Auk* 103(2):370–78 (1986).

993 Brigham, R.M. & Barclay, R.M.R., Lunar influence on foraging and nesting activity of common poorwills Phalaenoptilus nuttallii. *Auk* 109(2):315–20 (1992).

994 Grzimek, B., Eierlegende Säugetiere. *Grzimeks Tierleben* 10:43 (1967).

995 Sadleir, R.M.F.S. & Tyndale-Biscone, Ch., Photoperiod and the termination of embryonic diapause in the marsupial Macropus eugenii. *Biology of Reproduction* 16(5):605–8 (1977).

996 Erkert, H.G., Der Einfluß des Mondlichtes auf die Aktivitätsperiodik nachtaktiver Säugetiere. *Oecologia* 14(3):269–87 (1974).

997 Erkert, H.G., Beleuchtungsabhängiges Aktivitätsoptimum bei Nachtaffen (Aotus trivirgatus). *Folia Primatologica* 25:186–92 (1976).

998 Erkert, H.G., Lunar periodic variation of the phase angle difference in nocturnal animals under natural Zeitgeber conditions near the equator. *International Journal of Chronobiology* 4(2):125–38 (1976).

999 Erkert, H.G. & Groeber, J., Direct modulation of activity and body temperature of owl monkeys (Aotus lemurinus griseimembra) by low light intensities. *Folia Primatologica* 47(4):171–88 (1986).

1000 Morrison, D.W., Lunar phobia in a Neotropical fruit bat, Artibeus jamaicensis (Chiroptera: Phyllostomidae). *Animal Behaviour* 26:852–55 (1978).
1001 Reith, C.C., Insectivorous bats fly in shadows to avoid moonlight. *Journal of Mammalogy* 63:685–88 (1982).
1002 Daly, M. et al., Behavioural modulation of predation risk, moonlight avoidance, and crepuscular compensation in a nocturnal desert rodent, Dipodomys merriami. *Animal Behaviour* 44(1):1–9 (1992).
1003 Lockard, R.B. & Owings, D.H., Moon related surface activity of bannertail Dipodomys spectabilis and fresno Dipodomys nitratoides, kangaroo rats. *Animal Behaviour* 22(1):262–73 (1974).
1004 Lockard, R.B., Experimental inhibition of activity of kangaroo rats in the natural habitat by an artificial moon. *Journal of Comparative and Physiological Psychology* 89(3):263–66 (1975).
1005 Klinowska, M., Lunar rhythms in activity urinary volume and acidity in the golden hamster Mesocricetus auratus. *Journal of Interdisciplinary Cycle Research* 1(4):317–22 (1970).
1006 Klinowska, M., A comparison of the lunar and solar activity rhythms of the golden hamster Mesocricetus auratus. *Journal of Interdisciplinary Cycle Research* 3(2):145–50 (1972).
1007 Brown, F.A.jr. & Park, Y.H., Synodic monthly modulation of the diurnal rhythm of hamsters. *Proceedings of the Society for Experimental Biology and Medicine* 125:712–15 (1967).
1008 Jahoda, J.C., The effect of the lunar cycle on the activity pattern of Onychomys leucogaster breviauritus. *Journal of Mammology* 54(2):544–49 (1973).
1009 Price, M.V., Waser, N.M. & Bass, T.A., Effects of moon light in micro habitat use by desert rodents. *Journal of Mammology* 65(2):353–56 (1984).
1010 Stutz, A.M., Lunar day variations in spontaneous activity of the mongolian gerbil. *Biological Bulletin of the Marine Biological Laboratory Woods Hole* 146(3):415–23 (1974).
1011 Wolton, R.J., The activity of free ranging wood mice Apodemus sylvaticus. *Journal of Animal Ecology* 52(3):781–94 (1983).
1012 Truchan, L.C. & Boyer, S.D., Barometric pressure correlations with spontaneous motor activity of albino mice Mus musculus. *Physiological Zoology* (Chicago) 45(3):204–14 (1972).
1013 Brown, F.A.jr., Shriner, J. & Ralph, C.L., Solar and lunar rhythmicity in the rat in 'constant conditions' and the mechanism of physiological time measurement. *American Journal of Physiology* 184:491–96 (1956).
1014 Harrison, J.L., Breeding rhythms of Selangor rodents. *Bulletin of The Raffles Museum* (Singapore) 24:109–31 (1952).
1015 Harrison, J.L., The moonlight effect on rat breeding. *Bulletin of the Raffles Museum* (Singapore) 25:166–70 (1954).
1016 Harrison, J.L., Moonlight and the pregnancy of Malayan forest rats. *Nature* 173:1002 (1954).
1017 Alkon, P.U. & Saltz, D., Influence of season and moonlight on Indian crested porcupines Hystrix indica. *Journal of Mammalogy* 69(1):71–80 (1988).
1018 Nielsen, S.M., Fishing arctic foxes Alopex lagopus on a rocky island in West Greenland. *Polar Research* 9(2):211–13 (1991).
1019 Trillmich, F. & Mohren, W., Effects of the lunar cycle on the galapagos fur seal, Arctocephalus galapagoensis. *Oecologia* (Berlin) 48(1):85–92 (1981).
1020 Laws, R.M., Age determination of pinnipeds with special reference to growth layers in the teeth. *Zeitschrift für Säugetierkunde* 27:129–46 (1962).

1021 Rombeck, A.M., *Studien zur Ontogenie des Seehundes (Phoca vitulina L. 1758). Altersmerkmale von Seehunden unter besonderer Berücksichtigung der Jungtiere.* Thesis for first state exam, December 1988, Universität Dortmund.
1022 Laws, R.M., A new method of age determination for mammals. *Nature* 169:972 (1952).
1023 Watts, P., Possible lunar influence on hauling-out behavior by the Pacific harbor seal (Phoca vitulina Richardsi). *Marine Mammal Science* 9(1):68–76 (1993).
1024 Elton, C. & Nicholson, M., The ten-year cycle in numbers of the lynx in Canada. *Journal of Animal Ecology* 11(2):215–44 (1942).
1025 Murray, M.G., The rut of impala Aepyceros melampus. Aspects of seasonal mating under tropical conditions. *Zeitschrift für Tierpsychologie* 59(4):319–37 (1982).
1026 Roy, J.H.B. et al., Effect of season of the year and phase of the moon on puberty and on the occurrence of estrus and conception in dairy heifers reared on high planes of nutrition. *Animal Production* 31(1):13–26 (1980).
1027 Subramaniam, A. et al., Effect of lunar phases on variability of inseminations in cattle. Australian Veterinary Journal 68(2):71–72 (1991).
1028 Ramanathon, O., Light and sexual periodicity in Indian buffaloes. *Nature* 130:169–70 (1932).
1029 Eschenlohr, S., *Stellare, lunare und biologische Einflüsse auf Reproduktionsmerkmale bei Ziegen.* Gesamthochschule Kassel-Universität, Fachbereich: Landwirtschaft, internationale Agrarentwicklung und ökologische Umweltsicherung. Diplomarbeit, Witzenhausen 1995.
1030 Sinclair, A.R.E., Lunar cycle and timing of mating season in Serengeti Wildebeest. *Nature* 267(5614):832f (1977).
1031 Habermehl, K.-H., *Altersbestimmung bei Wild- und Pelztieren.* Hamburg (Paul Parey) 1985.
1032 Türcke, F. & Tomiczek, H., *Das Muffelwild. Naturgeschichte, Ökologie, Hege, Jagd.* Hamburg (Parey) 1982.
1033 Sands, J.M. & Miller, L.E., Effects of moon phase and other temporal variables on absenteeism. *Psychological Reports* 69(3 Part 1):959–62 (1991).
1034 Hicks-Caskey, W.E. & Potter, D.R., Effect of the full moon on a sample of developmentally delayed institutionalized women. *Perceptual and Motor Skills* 72(3 Part 2):1375–80 (1991).
1035 Hicks-Caskey, W.E. & Potter, D.R., Weekends and holidays and acting-out behaviour of developmentally delayed women: a reply to Dr. Mark Flynn. *Perceptual and Motor Skills* 74:344–46 (1992).
1036 Hejl, Z., Daily, lunar, yearly, and menstrual cycles and bacterial or viral infections in man. *Journal of Interdisciplinary Cycle Research* 8(3–4):250–53 (1977).
1037 Criss, T.B. & Marcum, J.P., A lunar effect on fertility. *Social Biology* 28(1–2):75–80 (1981).
1038 Gautherie, M. & Gros, Ch., Circadian rhythm alteration of skin temperature in breast cancer. *Chronobiologica* 4:1–17 (1977).
1039 Laue, H.B. von, Zeitphänomene der Krebskrankheit. *Deutsche Zeitschrift für Onkologie* 3:64–73 (1991).
1040 Simpson, H.W. & Griffiths, K., The diagnosis of Breast-Pre-Cancer by chonobra. *Chronobiology International* 6:355–93 (1989).
1041 Weckenmann, M., Stegmaier, J. & Rauch, E., On the spectrum of the reactive periods in patients with malignoma. Abstracts of the 7th Meeting of the European Society for Chronobiology, Marburg, Germany, May 30 – June 2, 1991. *Journal of Interdisciplinary Cycle Research* 22(2):202 (1991).

1042 Benvenuti, M. et al., Significant components in the circadian band infradian domain of serum iron and ferritin. *Chronobiologia* 10:390 (1983).
1043 Cutler, W.B. et al., Lunar influences on the reproductive cycle in women. *Human Biology* 59(6):959–72 (1987).
1044 Cutler, W.B. et al., Lunar influences on the reproductive cycle in women. 17th International Congress of the International Society of Psychoneuroendocrinology, Chapel Hill-Durham, North Carolina, June 28 – July 3, 1987. *Neuroendocrinology Letters* 9(3):200 (1987).
1045 Heckert, H., *Mondperiodische Schwankungen in der Sterberate einer Großstadtbevölkerung und Bemerkungen zu Problemen bei der Untersuchung synodischlunarer Einflüsse.* Dissertation, Hamburg 1956.
1046 Ossenkopp, K.P. & Ossenkopp, M.D., Self inflicted injuries and the lunar cycle. A preliminary report. *Journal of Interdisciplinary Cycle Research* 4(4):337–48 (1973).
1047 Buckley, N.A., Whyte, I.M. & Dawson, A.H., There are days... and moons. Self-poisoning is not lunacy. *The Medical Journal of Australia* 159(11–12):786–89 (1993).
1048 Alonso, Y., Geophysical variables and behaviour. LXXII: Barometric pressure, lunar cycle, and traffic accidents. *Perceptual and Motor Skills* 77(2):371–76 (1993).
1049 Kelly, I.W., Laverty, W.H. & Saklofske, D.H., Geophysical variables and behavior. LXIV: An empirical investigation of the relationship between worldwide automobile traffic disasters and lunar cycles: no relationship. *Psychological Reports* 67:987–94 (1990).
1050 Laverty, W.H. et al., Geophysical variables and behavior. LXVIII: Distal and lunar variables and traffic accidents in Saskatchewan 1984 to 1989. *Perceptual and Motor Skills* 74(2):483–88 (1992).
1051 Templer, D.I., Veleber, D.M. & Brooner, R.K., Geophysical variables and behavior, 6: Lunar phase and accident injuries: a difference between night and day. *Perceptual and Motor Skills* 55(1):280–82 (1982).
1052 Templer, D.I., Brooner, R.K. & Corgiat, M.D., Geophysical variables and behavior, 14: Lunar phase and crime: fact or artifact. *Perceptual and Motor Skills* 57(3 Part 1):993f (1983).
1053 DeCoursey, P.J. (ed.), *The Belle W. Baruch Library in Marine Sciences,* No.4. *Biological rhythms in the marine environment.* Papers presented at a symposium, held April 24–27, 1974, at the Belle W. Baruch Coastal Research Center, Georgetown, SC. University of South Carolina Press 1976.
1054 Enright, J.T., Orientation in time: endogenous clocks, in: Kinne, O. (ed.), *Marine Ecology* II, pp.917–45. London (Wiley) 1975.
1055 McDowall, R.M., Lunar rhythmus in aquatic animals, A general review. *Tuatara* 17(3):133–44 (1970).
1056 Mletzko, H.G. & Mletzko, J., *Biorhythmik – Elementareinführung in die Chronobiologie.* Wittenberg-Lutherstadt (Die Neue Brehm-Bücherei No.507) 1977.
1057 Moore-Ede, M.C., Sulzmann, F.M. & Fuller, Ch.A., *The Clocks that Time Us. Physiology of the circadian timing system.* Cambridge (Harvard University Press) 1982.
1058 Naylor, E. & Hartnoll, R.G. (eds.), *Cyclic phenomena in marine plants and animals.* Proceedings of the 13th Marine Biology Symposium, Isle of Man, 27 September – 4 October 1978. Oxford (Pergamon Press) 1979.
1059 Palmer, J.D., Tidal rhythms: the clock control of the rhythmic physiology of marine organisms. *Biological Reviews* 48:377–418 (1973).

1060 Palmer, J.D., *Biological clocks in marine organisms: the control of physiological and behavioral tidal rhythms.* London (Wiley) 1974.
1061 Palmer, J.D., Brown, F.A.jr. & Edmunds, L.N.jr., *An introduction to biological rhythms.* New York (Academic Press) 1976.
1062 Saunders, D.S., *An introduction to biological rhythms.* Glasgow (Blackies) 1977.
1063 Tomassen, G.J.M. et al. (eds.), *Geo-cosmic relations; the earth and its macroenvironment.* Proceedings of the First International Congress on Geo-cosmic Relations, organized by the Foundation for Study and Research of Environmental Factors (S.R.E.F.), Amsterdam, April 19–22, 1989. Wageningen (Pudoc) 1990.
1064 El-Saadany, G. & Abd-El-Fattah, M.L., Contributions to the ecological studies on the cotton pests in Egypt, Part 3: The effect of lunar phases on the nocturnal activity of certain lepidoptera. *Zeitschrift für Angewandte Entomologie* 79(1): 17–20 (1975).
1065 Nemec, S.J., Effect of lunar phases on light trap collections and populations of bollworm moths. *Journal of Economic Entomology* 64(4):860–64 (1971).
1066 Bowden, J., The significance of moon light in photoperiodic responses of insects. *Bulletin of Entomological Research* 62(4):605–12 (1973).
1067 Banerjee, T.C., Haque, N. & Mahapatra, A.K., Influence of rice crop climate and lunar cycles on the population patterns of Scirpophage nivella. *Insect Science and its Application* 7(5):593–98 (1986).
1068 Shrivastava, S.K., Shukla, B.C. & Shastri, A.S.R.A.S., Effects of lunar cycle on light trap catches of Spodoptera litura Fabricius. *Indian Journal of Agricultural Science* 57(2):117–19 (1987).
1069 Sternlicht, M., Effects of moon phases and of various trap colors on the capture of Prays citri males in sticky traps. *Phytoparasitica* 2(1):35–40 (1974).
1070 Danthanarayana, W., Diel and lunar flight periodicities in the light brown apple moth Epiphyas postvittana, Tortricidae, and their possible adaptive significance. *Australian Journal of Zoology* 24(1):65–73 (1976).
1071 Goodwin, S. & Danthanarayana, W., Flight activity of Plutella xylostella, Lepidoptera, Yponomeutidae. *Journal. Australian Entomological Society* 23(3): 235–40 (1984).
1072 Kavaliers, M. & Macvean, C., Effect of temperature and lunar phase on the phototactic responses of larvae of the wax moth Galleria mellonella, Lepidoptera, Pyralidae. *Entomologia Experimentalis et Applicata* 28(2):224–28 (1980).
1073 Fechter, H., Seelilien und Haarsterne, in: *Grzimeks Tierleben* Vol.3, p.297, Zürich 1970.
1074 Berghahn, R., Effects of tidal migration on the growth of 0-group-plaice (Pleuronectes platessa L.) in the North Frisian Wadden Sea. *Berichte der Deutschen Wissenschaftlichen Kommission für Meeresforschung* 31:209–26 (1987).
1075 Gibson, R.N., Tidal and circadian activity rhythms in juvenile plaice, Pleuronectes platessa. *Marine Biology* (Berlin) 22:379–86 (1973).
1076 Pilgrim, U. et al., Normal values of immunoglobulins in premature and in fullterm infants, calculated as percentiles. *Helvetica Paediatrica Acta* 30:121–34 (1975).
1077 Russel-Hunter, W.D., Apley, M.L. & Hunter, R.D., Early life history of Melampus and the significance of semilunar synchrony. *Biological Bulletin of the Marine Biological Laboratory Woods Hole* 143(3):623–56 (1972).
1078 Müller, A.H., *Lehrbuch der Paläozoologie,* Vol.2: *Invertebraten,* Part 1: Protozoa – Mollusca 1, pp.645ff; Anhang: Über rhythmische Wachstumsvorgänge bei Muscheln und anderen hartteilbildenden Invertebraten. Stuttgart (Fischer) 1993.

1079 Sinclair, A.R.E. et al., Can the solar cycle and climate synchronize the snowshoe hare cycle in Canada? Evidence from tree rings and ice cores. *The American Naturalist* 141:173–98 (1993).
1080 Stenseth, N.C. & Ims, R.A., Population dynamics of lemmings: temporal and spatial variation – an introduction, in: Stenseth, N.C. & Ims, R.A. (eds.), *The biology of Lemmings.* New York (Academic Press) 1993.
1081 Frank, St., *Das große Bilderlexikon der Fische.* Munich (Bertelsmann) 1975.
1082 Dircksen, R., *Das kleine Vogelbuch.* Gütersloh (Bertelsmann) 1952.
1083 Eisenhuth, A. (ed.), *Das Weltall im Bild.* Köln (Styria) 1971.
1084 Krümmel, O., *Handbuch der Ozeanographie,* Vol.2. Stuttgart 1911.
1085 Riedl, R., *Fauna und Flora des Mittelmeeres.* Hamburg (Parey) 1983.
1086 *Knaurs Pflanzenreich in Farben,* Vol.3: *Niedere Pflanzen.* Munich (Droemer, Knaur) 1967.
1087 Gutmann, W.F., *Meerestiere am Strand in Farben.* Ravensburg (Otto Maier) 1967.
1088 Haeckel, E., *Art Forms in Nature (100 Plates).* New York (Dover) 1974.
1089 De Haas, W. & Knorr, F., *Was lebt im Meer?* Stuttgart (Kosmos) 1971.
1090 Kaestner, A., *Lehrbuch der Speziellen Zoologie,* Vol.1 (Wirbellose): 3. Teil, B. Stuttgart (Fischer) 1973.
1091 Currie, R.G. & O'Brien, D.P., Deterministic signals in precipitation records from the American corn belt. *International Journal of Climatology* 10:179–89 (1990).
1092 Currie, R.G. & O'Brien, D.P., Periodic 18,6-year and cyclic 10 to 11 year signals in northeastern United States precipitation data. *International Journal of Climatology* 8:255–81 (1988).
1093 Chapman, F.R.S., The lunar tide in the atmosphere. *The Meteorological Magazine* 74:273–81 (1939).
1094 Adderley, E.E. & Bowen, E.G., Lunar component in precipitation data. *Science* 137:749f (1962).
1095 Bradley, D.A., Woddbury, M.A. & Brier, G.W., Lunar synodical period and widespread precipitation. *Science* 137:748f (1962).
1096 Bigg, E.K., A lunar influence on ice nucleus concentrations. *Nature* 197:172f (1963).
1097 Brier, G.W. & Bradley, D.A., The lunar synodical period and precipitation in the United States. *Journal of the Atmospheric Sciences* 21:386–95 (1964).
1098 Haurwitz, B., Atmospheric tides. *Science* 144:1415–22 (1964).
1099 Lund, I.A., Indications of a lunar synodical period in United States observations of sunshine. Journal of the Atmospheric Sciences 22:24–39 (1965).
1100 Lethbridge, M.D., Relationship between thunderstorm frequency and lunar phase and declination. *Journal of Geophysical Research* 75 (27):5149–54 (1976).
1101 Hanson, K., Maul, G.A. & McLeish, W., Precipitation and the lunar synodic cycle: phase progression across the United States. *Journal of Climate and Applied Meteorology* 26:1358–62 (1987).

Index of authors

The numbers refer to the bibliographic reference list, p.219

Abd-El-Fattah, M.L. 1064
Abe, N. 380, 381, 444, 445, 448, 449, 451
Abel, K.M. 331
Able, K.W. 952, 953
Abrami, G. 7
Adderley, E.E. 1094
Adhub, A.H.Y. Al- 660
Ahmed, S. 929
Aiyar, R. G. 527
Akselman, R. 298
Al-Adhub, A.H.Y. 660
Albertis, L.M. d 579
Aldrich, J.C. 665
Alheit, J. 617
Alkon, P.U. 1017
Alldredge, A.L. 587
Allen, R.L. 951
Allen, W.E. 326
Alonso, J.A. 976
Alonso, J.C. 976
Alonso, Y. 1048
Alpatov, A.M. 118
Altevogt, R. 718
Amelung, W. 266
Amirthalingam, C. 462
Andersson, S. 320
Apfelbaum, M. 209
Apley, M.L. 1077
Archibald, H.L. 12
Aristotle 1, 284
Armonies, W. 478
Arnold, G.P. 956, 957
Arudpragasam, K.D. 669, 670
Aschoff, J. 28, 89, 99, 132, 136, 210, 454, 459, 598
Augustine, Aurelius 285
Aveni, Anthony, F. 117
Azuma, N. 943

Babcock, R.C. 64, 368, 382, 403, 405, 818
Bach, E. 235

Balch, W.M. 299, 300
Balzer, I. 104, 113
Banerjee, T.C. 1067
Barclay, R.M.R. 993
Barnes, A. 764
Barnes, D.J. 401
Barnes, H.S. 385
Barnwell, F.H. 72, 114, 694, 701, 702, 720
Bass, T.A. 1009
Basset, Y. 759
Battle, H.J. 455, 456, 860
Beets, J.L. 231
Bell, L.J. 431
Benayahu, Y. 406
Bennett, F.D. 761
Bennett, M.F. 71, 348, 481, 567, 707, 710, 712
Benson, J.A. 604
Benvenuti, M. 1042
Beretta, G.A. 920
Bergh, O. 183
Berghahn, R. 1074
Berghuis, E.M. 187
Bergin, M.E. 705
Bergquist, P.R. 361
Berry, A.J. 443, 442
Berthold, P. 93
Bhatt, R.M. 766
Bider, J.R. 968
Biedermann, F. von 287
Bigg, E.K. 1096
Bijma, J. 353, 354
Birbaumer, N. 213
Birley, M.H. 769
Birukow, G. 199, 416
Biscone, Ch. Tyndale- 995
Bishop, J.M. 648
Blaber S.J.M. 962
Black, R. 391
Blaxter, J.H.S. 960
Blume, J. 671
Boeuf, G. 841

Bohn, G. 66, 369, 370, 407
Boisserée, S. 287
Boklage, C.E. 250
Booth, D.J. 920
Borgese, M.B. 169
Borst, L.B. 240
Bottjer, D.J. 19
Bouíe, A. 256
Bouíe, J. 256
Boulos, Z. 115
Boulter, D. 189
Bounias, M. 758
Bourget, E. 477
Bowden, J. 737, 751, 1066
Bowen, E.G. 1094
Bowman, M.J. 300
Boyer, S.D. 1012
Bracher, R. 297
Brack, V.jr. 739
Bradford, C.S. 874, 875
Bradley, D.A. 1095, 1097
Brady, J. 123
Braley, R.D. 479
Branch, G.M. 631
Brandt-Reges, I. 273
Braunroth, E. 344
Brawley, S.H. 319
Brawn, J.D. 928
Brett, W.J. 74, 446, 973
Brier, G.W. 1095, 1097
Brigham, R.M. 993
Brock, M.A. 364
Brock, V. 477
Brock, V.E. 903
Brockmann 573
Brooke, M. de L. 693
Brooks, G.W. 222
Brooner, R.K. 1051, 1052
Brosche, P. 54, 55
Brown, C.A. 820
Brown, E.S. 738, 740
Brown, F.A. 641
Brown, F.A.jr. 72, 73, 74, 80, 97, 98, 99, 100, 101, 102, 103, 105, 106, 107, 108, 109, 110, 317, 318, 345, 348, 351, 412, 413, 435, 446, 463, 707, 710, 711, 715, 964, 1007, 1013, 1061
Brown, F.M. 193, 194
Brown, F.R.S. Gilpin- 38
Brown, J.B. Gilpin- 38

Bruce, N.C. 190
Bruce, V.G. 190
Brunner, H. 236, 237
Bryan, J.H. 764
Buckley, N.A. 1047
Buddemeier, R.W. 399, 400
Bühler, G. 260
Bühler, W. 257
Bull, G. 367
Bulmer, M.G. 13
Bünning, E. 90, 146, 191, 671, 672
Burrows, M.T. 958
Burrows, W. 538
Byrnes, G. 230

Cailleux, A. 56
Cameron, A.M. 399
Camões, L. Vaz de 18
Campana, S.E. 954
Campbell, D.E. 231
Carandente, F. 270
Carpenter, G.H. 779
Carson, R.L. 806
Caskey, W.E. Hicks- 1034, 1035
Caspers, H. 33, 36, 539, 780
Castillo, A.R. 898
Catala-Stucki, I. 49
Chadee, D.D. 765, 768
Chakraverty, M. 355
Chandrashekaran, M.K. 664, 753
Chapin, J.P. 981, 982
Chapman, F.R.S. 1093
Charlwood, J.D. 767, 769
Chatterji, A. 684
Chatterton, T.D. 666
Chelazzi, G. 426, 428
Chen, T.H. 171
Chevallier, J.M. 56
Chevrel, R. 781
Chisholm, S.W. 174
Chornesky, E.A. 398
Chou, H.-M. 165
Christie, A.O. 327
Christy, J.H. 685, 704, 709
Church, B.M. 737
Church, G. 965, 969, 970, 972
Cicero, Marcus Tullius 2
Clark, F.N. 888
Clark, L.B. 531, 532
Clayton, M.N. 325
Cloudsley-Thompson, J.L. 248

INDEX OF AUTHORS

Cochran, J.K. 47
Cochran, R.C. 873
Coffroth, M.A. 395, 396, 397
Cohen, J.A. 573
Colin, P.L. 901
Collin, A. 540
Collins, D. 37
Conover, D.O. 893, 894
Conrad, K. 253
Corbet, P.S. 744
Corbet, S.A. 748
Cordiner, S. 938
Corgiat, M.D. 1052
Cornélissen, G. 52, 270, 274
Corney, B.G. 541
Cornuet, J.M. 758
Corpuz, G.C. 424
Costlow, J.D. 683
Coursey, P.J. De 129, 141, 150, 627, 697, 706, 959, 1053
Cowen, R.K. 945
Cox, P.A. 352
Crane, J. 695, 700
Crawshay, L.R. 4
Creese, R.G. 450
Creutzberg, F. 827
Crisp, D.J. 470, 471, 473, 474
Criss, T.B. 1037
Cristini, P. De 321
Cronin, T.W. 634, 636, 637
Crozier, W.Y. 420
Culver, R. 232
Currie, R.G. 1091, 1092
Cutler, W.B. 246, 1043, 1044

Daan S. 134, 135, 136
d Albertis, L.M. 579
Daly, M. 1002
Dantan, J.L. 518, 519
Danthanarayana, W. 1070, 1071
Darwin, Ch. 8
Davies, J.B. 770
Dawson, A.H. 1047
DeCoursey, P.J. 129, 141, 150, 627, 697, 706, 959, 1053
De Cristini, P. 321
De Greff, J.A. 346
De Groot, S.J. 960
De Haas, W. 1089
De Oliveira, D. 741
De Ruyck, A.M.C. 616

De Vries, M.C. 690
De Wilde, P.A.W.J. 187, 663
Defant, A. 21
Della Santina, P. 426, 427, 428
DeLong, E.F. 181
Denton, E.J. 38
Depledge, M.H. 682
Dewa, R.S. 722
Di Michele, L. 879, 880, 882
Dieckhues, B. 211
Dieleman, J. 593
Diels, H. 283
Dietrich, G. 22
Dircksen, R. 1082
Dittel, A.I. 588
Doguzhaeva, L. 46
Doherty, P.J. 906, 922
Doiron, N. 741
Donaldson, T.J. 900
Donn, T.E.jr. 484, 616
Dotan, A. 817
Douglass, J.K. 635
Dowse, H.B. 717
Dowse, H.O. 290
Dresler, A. 196, 197
Drew, E.A. 331
Drzewina, A. 662
Dufour, V. 942
Durchon, M. 507
Duviard, D. 749

Earnest, D.J. 136
Eaton, J.W. 302
Ede, M.C. Moore- 1057
Edmunds, L.N.jr. 188, 1061
Edwards, J.M. 602
Edwards, V.C. Wynne- 991
Eichler, H. 186
Eisenhuth, A. 1083
El-Saadany, G. 1064
Elmhirst, R. 365
Elton, C. 1024
Emata, A.C. 869
Endraß, U. 782
Enright, J.T. 67, 87, 129, 130, 137, 515, 597, 625, 628, 984, 1054
Epifanio, C.E. 588
Erdmann 834
Erez, J. 353
Erkert, H.G. 996, 997, 998, 999
Esaki, T. 773

Eschenlohr, S. 1029
Evans, J.W. 50
Evans, L.V. 327

Fage, L. 488, 487, 504, 505, 516
Farbridge, K.J. 844, 850, 851, 854
Farhang, S. Ross- 569
Farrow, G.E. 469
Fattah, M.L. Abd-El- 1064
Fauré-Fremiet, E. 301, 307, 356, 357
Featonby-Smith, B.C. 316
Fechter, H. 1073
Felder, D.L. 733, 734
Ferguson, D. 971
Fincham, A.A. 595, 606
Fingerman, M. 667, 703, 708, 889
Fish, J.D. 608, 609, 618
Fish, S. 618
Fitzgerald, G.J. 968
Flügel, W. 404
Foley, D.H. 764
Fong, P.P. 514
Fortlage, C. 224
Forward, R.B.jr. 605, 633, 634, 635, 636, 637, 688, 689
Foster, S.A. 916
Foster, W.A. 578
Fox, H.M. 810, 811, 812, 813, 814
Francé, R.H. 735
Frank, St. 1081
Franke, H.D. 498, 499, 500, 501, 502, 503
Frankenberg, E. 974
Fraser, C.M. 494
Freeland, R.O. 317, 348
Fremiet, E. Fauré- 301, 307, 356, 357
Frey, J. 233
Fricke, H.W. 917
Friedländer, B. 542, 543, 544, 545, 546, 547
Froelich, A. Szmant- 377, 378
Fromont, J. 361
Fryer, G. 583, 778
Fuhrman, J.A., 185
Fuller, C.A. 119, 1057
Funke, W. 425
Fuss, C.M.jr. 646

Gaertner, T. von 344
Galbraith, H. 979
Gallien, L. 510
Galloway, T.W. 490

Galzin, R. 942
Gamble, G.W. 408, 409
Ganey, J.L. 990
Garbarini, P. 565
Garcia, L.M.B. 899
Gärtner, H. Knörr- 255
Gautherie, M. 1038
Georgévitch, J. 517
Geppetti, L. 96
Gibson, R.N. 823, 934, 935, 936, 937, 939, 940, 959, 960, 1075
Giese, A.C. 564
Gilpin-Brown, F.R.S. 38
Gilpin-Brown, J.B. 38
Gittings, S.R. 386
Gliwicz, M.Z. 582
Glynn, P.W. 388, 402
Goethe, J.W.v. 282, 288, 289
Gompel, M. 371, 480
Goodenough, J.E. 414, 415
Goodwin, S. 1071
Goswami, S.C. 657
Goswami, U. 657
Grahame, J. 436
Grau, E.G. 847, 848, 849, 933
Grave, B.H. 421, 483
Gravier, Ch. 518, 519, 528, 548
Graviou, E. 341, 342, 343
Gray, J.E. 6
Gray, J.S. 475
Greeley, M.S.jr. 863, 872, 884, 885, 870, 871
Green, D.G. 926, 927
Greenspan, B.N. 716
Greenwald, L. 39
Greenwald, O.E. 39
Greenwood, J.G. 584, 585
Greff, J.A. De 346
Griffith, D.R.W. 656
Griffiths, K. 1040
Grobbelaar, N. 166
Groeber, J. 999
Groos, G.A. 136
Groot, S.J. De 960
Gros, Ch. 1038
Gruber, S.H. 820
Gruner, H.-E. 293
Grzimek, B. 292, 949, 994
Guillon, P. 241
Gunaga, V. 638
Gunn, J.S. 862

Gunther, E.R. 615
Gutmann, W.F. 1087
Gwinner, E. 94, 95

Haas, W. De 1089
Habermehl, K.-H. 1031
Haeckel, E. 1088
Hagen, H.-O. von 719
Hahn, K.O. 433
Hain, J.H.W. 830
Halberg, F. 102
Halberg, F. 52, 71, 84, 102, 270, 275
Hall, D.W. 771, 772
Hamner, K.C. 116
Hanson, K. 1101
Haque, N. 1067
Haraldstad, O. 826
Hardege, J.D. 509, 513
Hardeland, R. 104, 113
Hardy, A.C. 615, 836
Harkness, W.L. 989
Harrison, J.L. 1014, 1015, 1016
Harrison, P.L. 366
Hartland-Rowe, R. 745, 746
Hartmann, W. 212
Hartnoll, R.G. 154, 187, 302, 430, 593, 706, 792, 1058
Haseman, J.D. 440
Hashimoto, H. 798
Hastings, J.W. 101, 115
Hastings, M.H. 619, 620, 621
Hauenschild, C. 125, 126, 520, 549
Haurwitz, B. 1098
Haus, E. 214, 271
Hay, D.E. 833
Hayes, D.K. 131, 272, 717
Heales, D.S. 653
Heckert, H. 221
Heckert, H. 1045
Hefford, A.E. 858, 859
Heimbach, F. 152, 783, 790, 792
Hejl, Z. 1036
Hemleben, C. 353, 354
Hemmer, M.J. 895, 896
Hempelmann, Fr. 521
Henmi, Y. 692
Hensel, H. 263
Heraclites 283
Herbing, I. Hunt von 930
Herdman, E.C. 304
Herdman, W.A. 303

Hermans, C.O. 564
Herpin, R. 489, 497, 511, 512
Herrnkind, W.F. 648
Heslinga, G.A. 432
Hess, H.C. 925
Heusner, A.A. 628
Heyward, A. 363
Hicks-Caskey, W.E. 1034, 1035
Hidaka, T. 725, 726
Hildebrandt, G. 266, 264, 273, 281a
Hill, E. 496
Hillmann, A. 432
Hines, A.H. 883
Hines, M.N. 713, 714
Hodgson, W.L. 835
Hoerner, W. 262
Hoffman, K.S. 933
Hoffmann, K. 598
Hofman, P.J. 316
Hollenberg, G.J. 329
Hollwich, F. 211
Holmström, W.F. 611, 612, 613
Holtschmit, K.H. 655
Honegger, H.-W. 31, 698
Hopkins, C.L. 855, 856
Hoppe, W.F. 358, 359
Hora, S.L. 742
Horn, M.H. 937
Horst, G. van der 485
Horst, R. 550, 561
Hosemann, H. 258
Hostrander, L. 194
Hough, A.R. 589, 590, 592, 594
House, M.R. 469
Hövel, W.-T. 57
Hoyt, W.D. 310, 311
Hsiao, S.M. 68, 69, 864, 865, 867, 869
Huang, T.C. 167
Hughes, D.A. 643, 647
Hughes, W.W. 41
Hunt von Herbing, I. 930
Hunte, W. 930
Hunter, R.D. 1077
Hunter, W.D. Russel- 1077
Huntsman, A.G. 491

Idyll, C.P. 890
Iliffe, T.M. 809
Ims, R.A. 1080
Ito, R.Y. 390
Izuka, A. 506, 529

Jablonski, D. 19
Jacoby, C.A. 584, 585
Jahoda, J.C. 1008
Jens, G. 822
Johannes, R.E. 902
Johnson, K.G. 387
Jokiel, P.L. 389, 390
Jones, D.A. 622
Jones, D.S. 42, 51
Jones, E.W. Knight- 566
Jonsson, B. 826
Jonsson, P.R. 306
Jores, A. 219
Jorgensen, O. 305
Just, E.E. 508, 526
Justice, A. Wirz- 215

Kaestner, A. 293, 1090
Kahn, P.G.K. 40
Kaiser, H. 52
Kalle, K. 22
Kalvas, A. 320
Kautsky, L. 320
Kavaliers, M. 839, 1072
Keeble, F. 408, 409, 410
Keely, I.W. 232
Keener, P. 905
Keeton, T.W. 988
Keeton, W.T. 988
Kelly, I.W. 228, 229, 230, 233, 242, 1049
Kennedy, B. 805
Kenney, B.E. 635
Kerfoot, W.B. 754, 755
Kikuchi, T. 384, 614
King, J.M. 587
Kinne, O. 1054
Kinzie, R.A. 400
Kinzie, R.A.III 394
Kipp, F.A. 983
Kirby, C.F.jr. 250
Klages, L. 261
Klapow, L.A. 626, 627, 629
Klein, T. 217
Klein-Schwartz, W. 227
Kline, D.L. 771, 772
Klinowska, M. 1005, 1006
Kluge, H. 212
Kneib, R.T. 659, 877
Knight-Jones, E.W. 566
Knöpp, H. 824

Knorr, F. 1089
Knörr, K. 255
Knörr-Gärtner, H. 255
Kobayashi, N. 801
Kohlrausch, A. 198
Kojis, B.L. 376
Kolisko, E. 159
Kolisko, L. 158, 159
König, K. Schmidt- 988
Korall, H. 75
Korringa, P. 466, 522
Koskimies, J. 11
Koskinen, R. 30
Kosuge, T. 686, 687, 691
Kowalewski, M. 857
Krämer, A. 540, 551, 552, 553, 559
Krauss, W. 22
Kremer, B.P. 63, 411
Kruchinin, O.N. 897
Krümmel, O. 1084
Kubota, T. 819
Kuckuck, P. 309, 330
Kuznetsov, Y.U.A. 897
Kynard, B.E. 832, 893

Landman, N.H. 43, 47
Lang, H.J. 199, 200, 201, 202, 203, 204, 205, 206
Langman, J. 244
Larkin, T. 988
Larwood, G.P. 19
Laskar, J. 17
Lasker, H.R. 912
Laue, H.B. von 1039
Lausi, D. 321
Laval, R.K. 739
Laverty, W.H. 1049, 1050
Lavesack, M. 147
Law, S.P. 247
Laws, R.M. 1020, 1022
Leach, G.J. 879
Leatherland, J.F. 844, 850, 851, 854
Legendre, R. 487, 488, 504, 505, 523
Legg, C.J. 761
Lemmer, B. 268
Lenderking, R.E. 434
Lesauter, J. 138
Lessios, H.A. 802
Lethbridge, M.D. 1100
Lewis, J.F. 312
Lewis, R.D. 604

Lichatowich, T. 950
Lieber, A.L. 225, 226
Lillie, F.R. 508
Lillie, T.H. 771, 772
Lin, H.C. 699
Lindauer, M. 76, 77, 78
Lindeberg, B. 32
Little, C. 429
Liu, P.M. 390
Lobel, P.S. 909
Lockard, R.B. 1003, 1004
Lohmann, K. 634
Lohmann, K.J. 70
Loneragan, N.R. 653
Long, E.F. De 181
Lonne, O.J. 475
Loosanoff, V.L. 464, 465, 467
Lorenz, K. 223
Lough, J.M. 401
Lowe, M.E. 703
Lowe, R.H. 825
Loya, Y. 65
Lummel, L.A.E. van 493
Lund, I.A. 1099
Lutz, A. 774
Lutz, R.A. 53, 476
Lysaght, A.M. 439

McClay, Dr. 808
McCleave, J.D. 828
MacDonald, J.D. 554
MacDonald, W.W. 776
McDowall, R.M. 1055
McFarland, W.N. 907
McGaw, I.J. 678
McGehee, M.A. 904
MacGregor, R. III 870, 871, 872, 885
MacIntosh, W.C. 555
McKeown, B.A. 852
McLachlan, A. 485, 616
McLeish, W. 1101
McMullan, P.M. 665
McNeil, R. 980
Macvean, C. 1072
Mahapatra, A.K. 1067
Marchase, R. 808
Marcum, J.P. 1037
Marion, K.R. 871, 872, 885
Markert, B.J. 492
Markert, R.E. 492
Marraro, Ch. 941

Marsden, I.D. 722
Marsh, B.A. 631
Marshall, S.M. 393
Marte, C.L. 898
Martens, R. 242
Martin, A.W. 49
Martin, H. 75, 76, 78
Martin, S.J. 228
Martini, J. 720
Masayuki, S. 724
Maul, G.A. 1101
Maw, M.G. 350
May, D.I. 325
May, E. 416
Mayer, A.G. 533, 534, 535
Meier, A.H. 68, 69, 864, 865, 866, 867, 869
Menacker, W. 239
Menaker, A. 239
Menaker, M. 291
Mergenhagen, D. 120, 121
Mergenhagen, E. 120
Metcalfe, J.D. 956
Michele, L. Di 879, 880, 882
Middaugh, D.P. 891, 892, 895, 896
Miklas, J.J. 883
Mikulecky, M. 581
Mileikovsky, S.A. 437
Miles, L.E.M. 207, 208
Milia, A. Di 96
Miller, C.A. 866
Miller, L.E. 1033
Millet, B. 347
Mills, A. 609
Mills, A.M. 992
Milson, T.P. 977, 978
Milton, D.A. 962
Mitchell, S.A. 777
Mitsui, A. 168
Mletzko, H.G. 1056
Mletzko, J. 1056
Mobberly, W.C.jr. 703
Mohren, W. 1019
Mohssine, E.H. 758
Montfrans, J. van 668
Moore, H.B. 815
Moore, K. 252
Moore, P.G. 603
Moore-Ede, M.C. 1057
Moran, V.C. 752
Moreira, M.H. 683

Morgan, E. 124, 610, 611, 612, 613, 938, 939
Morrell, T.E. 989
Morrison, D.W. 1000
Moser, P.W. 574
Moyer, J.T. 923
Mozart, W.A. 286
Muench, K.A. 887
Mukhopadhyay, S. 750
Mukhopadhyaya, M.C. 736
Muller, K. 861
Müller, A. 743
Müller, A.H. 1078
Müller, D. 146, 313, 314, 671
Müller, E.F. 222
Murai, M. 686, 687
Murray, M.G. 1025

Nag, A. 763
Nakai, K. 915
Nascimento, I.A. 652
Natarajan, P. 649, 650
Nath, P. 763
Nayak, L.D. 586
Naylor, E. 123, 142, 143, 147, 153, 154, 187, 302, 427, 426, 589, 590, 592, 593, 594, 602, 617, 619, 622, 623, 624, 658, 660, 669, 670, 673, 674, 675, 676, 677, 678, 679, 680, 681, 706, 723, 792, 1058
Neelakantan, B. 638
Neelakantan, K. 638
Nelson, D.A. 987
Nemec, S.J. 1065
Neumann, D. 26, 27, 28, 31, 128, 144, 148, 149, 150, 151, 152, 784, 785, 786, 787, 788, 789, 790, 791, 792, 793, 794, 795
Newell, G.E. 438
Nicholson, M. 1024
Nielsen, S.M. 1018
Nishioka, R.S. 842, 846
Nojima, S. 384
Nomejko, Ch.A. 464, 465, 467
Northcott, S.J. 939
Norton, T.A. 324
Nowinszky, L. 156
Nursall, J.R. 941

O Brien, D.P. 1091, 1092
Ochi, H. 918

Oderda, G.M. 227
Oehmke, M. 756, 757
Oka, H. 797, 798
Okada, K. 562
Olcese, J. 81
O Leary, J.A. 832
Oliveira, D. De 741
Olsen, G.J. 182
Omori, K. 614
Orth, R.J. 668
Orton, J.H. 468
Osawa, K. 530
Osgood, K.E. 883
Osley, M. 240
Ossenkopp, K.P. 1046
Ossenkopp, M.D. 1046
Otubusin, S.O. 961
Oudemans, A.C. 580
Owings, D.H. 1003

Palincsar, J.S. 482
Palmén, E. 32
Palmer, J.D. 25, 88, 101, 111, 131, 133, 139, 140, 290, 308, 717, 732, 1059, 1060, 1061
Pandian, R.S. 753
Pang, K. 60
Panikkar, N. K. 527
Panzram, H. 349
Papi, F. 599, 600, 601
Pardi, L. 599, 600, 601
Park, Y.H. 413, 1007
Parsons, G.J. 461
Parthier, P. 189
Partridge, J.C. 429
Patnaik, E. 586
Patra, A.K. 586
Patton, D.G. 571
Paula, J. 632
Pauly, J.E. 71, 84, 102, 131, 717
Pearse, J.S. 564, 803, 804, 805, 807, 809
Pentz, U. von 220
Persson, B. 762
Peters, E.C. 398
Petersen, C.W. 924, 925, 928
Peterson, W.T. 300
Pevet P. 128
Pflüger, W. 29
Piéron, H. 370, 372, 373, 374, 375
Pilgrim, U. 1076

Pinder, W. 279
Pitcher, C.R. 661
Pittendrigh, C.S. 134, 135
Plinius Secundus Cajus 3
Pochobradsky, J. 249
Pompea, S.M. 40
Poole, S.J. 978
Poovachiranon, S. 686, 687
Potter, D.R. 1034, 1035
Potts, F.A. 495
Powell, Th. 556
Preece, G.S. 607
Price, A.R.G. 654
Price, C.H. 452, 453
Price, M.V. 1009
Proctor, L.M. 185
Prosser, C. L. 115
Provost, M.W. 775
Prunet, P. 841
Purkinje, J.E. 195

Quartier, A.A. 845
Queiroga, H. 683

Ralph, C.L. 317, 568, 711, 1013
Ramanathon, O. 1028
Randall, J.E. 903
Ranzi, S. 524, 525
Rao, K.P. 457, 458
Rauch, E. 1041
Ray, H. 355
Raynal, D.M. 208
Reaka, M.L. 591
Reges, I. Brandt- 273
Reichert, M.J.M. 358
Reid, D.G. 143, 153, 623, 624, 681
Reinberg, A.E. 269
Reiter, R.J. 131, 717
Reith, C.C. 1001
Remmert, H. 35, 985
Render, J.H. 951
Rensing, L. 91, 192, 267
Reuss, S. 81
Reutter, M. 377, 378
Rhoads, D.C. 53, 476
Richardson, C.A. 470, 471, 472, 473, 474
Richkus, W.A. 831
Richmond, R.H. 389
Riedl, R. 1085
Rietveld, W.J. 214

Riggs, L. 377, 378
Robertson, D.R. 911, 926, 927, 928, 948, 966, 967
Roch, F. 486
Rochard, J.B.A. 978
Rodriguez, G. 658
Rombeck, A.M. 1021
Romero, J.M. 655
Rosenberg, G.D. 400
Roßlenbroich, B. 238
Ross, B. 442
Ross, R.M. 919, 932
Ross-Farhang, S. 569
Rossiter, A. 913
Röthig, P. 261
Rotton, J. 229, 232, 233
Round, F.E. 122a, 308
Rounds, H.D. 332, 333, 334, 335, 336, 337, 338
Rountree, R.A. 952
Rowe, R. Hartland- 745, 746
Roy, J.H.B. 1026
Rudder, B. de 243
Rudloe, A. 572, 575, 576, 577
Rumphius, G.E. 34
Runcorn, S.K. 44, 400
Runham, N.W. 470, 471, 473, 474
Rusak, B. 115
Russel-Hunter, W.D. 1077
Ruyck, A.M.C. De 616
Rye, D.M. 47
Ryer, C.H. 668
Ryland, J.S. 403

Saadany, G. El- 1064
Sadleir, R.M.F.S. 995
Sadler, W.A. 855
Sadovy, Y. 904
Sager, G. 23, 58
Saha, S.K. 736
Saigusa, M. 127, 725, 726, 727, 728, 729, 730, 731
Saito, Y.Z. 276
Saklofske, D.H. 228, 242, 1049
Saltz, D. 1017
Sandeen, M.I. 108, 318, 435, 447, 711
Sands, J.M. 1033
Santina, P. Della 426, 427, 428
Santini, G. 428
Sarano, F. 362
Saunders, D.S. 1062

Saunders, W.B. 45
Savage, R.E. 835, 836
Sawara, Y. 943, 944
Schad, W. 20, 265
Scheving, L.E. 71, 84, 102
Schluck, L. 235
Schmidt, H. 245
Schmidt, R.F. 213
Schmidt-König, K. 988
Schnakenbeck, 837
Schroeder, P.C. 564
Schubert, G.H. 280, 281
Schwanck, E. 914
Schwartz, W. Klein- 227
Schweig, G. 218
Schweiger, H.G. 189
Scott, J.W. 563
Scow, K.M. 80
Scrimgeour, J.B. 256
Sellick, R.D. 748
Semm, P. 81
Semmens, E.S. 157
Severinghaus, L.L. 699
Shapiro, D.Y. 901, 904
Shastri, A.S.R.A.S. 1068
Sherr, E.B. 173
Sherrill, M.T. 891
Sheshappa, D.S. 929
Shlesinger, Y. 65
Shriner, J. 710, 1013
Shrivastava, S.K. 1068
Shukla, B.C. 1068
Siddorn, J.W. 740
Siedler, G. 22
Siewing, R. 360
Siivonen, L. 9, 10, 11
Silver, R. 138
Simpson, C.J. 368
Simpson, H.W. 1040
Simpson, P. 302
Sinclair, A.R.E. 1030, 1079
Singh, B.P. 155
Sitte, P. 164
Skewes, T.D. 661
Smith, B.C. Featonby- 316
Smith, G.C. 986
Smith, G.M. 328
Smith, J.E. 438
Smith, S.R. 385
Soong, K. 379
Sorokin, M.A. 897

Spieß, H. 160, 161, 339, 340
Sponaugle, S. 945
Spruyt, E. 346
Staden, J. van 316
Stair, J.B. 5, 6
Staples, D.J. 640, 651
Starck, D. 293
Stark, D. 254
Stegmaier, J. 1041
Steiner, R. 278
Stenseth, N.C. 1080
Stephens, G.C. 435, 447
Stephenson, A. 417
Stephenson, T.A. 393
Sternlicht, M. 1069
Stifter, A. 16
Stimson, J.S. 392
Stoddart, J.A. 391
Storch, V. 294
Stradling, D.J. 761
Strawn, K. 868
Stucki, I. Catala- 49
Sturmwasser, F. 454
Stutz, A.M. 1010
Subramaniam, A. 1027
Suessenguth, K. 163
Sulzmann, F.M. 122, 1057
Summerville, D. 240
Sündermann, J. 54, 55
Swann, J. 136
Sweeney, B.M. 92, 169
Sweeting, R.M. 852
Szedlmayer, S.T. 953
Szmant, A.M. 383
Szmant-Froelich, A. 377, 378

Tahara, M. 322, 323
Tanaka, M. 614
Tang, S.-F. 460
Tankersley, R.A. 688, 689
Taylor, A.C. 680
Taylor, G.J. 61
Taylor, L.R. 738
Taylor, M.H. 874, 875, 876, 878, 879, 880, 881, 882
Taylor, P.D. 19
Taylor, R.A.J. 760
Teagle, L. 429
Templer, D.I. 234, 1051, 1052
Tennent, D.H. 816
Thamdrup, H. 441

INDEX OF AUTHORS

Thibault, M. 980
Thienemann, A. 796
Thompson, I. 42
Thompson, J.L. Cloudsley- 248
Thompson, L.M. 14
Thomson, D.A. 887
Thorpe, J.E. 823, 840, 960
Thorrold, S.R. 838, 946
Thresher, R.E. 821, 910
Tikasingh, E.S. 768
Tjønneland, A. 747
Todd, P.R. 829
Tokunaga, M. 773
Toledo, J.D. 898
Tomassen, G.J.M. 86, 1063
Tomiczek, H. 1032
Touitou, Y. 214, 271
Treadwell, A.L. 536, 537
Trendall, J.T. 661
Trillmich, F. 1019
Truchan, L.C. 1012
Truschel, M. 808
Tsuji, F.I. 496
Türcke, F. 1032
Turek, F.W. 136
Tyndale-Biscone, Ch. 995

Urbantat, V. 259

Vaas, R. 62
Vance, D.J. 639, 640, 651, 653
Vanderplank, F.L. 799
Veiga, J.P. 976
Veleber, D.M. 234, 1051
Ventsch, C.M. 300
Verbelen, J.P. 346
Verheijen, F.J. 975
Vertrees, N.J. 492
Victor, B.C. 926, 927, 931
Vielhaben, V. 145
Vines, R.G. 15
Volestad, L.A. 826
Voss, J. 947
Vries, M.C. De 690

Waas, B.P. 868
Wagner, G.F. 852
Walker, B.W. 886
Walker, J.C.G. 59
Walker, T.J. 800
Walsby, A.E. 170

Ward, P. 39, 48
Ward, P.D. 37, 45, 49
Ward, R.R. 112
Warman, C.G. 624, 681
Waser, N.M. 1009
Wasserman, G.S. 570, 571
Waterman, T.H. 85
Watts, P. 1023
Webb, H.M. 72, 74, 86, 107, 141, 318, 446, 447, 707, 715
Webb, P.W. 956
Weckenmann, M. 1041
Weeks, J.M. 603
Weiler, D. 901
Welch, P.S. 490
Wellnitz, K. 354
Welsch, U., 294
Wendt, H.W. 277
Werner, Y.L. 974
Westermann, G.E.G. 37
Wever, R. 82, 83, 84, 210, 291
Wheeler, J.F.G. 641, 642
Whitmee, S.J. 557
Whyte, I.M. 1047
Wickham, D.A. 644, 645
Wieser, W. 630
Wigglesworth, J.M. 656
Wilde, P.A.W.J. De 187, 663
Wildish, F.J. 596
Williams, A.H. 921
Williams, B.G. 142, 666, 677, 679, 723
Williams, C.B. 155
Williams, J.A. 154
Williams, J.L. 315
Willis, B.L. 368
Willoughby, N.G. 748
Willows, A.O.D. 70
Wilson, J.M. 866
Wilson, M.A. 208
Wiltschko, R. 79
Wiltschko, W. 79
Winans, G.A. 846
Wing, L.W. 982
Winn, H.E. 831
Winn, L.K. 831
Wippelhauser, G.S. 828
Wirz-Justice, A. 215
Witkowski, A. 857
Woddbury, M.A. 1095
Wolf, W. 98
Wolfrath, B. 721

Wollnik, F. 216
Wolton, R.J. 1011
Wood, F.J. 24
Woodworth, W. 558, 559, 560
Wooldridge, T. 485
Worthmann, H.O. 908
Wright, J.R. 430
Wyers, S.C. 385
Wynne-Edwards, V.C. 991

Yahner, R.H. 989
Yamahira, K. 963
Yamauchi, K. 853
Yeemin, T. 384

Yoshioka, E. 418, 419
Youngson, A.F. 843
Youthed, G.J. 752

Zahnle, K.J. 59
Zann, L.P. 422, 423
Zemek, R. 581
Zerger, U. 339
Ziac, D.C. 251
Zimmermann, T.L. 733, 734
Zincone, L.H. 250
Zucker, N. 696, 697
Zürcher, P.E. 162

Index of species

Italic numbers refer to the species catalogue, other numbers to the main text.

Abudefduf saxatilis (sergeant major) 203, 204
— *troschelii* (Pacific sergeant fish) 203, 204
Acanthopleura japonica 166
Acanthozostera gemmata 166
Acanthurus bahianus 206
— *chirurgus* 206
— *coeruleus* 206
— *triostegus sandvicensis* 207
Achillea spec. (yarrow) 111, *159*
Acropora cuneata 163
— *palifera* 163
Actinia equina (sea or beadlet anemone) 95, *163*
Adinia xenica 199
Aegopodium podagraria (ground elder) 10, *160*
Aepyceros melampus (impala) 215
Agrostemma githago (corn cockle) *160*
— *ipsilon* (cutworm) *191*
algae, luminous 52, *157*
Allium porrum (leek or porrum) 111, *159*
— *ursinum* (wild garlic) 11
Alopex lagopus (Arctic fox) 214
Alosa aestivalis (blueback herring) *196*
Alosa sapidissina (American shad) *196*
Amphidinium herdmaniae (= *operculatum*) *157*
Amphiprion bicinctus 203
— *clarkii* (tropical anemone fish) 203
— *melanopus* (anemonefish) 204
Amphitrite ornata 174
Anchistoides antiguensis 181
Anemone nemorosa (wood anemone) 10, *160*
— *viridis* (wood anemone) 83, *160*
anemone, sea or beadlet 95, *163*
—, snakelocks 83
—, wood 10, *160*
anemone fish *204*
—, tropical 203

Anemonia viridis (snakelocks anemone) 83
Anguilla anguilla (fresh-water eel) 12, *195*
— *australis* (shortfin) *196*
— *dieffenbachii* *196*
— *rostrata* (American eel) *196*
Anopheles annulipes *191*
— *bellator* Knab *191*
— *culicifacies* *191*
— *farauti* *191*
ant-eater, Australian spiny 211
Aotus lemurinus griseinembra (night ape) 212
— *trivirgatus griseinembra* 110, *212*
ape, night 212
Apis mellifera carnica (Carnica honey bee) *189*
— — *intermissa* (Moroccan honey bee) *189*
— — *mellifera* (worker honey bee) 11, *190*
Aplysia californica (Californian sea hare) *169*
Apodemus sylvaticus (wood mouse) 213
Archaeomysis maculata 178
Arctocephalus galapagoensis (Galápagos fur seal) 214
Arenicola cristata 56, *174*
— *marina* (lugworm) 56, *174*
Armigeres subalbatus 189
Artibeus jamaycensis (Jamaican fruit bat) 212
— *lituratus* 109f, *212*
Athecate hydroids 168
Athripsodes stigma 108, *190*
— *ugandanus* 108, *190*
auklet, Cassin's 210
Austrocochlea obtusa 167
Avena spec. (oats) 111, *160*

barley 111, *160*
—, pearl *160*

bass, sea *201*
bat, Egyptian fruit 110, *212*
—, Jamaican fruit *212*
Bathyporeia pelagica 179
— pilosa 179
Bdella interrupta 176
bean, common 11, 111, *159f*
—, broad 11, *160*
bee, Carnica honey *189*
—, Moroccan honey *189*
—, North American *189*
—, worker honey 11, *190*
Bembicium nanum 167
Biddulphia aurita 158
bladderwrack 59, 98f, *158*
Blennius (= Lipophrys) pholis (shanny) *205*
bluehead wrasse *205*
Bohadschia argus 194
bollworm moth *190*
Bombycilla garrulus (waxwing) 17f, *211*
Bos prinigenius taurus (domestic cattle) *215*
Brassica oleracea (white cabbage) *159*
brook lamprey (European palolo worm) 70, *173*
Bubalus arnee (Indian buffalo) 109, *215*
Bubo virginianus (great horned owl) *211*
buffalo, Indian 109, *215*
Bufo americanus (American toad) *208*
— biporcatus 208
— fowleri 17, *208*
— melanostictus (black-spined toad) 17, *208*
burbot (eel pout) *199*
butterfish, gulf *207*
butterflyfish *203*

Callinectes sapidus (blue crab) *183*
Callista (= Cytherea) chione (Venus shell) *170*
Campanula rapunculoides (creeping bellflower) 10f, *160*
Campanularia flexuosa 162
Cancer novaezelandiae 183
— pagurus (rock-dwelling crab) *183*
candytuft, rocket *160*
capercaillie 17, *209*
Capra hircus (= aegagrus) (goat) *215*
Caprinulgus europaeus (nightjar) *211*

Carcinus maenas (shore crab) *183*
Cardium (= Cerastorma) edule (common cockle) 49, *170*
carrot 14, *159*
—, wild 14, *160*
Cataleptodius taboganus 184
caterpillar, tobacco *190*
Catostomus commersonii (white sucker) *197*
Cellana exarata 166
Centrechinus (= Diadema) setosum (Indo-Pacific diadem sea urchin) 9, *194*
Centropyge potteri 202
Centrostephanus coronatus 194
Cephalopholis boenack 202
Cerastorma (= Cardium) edule (common cockle) 49, *170*
Ceratocephale osawai (= Tylorhynchus chinesis) (Japanese palolo worm, 'Itome') *173*
Chaenogobius castaneus (biringo-goby) *206*
Chaetodon capistratus (butterflyfish) *203*
Chaetopleura apiculata 166
Chaoborus spec. 192
— anomalus 192
— edulis 192
Charadrius wilsonia cinnamominus (thick-billed plover) *210*
Charybdis cruciata 184
Chasmichthys gulosus 206
Chironomus (= Nilodorum) brevibucca 192
Chiton tuberculatus (West Indian chiton) *166*
Chone teres 174
Chromis cyanea 204
— multilineata 204
Chromulina psammobia (lumnious algae) 52, *157*
Cittarium pica (West Indian topshell) *167*
clam, soft-shelled 47, 49, *171*
Clavularia hamra 165
Clibanarius misanthropus 183
Clinotanypus claripennis 108, *192*
Clunio marinus (marine midge) 59–62, 99, *193*
— pacificus 193

Clupea harengus (Atlantic herring) 47, *196*
— *pallasii* (Pacific herring) *196*
Cnidaria 82, *162*
cockle, common 49, *170*
cockroach, American 111, *188*
Coenobita clypeatus *183*
Coleus blumei (common coleus) 111, *159*
Columba livia (homing pigeons) *211*
Comanthus japonicus (Japanese feather star) *193*
comfrey, tuberous 11, *160*
Conchophthirius lamellidens *162*
Connochaetes taurinus (brindled gnu) 109, *215*
Convoluta roscoffensis 54f, *165*
coral 10, 19, 63, 66, 79, 81-85, *163*
coral, blue *165*
—, sea anemone-type mushroom *164*
corn cockle *160*
Corophium arenarium *179*
— *volutator* (mud shrimp) *179*
Corydalis cava (bulbous corydalis) 10, *160*
Coryphoblennius galerita *205*
Coryphopterus glaucofraenum *206*
crab, Atlantic *175*
—, blue *183*
—, rock-dwelling *183*
—, shore *183*
crane, wintering common 110, *209*
Crangon crangon (brown shrimp) *182*
Crassostrea virginica (American oyster) *170*
cress, garden *161*
Culex (= Melanoconion) candelli *191*
— *portesi* *192*
— *taeniopus* *192*
Culicoides spec. *192*
— *peliliouensis* *192*
Cumingia tellinoides *171*
cutworm *191*
Cyclograpsus lavauxi *186*
Cylindrotheca signata *158*
Cytherea (= Callista) chione (Venus shell) *170*

damselfish, Caribbean *204*
—, threespot *204*
—, tropical *204*

Daucus carota (carrot) 14, *159, 160*
Decapoda *180*
Delphinium consolida (larkspur) 111, *159f*
Derbesia (= Halicystis) ovalis 56, *159*
Diadema (= Centrechinus) setosum (Indo-Pacific diadem sea urchin) 9, *194*
— *antillarum* (long-spined sea urchin) *194*
Diatomeae 50–53, *157*
Dictyota ciliolata *158*
— *dentata* *158*
— *dichotoma* 55f, 105, *158*
Diemictylus (= Triturus) viridescens (red spotted newt) 111, *208*
Diplonychus nepoides *189*
Diploria strigosa *164*
Dipodomys merriami (Merriam's kangaroo rat) *212*
— *spectabilis* (bannertail kangaroo rat) 109, *212*
Donax serra (white mussel) *171*
Dugesia dorotocephala (black planaria) *165*

earthworm 64, *175*
Ecklonia maxina *158*
eel, American *196*
—, fresh-water 12, *195*
—, shortfin *196*
eel pout (burbot) *199*
elder, ground 10, *160*
elm, American 111, *159*
Emerita analoga *183*
— *asiatica* *183*
Enchelyopus cinbrius (four-bearded rockling) *199*
Encrasicholina devisi *197*
— *heterolobus* *197*
Enhalus acoroides *161*
Ensis directus *171*
Enteromorpha intestinalis (gut weed) 56, *159*
Epheron (= Polymitarcis) virgo *188*
Epinephelus guttatus *202*
— *merra* *202*
— *striatus* *202*
Epiphyas postvittana *190*
Euapta godeffroyi *194*
Eucidaris tribuloides *193*
Euglena limosa (= obtusa) 52, *157*

Eulalia punctifera 171
Eunice (= *Leodice*) *fucata* (Atlantic palolo worm) 70, *173*
— *harassi* (brook lamprey, European palolo worm) 70, *173*
— *viridis* (Pacific palolo worm) 10, 63–70, 84f, *173*
Euphasiopteryx ochracea 193
Euphausia frigida 180
— *superba* 180
Euplanaria gonocephala (river planaria) 11, *166*
Eupomacentrus planifrons (threespot damselfish) 204
Eurydice longicornis 179
— *pulchra* 179
Eurytemora affinis 177
Eusyllis blomstrandi 171
Excirolana chiltoni 92, 94, 100, 103, 105, *180*
— *natalensis* 179

Fandulus grandis (gulf killifish) 199
— *heteroclitus* (common mummichog) 200
— *pulvereus* 199
— *sinilis* (longnose killifish) 200
Favia fragum 163
fish 195
—, reef-inhabiting 195
flounder, starry 207
—, summer 207
Forsythia spec. (forsythia) 111, *159*
fox, Arctic 214
frog, leopard 110, *208*
Fucus spec. 59, *158*
— *ceranoides* 158
— *vesiculosus* (bladderwrack) 98f, *158*
— *virsoides* 158
Fungia actiniformis var. palawensis (sea anemone-type mushroom coral) 164

Galaxias attenuatus (= *maculatus*) (New Zealand whitebait) 199
Galleria mellonella (wax moth) 191
Gammarus chevreuxi 178
— *zaddachi* 178
garlic, wild 11
gecko, house 209
Geranium spec. (wild geranium) 111, *159*
gerbil, Mongolian 213

Globigerinella siphonifera 161
Globigerinoides sacculifer 161
Gloligerinoides ruber 161
Glossina pallidipes (tsetse fly) 193
Gnatholepis thompsoni 206
gnu, brindled 109, *215*
goat 215
goby, biringo- 206
—, numachichibu- 206
Goniastrea aspera 163
Gonodactylus chiragra 177
— *falcatus* 177
— *graphurus* 177
— *zacae* 177
Gonyaulax excavata 157
grape vine 11, *160*
grayling, European 198
grenadier, blue 199
grouse, black 17, *209*
grunion 71, *201*
—, California 201
grunts, French 202
Grus grus (wintering common crane) 110, *209*
gulf weed 158
guppy 15f, 120, *200*
gut weed 56, *159*
Gymnodinium spec. 157

Haemulon flavolineatum (French grunts) 202
Halicystis (= *Derbesia*) *ovalis* 56, *159*
Halineda discordea 159
— *hederacea* 159
— *tuna* 159
Haliotis tuberculata (tuberculated sea bear) 166
Halodule pinifolia 161
hamster, golden 14, *213*
Hantzschia virgata (= *amphioxys*) 52, *157*
Haptosquilla glyptocercus 177
hare, snowshoe 17, *214*
Hediste diversicolor 171
Helianthus annuus (sunflower) 160
Helice crassa 94, 101, 103, *186*
Heliopora coerulea (blue coral) 165
Heliothis zea (bollworm moth) 190
Hemicentrotus pulcherrinus 193
Hemigrapsus edwardsi 186
Herklotsichthys castelnaui 196

INDEX OF SPECIES

herring, Atlantic 47, *196*
—, blueback *196*
—, Pacific *196*
Heterocentrotus mammillatus 194
Hexagenia bilineata 188
Holacanthus ciliaris 203
Homo sapiens 216–18
Hordeum spec. (barley) 111, *160*
— *distichon* (pearl barley) *160*
Hubbsiella sardina (sardine silverside) 71, *201*
Hystrix indica 214

Iberis amara (rocket candytuft) *160*
Ilyoplax delsmani 184
— *gangetica* 184
— *orientalis* 184
impala 215
insects *187*
—, night *188*
Itome, Japanese palolo worm *173*

Japanese feather star *193*

killifish, gulf *199*
—, longnose *200*
Kyanassa (= *Nassarius*) *obsoletus* (mud dog whelk) 59, *168*

Lactuca sativa capitata (head lettuce) 111, *159*
Laomedea (= *Obelia*) *geniculata* 56, 58, *162*
lapwing 110, *209*
larkspur 111, *159f*
Lates calcarifer (sea bass) *201*
leafhoppers, rice green *189*
Lebistes (= *Poecilia*) *reticulatus* (guppy) 15f, 120, *200*
leek 111, *159*
lemming, Norway 17, *213*
Lemmus lemmus (Norway lemming) 17, *213*
lemon balm 111, *159*
Leodice (= *Eunice*) *fucata* (Atlantic palolo worm) 70, *173*
Lepidium sativum (garden cress) *161*
Lepidoptera 190
Leptonereis glauca 172
Lepus americanus (snowshoe hare) 17f, *214*

lettuce 111, *159*
Leuresthes sardina (California grunion) *201*
— *tenuis* (grunion) 71, *201*
Levisticum officinale (lovage) 111, *159*
Lilium tigrinum (tiger lily) 111, *159*
Linnothrissa miodon (Tanganyikan sardine) *197*
Linulus polyphemus (Atlantic crab) *175*
Littoraria strigata 167f
Littorina angulifera 167
— *littorea* (common periwinkle) 51f, *167*
— *neritoides* (European rock periwinkle) 51f, *167*
— *obtusata* (flat periwinkle) 51f, *167*
— *rudis* (rough periwinkle) *167*
— *saxatilis* (rough periwinkle) 51f, *167*
Littorinopsis (= *Melarhaphe*) *scabra* 168
Lota lota (eel pout, burbot) *199*
lovage 111, *159*
lugworm 56, *174*
Lumbriconereis spec. 174
— *sphaerocephale* 173
Lumbricus terrestris (earthworm) 64, *175*
Lutjanus fulvus (= *variegiensis*) *202*
Lycopersicum esculentum (tomato) *160*
Lynx canadensis (Canadian lynx) *214*
Lyrurus tetrix (black grouse) 17, *209*
Lysidice fallax 174
Lysidice oele (wawo worm) 63, *174*
Lytechinus (= *Toxopneusthes*) *variegatus* (West Indian white sea urchin) *194*

mackerel, tuna-like *206*
Macoma balthica (Baltic macoma) 50, 52, *170*
Macrophthalamus boteltobagoe 184
— *hirtipes* 93f, 101–3, *185*
— *japonicus* 185
Macropus eugenii (tammar wallaby) 109, *211*
Macruronus novaezelandiae (blue grenadier) *199*
Madreporaria 82f, *163*
Maesopsis eminii 160
maize 111, *160*
Manicina areolata 164
marine midge 59–62, 99, *193*

Marinogammarus marinus 178
mayfly 107f, 112, *188*
—, tropical 108, *188*
Melampus bidentatus 168
Melanerita atramentosa 166
Melanoconion (= *Culex*)
Melarhaphe (= *Littorinopsis*) *scabra* 168
Melissa officinalis (lemon balm) 111, *159*
Menidia beryllina (inland silverside) *201*
— *menidia* (Atlantic silverside) *201*
— *peninsulae* (tidewater silverside) *201*
Mercenaria (= *Venus*) *mercenaria* (northern quahog) *170*
Meriones unguiculatus (Mongolian gerbil) *213*
Mesocricetus auratus (golden hamster) 14, *213*
Metapenaeus 182
— *dobsoni* 181
— *endeavouri* 181
Microspathodon bairdi 204
— *chrysurus* 204
— *chrysurus* (Caribbean damselfish) 204
— *dorsalis* 204
midge, marine 59–62, 99, *193*
millipede, grassland *187*
Montastrea annularis 164
— *cavernosa* 164
— *valenciennesi* 164
Morula marginalba 168
moth, bollworm *190*
—, wax *191*
mouflon, European 215
mouse, grasshopper *213*
—, white 14, *213*
—, wood *213*
mud shrimp *179*
Mugilidae 205
mummichog, common 200
Mus musculus (white mouse) 14, *213*
mussel, Californian blue *169*
—, common 47, 49, *169*
—, white *171*
Mya arenaria (soft-shelled clam) 47, 49, *171*
Mycteroperca microlepis 202
Myotis yumanensis 212
Myrmeleon obscurus 189

Mytilus californicus (Californian blue mussel) *169*
— *edulis* (common mussel) 47, 49, *169*

Naesa bidentata 180
Nassarius (= *Kyanassa*) *obsoletus* (mud dog whelk) 59, *168*
Navicula ammophila 158
Negaprion brevirostris (lemon shark) *195*
Nematodinium spec. 157
Nemorma tingitana 158
Neofibularia nolitangere 162
Neolamprologus moorii 203
Neomysis integer 178
Nephotettix spec. (rice green leafhoppers) *189*
— *nigropictus* 189
— *virescens* 189
Nereis irrorata 172
— *japonica* (Japanese palolo worm) *172*
— *linbata* (= *succinea*) *172*
— *longissina* 172
— *pelagica* 172
newt, red spotted 111, *208*
nightjar *211*
Nitzschia colsterium 158
Noctuidae 191
Nodilittorina pyramidalis 168

oats 111, *160*
Obelia (= *Laomedea*) *geniculata* 56, 58, *162*
Ocypode ceratophthalmus 185
Odontosyllis spec. 10, 63, *171*
— *enopla* 171
— *gibba* 171
— *hyalina* 171
— *phosphorea* 172
Oligoneuria rhenana 188
Oncorhynchus kisutch (silver salmon) *198*
— *masou* (Masu salmon) *198*
— *mykiss* 198
— *tschawytscha* (Chinook salmon) *198*
Onychomys leucogaster (grasshopper mouse) *213*
Ophioblennius atlanticus 205
Orchestia cavinana (shorehopper) 51, *178*
— *mediterranea* (shorehopper) 51, *178*

INDEX OF SPECIES

Orchestoidea corniculata 178
Orthomorpha coarctata (grassland millipede) 187
Ostrea edulis (common, edible, table or European oyster) 10, 58, 170
Otaria bryonia (maned seal) 214
Ovis ammon musinon (European mouflon) 215
owl, great horned 211
—, spotted 211
Oxyrrhis spec. 157
oyster, American 170
—, common, edible, table or European 10, 58, 170
Palaemon elegans (common prawn) 58, 182
— *serratus* (common prawn) 58, 182
Palaemonetes pugio 182
Palingenia robusta 188
palolo worm, Atlantic 70, 173
—, European (brook lamprey) 70, 173
—, Itome (Japanese) 172, 173
—, Pacific 10, 63–70, 84f, 173
Panaeus 182
— *esculentus* 181
Panulirus ornatus 183
Paracentrotus lividus 194
Paralichthys dentatus summer flounder 207
Parapeneopsis 182
Parathemisto gaudichaudii 179
Patella vulgata (common limpet) 52, 166
pea 111, 159
Pecten maximus (giant scallop) 58, 169
— *opercularis* (small scallop) 58, 170
Pediastrum boryanum 159
— *duplex* 159
Penaeopsis goodei 181
— *smithi* 181
Penaeus spec. (penaeid shrimps) 182
— *duorarum* 181
— *indicus* (prawns) 181
— *merguiensis* 181, 182
— *monodon* 181
— *schmitti* 182
— *semisulcatus* (tiger prawns) 182
— *stylirostris* 182
— *vannamei* 182
Peprilus burti (gulf butterfish) 207

Perinereis cultrifera 172
— *marioni* 173
— *nuntia var. brevicirrus* 173
Periplaneta americana (American cockroach) 111, *188*
periwinkle, common 51f, *167*
—, European rock 51f, *167*
—, flat 51f, *167*
—, rough 51f, *167*
pescada 202
Pharbites hispida 160
Phaseolus spec. (bean) 11, 111, *159*
— *vulgaris* (common bean) 11, 111, *159, 160*
Philodendron sagittifolium (arrow-leafed philodendron) 111, *159*
Phoca vitulina (common seal) 47, *214*
Phyllostomus hastatus 110, *212*
pigeons, homing *211*
Pionosyllis lamelligera 172
Pisum sativum (pea) 111, *159*
Placopecten magellanicus (giant scallop) *170*
Plagioscion monti 202
— *squamosissimus* (pescada) 202
plaice 207
planaria, black 165
—, river 11, *166*
plankton 83, 113, *176*
Platichthys (= Pleuronectes) platessa (plaice) 207
— *stellatus* starry flounder 207
Platynereis spec. 173
— *bicanaliculata* 173
— *dumerilii* 99, *173*
— *megalops* 173
Plectropomus 202
Pleuronectes (= Platichthys) platessa (plaice) 207
Pleurosigma aestuarii 52, *158*
plover, thick-billed *210*
Plutella xylostella 191
Pocillopora bulbosa 164
— *damicornis* 164
— *elegans* 164
— *eydouxi* 165
— *verrucosa* 165
Poecilia (= Lebistes) reticulatus (guppy) 15f, 120, *200*
Polycheira rufescens 195
Polykrikos spec. 157

Polymitarcis (= *Epheron*) *virgo* 188
Polyophthalmus pictus 174
Pomacentrus flavicauda (tropical damselfish) 204
— *nagasakiensis* 204
— *wardi* (tropical damselfish) 204
Pontogeloides latipes 179
Porites spec. 64, *165*
— *astreoides* 165
— *furcata* 165
— *lobata* 165
porrum 159
Portunus pelagicus 184
— *sanguinolentus* 184
potato 14, *161*
Povilla adusta (tropical mayfly) 108, *188*
prawn, common 58, *182*
Prays citri 190
Prionospio cirrifera 174
— *malmgreni* 174
Protopalythoa spec. 165
Pseudemys scripta (lettered terrapin) 110, *209*
Pseudoquilla ciliata 177
Psorophora confinnis 192
Ptychoramphus aleuticus (Cassin's auklet) 210
Ptyodactylus hasselquistii guttatus (house gecko) 209

quahog, northern *170*

Rana cancrivora 17, *208*
— *pipiens* (leopard frog) 110, *208*
rat, bannertail kangaroo 109, *212*
— house *213*
— Merriam's kangaroo *212*
Rattus mülleri 213
— *rajah* 213
— *rattus* (house rat) *213*
— *rattus jalorensis* 213
— *sabanus* 213
Retina plicata 166
Rhithropanopeus harrisii 184
Rhombus (= *Scophthalmus*) *maxinus* (turbot) 207
rocket candytuft *160*
rockling, four-bearded *199*
Rousettus aegyptiacus (Egyptian fruit bat) 110, *212*

Salmo gairdneri (rainbow trout) *197*
— *salar* (Atlantic salmon) 18, *197*
— *trutta lacustris* (lake trout) *197*
salmon, Atlantic 18, *197*
—, Chinook *198*
—, Masu *198*
—, silver *198*
Salpa fusiformis 195
sand hopper *178*
sardine, Tanganyikan *197*
Sargassum enerve (gulf weed) *158*
— *muticum* 158
— *vestitum* 159
scallop, giant 58, *170*
—, small 58, *170*
Scenedesmus quadricauda 159
Schöngastia (= *Thrombidium*) *vandersandei* 176
Scirpophaga nirella 190
Scoliopleura latestriata 158
Scomber colias (tuna-like mackerel) 206
Scophthalmus (= *Rhombus*) *maxinus* (turbot) 207
sea hare, Californian *169*
seal, common 47, *214*
—, Galápagos fur *214*
—, maned *214*
Sebastodes (= *Sebastes*) *taczanowskii* 201
sergeant fish, Pacific *203*
sergeant major *203*
Sesarma spec. 187
— *haematocheir* 99, *186, 187*
— *intermedium* 186
— *reticulatum* 187
— *rotandata* 187
shad, American *196*
shanny *205*
shark, lemon *195*
shipworm, Atlantic *171*
shorehopper 51, *178*
shrimp, brown *182*
—, mud *179*
Siganus rivulatus 207
silverside, Atlantic *201*
—, inland *201*
—, sardine 71, *201*
—, tidewater *201*
Siphonaria atra (Athecate hydroids) *168*
— *denticulata* 168
— *japonica* (Athecate hydroids) *168*

INDEX OF SPECIES

— *sipho* (Athecate hydroids) *168*
— *virgulata* *168*
Soja hispida (soya bean) *160*
Solanum lycopersicum (tomato) 111, *159*
— *tuberosum* (potato) 14, *161*
Sphecodogastra texana (North American bee) *189*
Sphingidae *191*
Spirorbis borealis *174*
Spodoptera litura (tobacco caterpillar) *190*
Spratelloides delicatulus *197*
— *gracilis* *197*
— *lewisi* *197*
Staurocephalus rudolphii *174*
Stauroneis salina *158*
Stegastes acapulcoensis *204*
— *diencaeus* *204*
— *dorsopunicans* *204*
— *leucostictus* *204*
— *partitus* *204*
— *planifrons* *204*
— *variabilis* *204*
Sterna fuscata (sooty tern) *210*
— *sumatrana* (black-naped tern) *210*
Stichopus chloronotus *194*
— *variegatus* *194*
Strix occidentalis (spotted owl) *211*
Strombidium oculatum *162*
sucker, white *197*
sunflower *160*
Surirella gemma *158*
Syllis amica *172*
Symphytum tuberosum (tuberous comfrey) 11, *160*
Synchelidium spec. 88, 103, *178*

Tachyglossus aculeatus (Australian spiny anteater) *211*
Takijugu niphobles *208*
Talitrus saltator (sand hopper) *178*
Talorchestia quoyana *178*
Tanytarsus balteatus 108, *193*
Tectus (= *Trochus*) *niloticus* (giant topshell) *167*
Teredo pedicellata (Atlantic shipworm) *171*
tern, black-naped *210*
—, sooty *210*
terrapin, lettered 110, *209*

Tetrao urogallus (capercaillie) 17, *209*
Thalassoma bisfasciatum (bluehead wrasse) *205*
— *duperrey* *205*
Thrombidium (= *Schöngastia*) *vandersandei* *176*
Thymallus thymallus (European grayling) *198*
Thysanoessa spec. *180*
tiger lily 111, *159*
Tilapia mariae *203*
toad, American *208*
—, black-spined 17, *208*
tobacco caterpillar *190*
tomato 111, *159*, *160*
Tribus lamprologini *203*
Tridacna derasa *170*
— *gigas* *170*
Tridentiger brevispinis (numachichibu-goby) *206*
Triticum spec. (wheat) 111, *160*
— *vulgare* (common wheat) *160*
Tritonia diomedea 90f, *169*
Triturus (= *Diemictylus*) *viridescens* (red spotted newt) 111, *208*
Trochus (= *Tectus*) *niloticus* (giant topshell) *167*
Tropidoneis vitrae *158*
trout, lake *197*
—, rainbow *197*
tsetse fly *193*
tuberculated sea bear *166*
turbot *207*
Tylorhynchus chinesis (= *Ceratocephale osawai*) (Japanese palolo worm, 'Itome') *173*
Tylos granulatus *180*
Typhlodromus pyri *176*
Typosyllis prolifera 99, *172*

Uca spec. 96, 103
— *beebei* *185*
— *crenulata* *185*
— *lactea lactea* *185*
— *latinanus* *185*
— *maracoani* *185*
— *minax* 184, *185*
— *mordax* *185*
— *musica terpsichores* *185*
— *princeps* *185*
— *pugilator* 94, *184*, *185*, *186*

— *pugnax* 94, *184, 185, 186*
— *speciosa* *186*
— *stylifera* *185*
— *tangeri* *186*
Ulmus americana (American elm) 111, *159*
Ulva angusta *159*
— *linza* *159*
— *lobata* *159*
— *stenophylla* *159*
— *taeniata* *159*
urchin, Indo-Pacific diadem sea 9, *194*
—, long-spined sea *194*
—, West Indian white sea *194*
Urosalpinx cinerea 168

Vanellus vanellus (lapwing) 110, *209*
Venus (= Mercenaria) mercenaria (northern quahog) *170*
Venus shell *170*
Vicia faba (broad bean) 11, *160*

vine, grape 11, *160*
Vitis vinifera (grape vine) 11, *160*

wallaby, tammor 109, *211*
Warnowia rubescens *157*
wawo worm 63, *174*
wax moth *191*
waxwing 17f, *211*
wheat 111, *160*
whelk, mud dog 59, *168*
whitebait, New Zealand *199*

Xantho floridus *184*
Xanthodius sternberghii *184*
Xestospongia testudinaria *162*

yarrow 111, *159*
Yoldia sapotilla *169*

Zea mais (maize) 111, *160*
Zooplankton *176*

Subject index

abiotic systems 20
absenteeism 216
accidents, *see* traffic accidents 218
aggression in women 216
amphidromic point 44
— system 44
amplitude 87
annular eclipse 27
anomalistic (apsidal) month 27, 34
aphelion 27
apogee 27
apsidal (anomalistic) month 27
apsides, line of 28

bacterial and viral inflammation 216
biographical rhythms 137
biological rhythms 134
births 126f, 129f, 216

cancer 216
Capture Hypothesis 80
captured rotation 29
chronobiology 7, 19
chronopharmacology 19
chronophysiology 19
Co-Accretion Hypothesis 81
colour sensitivity 117, 119, 217
conjunction 25
cosmic (Platonic) year 30
cotidal lines 43, 45

daily inequality 38
Darwin, George H. 80
deaths 126f, 217
draconitic (nodal) month 26, 34

eclampsia 126, 217
eclipse 25–27
—, annular 27
—, total 27
ecliptic 25
endogenous (endo-) rhythms 92, 94f, 97
endogenous organisms 87

Euler, Leonhard 77
Euler's number 77
exogenous (exo-) rhythms 95, 97
exogenous organisms 87
exogenous-endogenous rhythms 95

Fission Hypothesis 80
frequency 87

gravitational 42
Great Barrier Reef 81

Heisenberg, Werner 150
horizon 33
human organism 135
human rhythms 144

illness 138
Impact Hypothesis 81

lagging of tides 36
logarithms 76
lunar nodes 25
lunar periodicity 20
lunar phases 23
lunar rhythms 20
lunation 23
lundian rhythms 20

magnetic field 89–92
malaria 217
menstruation 217
Metonic cycle 34
monthly inequality 38
Moon, distance of 79

nadir 33
neap tide 38
nodal (draconitic) month 26
nodes, lunar 25
nutation cycle/period 26, 34

opposition 25

Palolo swarming 68
perigee 27
perihelion 27
period 87
Platonic (cosmic) year 30, 140
pneumonia 126, 217
poisoning 217
port time 36
precession 140
Purkinje phenomenon 117

quadrature 25

reaction norms 97
regression 35
rhythms of the moon 20
rise of tide 37

Saros cycle 27, 34
self-inflicted poisoning 217
(semi) solunar 20
semi-lunar 20
seven-year-rhythm 139
sex-determining 129
sidereal month 24, 34
sidereal year 24
sleeping/waking rhythm 121f, 136, 218
sol-semi-lunar 20
solunar 20
spray zone 49
spring tide 38
suicide 125
synchronous rotation 29
synodic month 20, 23, 34
syzygy 25

tidal amplitude 37
tidal curves 40
tidal forms 37
tidal movements 36
tidal range 37
tidal rhythms 20
tidal stream 42
tide 36–38
—, fall of 37
—, neap 38
—, rise of 37
—, spring 38
—, lagging of 36
timers 97, 99
total eclipse 27
traffic accidents 218
transgression 35
triggers 97
tropical month 30, 34
tropical year 30
tropics 33

uncertainty principle 150
uric acid 123–25
uricosuria 218

Van't Hoff's rule 96
violence 218

wisdom tooth 139

Zeitgestalt 105, 143
zenith 33

Eco-Geography

What we see when we look at landscapes

Andreas Suchantke

What do we really see when we look at a landscape? In detailed and telling observations Andreas Suchantke describes some of the most fascinating landscapes on Earth: encompassing the savannahs of East Africa, the rainforests of South America and Africa, the unique islands of New Zealand, the Great Rift Valley of Africa, and the Middle East.

He brings us to landscapes that have been severely damaged by human activity, and others — such as the island of Sri Lanka — where nature and human culture have been brought into paradisal harmony. His beautiful descriptions and illustrations alone are worth the journey, but these essays are more than great nature and ecology writing. Suchantke's real interest is a new way of seeing the physical landscape. This approach is not then just analysed objectively, but recreated imaginatively, with nature experienced as a form of meaning, a language. As Suchantke shows us, the quality of our relationship with nature is determined by how well we understand this language.

Eco-Geography is a ray of hope, to help us achieve the balance between the materialism of modern science, and the current philosophy of despair that sees human beings as destroyers of nature. It shows our potential to develop sensibilities that meet the needs of the planet, and to form a true nurturing partnership between nature and human culture.

Andreas Suchantke was born in Basel, Switzerland in 1933. Educated in zoology and botany at the universities of Munich and Basel, he taught natural sciences for many years in the Waldorf School in Zurich. As a freelance ecologist he worked extensively in Israel, in co-operation with the Society for the Protection of Nature.

Floris Books

Goethe on Science

An anthology of Goethe's scientific writings
Edited by Jeremy Naydler

Johann Wolfgang von Goethe (1749–1832) ranks with Shakespeare as a European man of letters, playwright and poet. But he himself considered his scientific work to be far more important than all his other achievements. In the twentieth century his ideas have been given special attention by scientists such as Adolf Portmann and Werner Heisenberg.

Jeremy Naydler provides a systematic arrangement of extracts from Goethe's major scientific works to provide a clear picture of his fundamentally different approach to scientific study of the natural world. According to Goethe, our deepest knowledge of phenomena can arise only from a contemplative relationship with nature, in which our feelings of awe and wonder are intrinsic. As conceived by him, science is as much a path of inner development as it is a way to accumulating knowledge.

From a Goethean standpoint, our modern ecological crisis is a crisis of relationship to nature. Goethe shows us a path of sensitive science which holds the potential for healing both nature and ourselves.

Jeremy Naydler is a philosopher, cultural historian and gardener, living in Oxford. He is author of *Temple of the Cosmos* and of *How Caterpillars acquire Wings*.

Floris Books

The Wholeness of Nature

Goethe's Way of Science

Henri Bortoft

The scientific work of Johann Wolfgang von Goethe represents a style of learning and understanding which is widely ignored today. The approach of modern science is largely detached, intellectual and analytical, and it is increasingly recognized that many of our contemporary problems stem from the resulting divorce from nature.

By contrast Goethe's way of science pursued understanding through the experience of the 'authentic wholeness' of what we observe. Working with the intuitive mode of consciousness, Goethe aimed at an encounter with the whole phenomenon in its relationship with the observer. In his way of seeing, rather than dividing merely in order to categorize, we should investigate the parts of an object in order to reveal the true nature of the whole.

Henri Bortoft examines the phenomenological and cultural roots of Goethe's way of science. He argues that Goethe's insights, far from belonging to the past, represent the foundation for a future science. This new science of nature, involving other human faculties besides the analytical mind, can provide understanding and explanation in a way which our present scientific attitudes, and the culture they serve, desperately lack.

Henri Bortoft has taught physics and the philosophy of science for most of his career. His postgraduate research was under David Bohm and Basil Hiley at Birkbeck College, London. He now lectures and gives seminars on Goethean science.

Floris Books